ABBREVIATIONS

a.c.	alternating current, or air cooled
A.C.G.I.	Associate of the City and Guilds of London Institute
A.E.S.D.	Association of Engineering and Shipbuilding Draughtsmen
A.E.U.	Amalgamated Engineering Union
A.F.R.Ae.S.	Associate Fellow of the Royal Aeronautical Society
A.I.A.A.	American Institute of Aeronautics and Astronautics
A.I.D.	Aeronautical Inspection Directorate
a.h.p.	actual horsepower
A.M.I.Mar.E.	Associate Member of the Institute of Marine Engineers
A.S.	Academy of Science
A.S.A.	American Standards Association
A.Sc.	Associate in Science
A.S.C.E.A.	American Society of Civil Engineers and Architects
A.S.E.E.	Association of Supervisory and Executive Engineers
A.S.M.E.	American Society of Mechanical Engineers
B.A.	Bachelor of Arts; British Association
B. & S.	Brown and Sharpe
b.d.c.	bottom dead centre
B.Eng.	Bachelor of Engineering
B.G.	Birmingham gauge
b.h.p.	brake horsepower
B.Sc.	Bachelor of Science
B.S.F.	British Standard Fine
B.S.I.	British Standards Institution
B.S.P.	British Standard Pipe
Btu	British thermal unit
°C	degree Celsius (Centigrade)
car.	carat
c.e.	chief engineer, civil engineer
C.Eng.	Chartered Engineer
c.g.	centre of gravity
c.g.s.	centimetre-gramme-second
c.i.	cast iron; compression ignition

ABBREVIATIONS (*Continued*)

cir.	circumference
c/l	centre line
cm	centimetre
c. of t.	centre of thrust
c.p.	candle power; centre of pressure
c.s.	cast steel; carbon steel
csk.	countersunk; countersink
d.c.	direct current
dia.	diameter
D.Eng.	Doctor of Engineering
d.o.	drawing office
drg.	drawing
D.Sc.	Doctor of Science
D.T.D.	Directorate of Technical Development
e.h.p.	effective horsepower
e.m.f.	electromotive force
°F	degree Fahrenheit
F.A.I.A.A.	Fellow of the American Institute of Aeronautics and Astronautics
F.C.G.I.	Fellow of the City and Guilds of London Institute
F.I.C.E.	Fellow of the Institution of Civil Engineers
F.I.E.E.	Fellow of the Institution of Electrical Engineers
F.I.Mech.E.	Fellow of the Institution of Mechanical Engineers
F.I.Min.E.	Fellow of the Institution of Mining Engineers
F.I.Prod.E.	Fellow of the Institution of Production Engineers
F.R.Ae.S.	Fellow of the Royal Aeronautical Society
F.R.S.	Fellow of the Royal Society
F.R.S.A.	Fellow of the Royal Society of Arts
g	gramme, value of gravity (see page 46)
g	gramme
gal	gallon
hp	horsepower
Hz	hertz

ABBREVIATIONS *(Continued)*

I.C.E.	Institution of Civil Engineers
i.c.e.	internal combustion engine
I.E.E.	Institution of Electrical Engineers
I.E.I.	Institution of Engineering Inspection
i.h.p.	indicated horsepower
I.Mar.E.	Institute of Marine Engineers
I.Mech.E.	Institution of Mechanical Engineers
I.Min.E.	Institution of Mining Engineers
in	inch
I.Prod.E.	Institution of Production Engineers
I.S.I.	Iron and Steel Institute
I.W.G.	Imperial Wire Gauge
J.I.E.	Junior Institution of Engineers
K	degrees Kelvin
k	kilo (prefix)
lb	pound
l.h.	left hand
m	metre; milli (prefix)
M.A.	Master of Arts
m/c	machine
M.C.E.	Master of Civil Engineering
m.e.p.	mean effective pressure
M.I.C.E.	Member of the Institution of Civil Engineers
M.I.E.E.	Member of the Institution of Electrical Engineers
M.I.Mar.E.	Member of the Institute of Marine Engineers
M.I.Mech.E.	Member of the Institution of Mechanical Engineers
M.I.Min.E.	Member of the Institution of Mining Engineers
M.I.Prod.E.	Member of the Institution of Production Engineers
MKS	metre-kilogramme-second
M.M.E.	Master of Mechanical Engineering
m.m.f.	magnetomotive force
mol. wt.	molecular weight
m.s.	mild steel
M.Sc.	Master of Science

ABBREVIATIONS *(Continued)*

p.d. potential difference
Ph.D. Doctor of Philosophy
p.w.g. pinion wire gauge

rad. radius, radian
R.Ae.C. Royal Aero Club
R.Ae.S. Royal Aeronautical Society
r.h. right hand
r.m.s. root mean square
r.p.h. revolutions per hour
r.p.m. revolutions per minute
R.S.A. Royal Society of Arts

S.A.E. Society of Automobile Engineers (U.S.)
S.I. Système International (International System) of units
sp. g. specific gravity
sp. ht. specific heat
s.t.p. standard temperature and pressure
s.v. slide valve; side valve
s.w.g. standard wire gauge

t.d.c. top dead centre
temp. temperature
t.p.i. threads per inch
t.p.l. threads per length

U.S.S. United States Standard

w.i. wrought iron
wt. weight

MATHEMATICAL SIGNS AND SYMBOLS

$=$	equal to
$\neq$	not equal to
$\equiv$	identical with
$\triangleq$	corresponds to
$\triangleq$ or $\approx$	approximately equal to
$\rightarrow$	approaches
$\simeq$	asymptotic to
$\propto$	proportional to
∞	infinity
$<$	less than
$>$	greater than
$\leqslant$	less than or equal to
$\geqslant$	greater than or equal to
$\ll$	much less than
$\gg$	much greater than
$+$	plus
$-$	minus
$\times$	multiply
$\div$	divide

$\frac{a}{b}$ or a/b	a divided by b
$\lvert a\rvert$	modulus of a
a^n	a raised to the power n
$a^{\frac{1}{2}}$ or $\sqrt{a}$	square root of a
$a^{1/n}$ or $\sqrt[n]{a}$	nth root of a
$\bar{a}$	mean value of a
$p!$	factorial p $(= 1 \times 2 \times 3 \times \ldots \times p)$
$\therefore$	therefore
$\because$	because
Σ	sum
Π	product
$\int$	integral sign
$\log_a x$	logarithm to the base a of x
$^\circ$	degree
$'$	second

THE GREEK ALPHABET

Α	α	alpha	Ν	ν	nu
Β	β	beta	Ξ	ξ	xi
Γ	γ	gamma	Ο	o	omicron
Δ	δ	delta	Π	π	pi
Ε	ϵ	epsilon	Ρ	ρ	rho
Ζ	ζ	zeta	Σ	σ ς	sigma
Η	η	eta	Τ	τ	tau
Θ	θ ϑ	theta	Υ	υ	upsilon
Ι	ι	iota	Φ	ϕ	phi
Κ	κ	kappa	Χ	χ	chi
Λ	λ	lambda	Ψ	ψ	psi
Μ	μ	mu	Ω	ω	omega

INTRODUCTION TO SI UNITS

This edition of *Newnes Engineer's Pocket Book* is based on the International System of units, which will become standard throughout Europe and the U.K., and which will replace the Imperial units presently used in engineering practice, such as the inch, pound, etc. The system, whose full title is "Système International d' Unités", is a rationalised and coherent system, based on the four units, metre, kilogramme, second and ampere, with the two supplementary units, the kelvin as the unit of temperature, and the candela as the unit of luminous intensity. After several international meetings over the years, the system of units was formally adopted and named at the eleventh meeting of the Conférence Générale des Poids et Mesures (CGPM) in 1960. All the important matters in this Pocket Book are given in these international units, which are described in detail, but as it takes time for changes of this kind to be thoroughly assimilated a few conversion tables between the old and the new units are included. The SI system based on the metre, kilogramme and second (MKS) also replaces the centimetre-gramme-second (c.g.s.) system.

In most tables there are preferred sizes of screws etc., and in some cases there are first, second and third choices. Ultimately it is intended that new designs should use only preferred sizes (first choice) and other sizes are called non-preferred; the latter may be convenient to use before metrication is fully adopted. In many tables only the first choices are given.

BASE UNITS

The unit of length is the metre (m)
The unit of mass is the kilogramme (kg)
The unit of time is the second (s)
The unit of electric current is the ampere (A)
The unit of thermodynamic temperature is the kelvin (K)
The unit of luminous intensity is the candela (cd)
The unit of the amount of substance is the mole (mol)

SUPPLEMENTARY UNITS

The unit of plane angle is the radian (rad)
The unit of solid angle is the steradian (sr)

DERIVED UNITS

Some of the derived units are given special names and these are listed below:

Quantity	*Special Name*	*Units*
Electric capacitance	farad (F)	As/V
Electric resistance	ohm (Ω)	V/A
Electromotive force, potential difference	volt (V)	W/A
Energy, work	joule (J)	Nm
Force	newton (N)	kgm/s^2
Frequency	hertz (Hz)	1/s
Illumination	lux (lx)	lm/m^2
Inductance	henry (H)	Vs/A
Luminous flux	lumen (lm)	cd sr
Magnetic flux	weber (Wb)	Vs
Magnetic flux density	tesla (T)	Wb/m^2
Power	watt (W)	J/s
Quantity of electricity	coulomb (C)	As

Other derived units are:

Quantity	*Units*
Acceleration	m/s^2
Angular acceleration	rad/s^2
Angular velocity	rad/s
Area	m^2
Density	kg/m^3
Diffusion coefficient	m^2/s
Dynamic viscosity	Ns/m^2
Electric field strength	V/m
Frequency	l/s
Kinematic viscosity	m^2/s
Luminance	cd/m^2
Magnetic field strength	A/m
Pressure or stress	N/m^2
Surface tension	N/m
Thermal conductivity	W/mK
Velocity	m/s
Volume	m^3

MULTIPLES

Multiples of units may be denoted by various prefixes as shown in the table.

Factor	*Prefix and Symbol*
10^{12}	tera (T)
10^{9}	giga (G)
10^{6}	mega (M)
10^{3}	kilo (k)
10^{2}	hecto (h)
10	deca (da)
10^{-1}	deci (d)
10^{-2}	centi (c)
10^{-3}	milli (m)
10^{-6}	micro (μ)
10^{-9}	nano (n)
10^{-12}	pico (p)
10^{-15}	femto (f)
10^{-18}	atto (a)

EXAMPLES

km = kilometre = 10^{3} metres
mm = millimetre = 10^{-3} metres
mA = millampere = 10^{-3} amperes
μs = microsecond = 10^{-6} seconds

Indices not included in the table, e.g. 10^{-8}, are not recommended for use.

DEFINITIONS OF UNITS

BASE UNITS

Metre (m)—The metre is the length equal to 1 650 763·73 wavelengths in vacuum of the radiation corresponding to the transition between the levels $2p_{10}$ and $5d_5$ of the krypton-86 atom.

Kilogramme (kg)—The kilogramme is the unit of mass; it is equal to the mass of the international prototype of the kilogramme.

Second (s)—The second is the duration of 9 192 631 770 periods of the radiation corresponding to the transition between the two hyperfine levels of the ground state of the caesium-133 atom.

Ampere (A)—The ampere is that constant current which, if maintained in two straight parallel conductors of infinite length, of negligible circular cross-section, and placed 1 m apart in vacuum, would produce between these conductors a force equal to 2×10^{-7} newton per metre of length.

Kelvin (K)—The kelvin of thermodynamic temperature is the fraction 1/273·16 of the thermodynamic temperature of the triple point of water.

Candela (cd)—The candela is the luminous intensity, in the perpendicular direction, of a surface of 1/600 000 m^2 of a black body at the temperature of freezing platinum under a pressure of 101 325 N/m^2.

Mole (mol)—The mole is the amount of substance of a system which contains as many elementary entities* as there are atoms in 0·012 kg of carbon 12.

These definitions have been formulated by the advisory committee on units of the International Committee of Weights & Measures.

*The elementary entities should be specified and can be atoms, molecules, ions, electrons, other particles or specified groups of such particles.

OTHER UNITS

Bar: a unit of pressure = 10^5 N/m^2.

Coulomb (C): the unit of electric charge; it is the quantity of electricity transported in one second by a current of one ampere.

Electron volt (eV): a unit of energy used in nuclear physics = $1{\cdot}60206 \times 10^{-19}$J.

Degree (°): a unit of plane angle = $\pi/180$ rad.

Farad (F): the unit of electric capacitance; it is the capacitance of a capacitor between the plates of which there is a potential difference of one volt when charged with one coulomb.

Henry (H): the unit of electric inductance; it is the inductance of a closed electrical circuit in which an e.m.f. of one volt is produced when the current varies uniformly at the rate of one ampere per second.

Hertz (Hz): the unit of frequency; it is equal to one cycle per second.

Joule (J): the unit of energy or work; it is the work done when a force of one newton acts over a distance of one metre.

Kilowatt-hour (kWh): a unit of energy; it is the energy expended when a power of 1000 watts is supplied for one hour.

Knot (kn): a unit of speed; it is equal to one nautical mile per hour.

Litre (l): a unit of volume equal to one cubic decimetre.

Lumen (lm): the luminous flux emitted within a unit solid angle of one steradian by a point source having a uniform intensity of one candela.

Lux (lx): the unit of illumination = 1 lm/m^2.

Newton (N): the force required to accelerate a mass of one kilogramme at one metre per second squared.

DEFINITIONS OF UNITS *(Continued)*

Ohm (Ω): the resistance between two points on a conductor at a potential difference of one volt when a current of one ampere is flowing.

Pascal (Pa): a unit of pressure = $1\ N/m^2$.

Poise (P): a unit of dynamic viscosity, usually quoted in centipoise (cP); $1cP = 10^{-3}\ Ns/m^2$.

Radian (rad): the angle subtended at the centre of a circle by an arc whose length is equal to the radius of the circle.

Siemens (S): the unit of conductance = $1/\Omega$.

Steradian (sr): the solid angle subtended at the centre of a sphere by a cap with an area equal to the radius squared.

Stoke (St): a unit of kinematic viscosity, usually quoted in centistokes (cSt); $1\ cSt = 10^{-6} m^2/s$.

Tesla (T): the unit of flux density = $1\ Wb/m^2$.

Tonne (t): a unit of mass = 10^3 kg.

Volt (V): the unit of e.m.f. and potential difference; it is the difference in potential between two points on a conductor carrying a constant current of one ampere when the power dissipated is one watt.

Watt (W): the unit of power = 1 J/s.

Weber (Wb): the unit of magnetic flux; it is the flux which, linking a coil of one turn, produces in it an e.m.f. of one volt as it is reduced to zero at a uniform rate in one second.

PHYSICAL QUANTITIES EXPRESSED DIMENSIONALLY

Length: metre [L]
Mass: kilogramme [M]
Time: second [T]
Quantity of electricity: coulomb [Q]
Area: square metre [L^2]
Volume: cubic metre [L^3]

PHYSICAL QUANTITIES *(Continued)*

Velocity: metre per second $[LT^{-1}]$
Acceleration: metre per second per second $[LT^{-2}]$
Mass density: kilogramme per cubic metre $[ML^{-3}]$
Force: newton $[MLT^{-2}]$
Pressure: newton per square metre $[ML^{-1}T^{-2}]$
Work: joule $[ML^{2}T^{-2}]$
Power: watt $[ML^{2}T^{-3}]$
Electric current: ampere $[QT^{-1}]$
Voltage: volt $[ML^{2}T^{-2}Q^{-1}]$
Electric resistance: ohm $[ML^{2}T^{-1}Q^{-2}]$
Electric conductance: siemens $[M^{-1}L^{-2}TQ^{2}]$
Inductance: henry $[ML^{2}Q^{-2}]$
Capacitance: farad $[M^{-1}L^{-2}T^{2}Q^{2}]$
Current density: ampere per square metre $[L^{-2}T^{-1}Q]$
Electric field strength: volt per metre $[MLT^{-2}Q^{-1}]$
Magnetic flux: weber $[ML^{2}T^{-1}Q^{-1}]$
Magnetic flux density: weber per square metre $[MT^{-1}Q^{-1}]$

SI EQUIVALENTS OF SOME COMMON IMPERIAL UNITS

1 in	= 25·4 mm
1 ft	= 304·8 mm
1 mile	= 1·609344 km
1 in^2	= 645·16 mm^2
1 ft^2	= 0·092903 m^2
1 $mile^2$	= 2·58999 km^2
1 acre	= 4046·86 m^2
1 in^3	= 16387·1 mm^3
1 ft^3	= 0·0283168 m^3
1 gal	= 4·54609 dm^3
1 ft/s	= 0·3048 m/s
1 mile/h	= 0·44704 m/s
1 mile/h	= 1·60934 km/h
1 lb	= 0·45359237 kg
1 ton	= 1016·05 kg
1 lb/ft^3	= 16·0185 kg/m^3
1 pdl	= 0·138255N
1 lbf	= 4·44822N
1 tonf	= 9·96402 kN
1 ft pdl	= 0·042140J

1 ft lbf	= 1·35582J
1 cal	= 4·1868J
1 Btu	= 1·05506kJ
1 lbf/in^2	= 6·89476kN/m^2
1 tonf/ft^2	= 107·252 kN/m^2
1 tonf/in^2	= 15·4443 MN/m^2
1 hp	= 746 W
1 Celsius unit	= 1 Kelvin unit
1 Fahrenheit unit	= 1 Rankine unit
1 Rankine unit	= 5/9 Kelvin unit

BS CONVERSION SLIDE

The British Standards Institution publishes a conversion slide which facilitates rapid conversion between corresponding values in SI and Imperial units.

METRIC CONVERSION FACTORS

To convert—

Millimetres to inches .	× .03937 or ÷ 25.4
Centimetres to inches .	× .3937 or ÷ 2.54
Metres to inches . .	× 39.37
Metres to feet . .	× 3.281
Metres to yards . .	× 1.094
Metres per second to feet per minute . .	× 197
Kilometres to miles .	× .6214 or ÷ 1.6093
Kilometres to feet .	× 3,280.864
Square millimetres to square inches .	× .00155 or ÷ 645.1
Square centimetres to square inches. .	× .155 or ÷ 6.451
Square metres to square feet . . .	× 10.764
Square metres to square yards . . .	× 1.2

METRIC CONVERSION FACTORS

To convert—

Square kilometres to acres	× 247.1
Hectares to acres . .	× 2.471
Cubic centimetres to cubic inches . . .	× .06 or ÷ 16.383
Cubic metres to cubic feet .	× 35.315
Cubic metres to cubic yards	× 1.308
Litres to cubic inches .	× 61.022
Litres to gallons . .	× .21998 or ÷ 4.545
Litres to cubic feet . .	÷ 28.316
Hectolitres to cubic feet .	× 3.531
Hectolitres to cubic yards .	× .131
Grammes to ounces (avoirdupois) . .	× .035 or ÷ 28.35
Grammes per cubic cm. to lb. per cubic inch .	÷ 27.7
Joules to foot-lb. . .	× .7373
Kilogrammes to oz. . .	× 35.3
Kilogrammes to lb. . .	× 2.2046
Kilogrammes to tons .	× .001
Kilogrammes per sq. cm. to lb. per sq. inch . .	× 14.22
Kilogramme-metres to foot-lb.	× 7.233
Kilogramme per metre to lb. per foot . . .	× .672
Kilogramme per cubic metre to lb. per cubic foot .	× .062
Kilogramme per cheval-vapeur to lb. per h.p. .	× 2.235
Kilowatts to h.p. . .	× 1.34
Watts to h.p. . . .	÷ 746
Watts to foot-lb. per second	× .7373
Cheval-vapeur to h.p. .	× .9863
Gallons of water to lb. .	× 10
Atmospheres to lb. per sq. inch . . .	× 14.7

Equivalents of Imperial and Metric Weights and Measures

Linear Measure

Imperial		
1 Inch	=	25.400 Millimetres.
1 Foot	=	0.30480 Metre.
1 Yard		**0.914399 Metre.**
1 Fathom	=	1.8288 Metres.
1 Pole	=	5.0292 ,,
1 Chain	=	20.1168 ,,
1 Furlong	=	201.168 ,,
1 Mile	=	1.6093 Kilometres.

Metric		
1 Millimetre (mm.) (1/1000 m.)	=	0.03937 Inch.
1 Centimetre (1/100 m.)	=	0.3937 Inch.
1 Decimetre (1/10 m.)	=	3.937 Inches.
1 Metre (m.)	=	**39.370113 Inches.** **3.280843 Feet.** **1.0936143 Yards.**
1 Decametre (10 m.)	=	10.936 Yards.
1 Hectometre	=	0.0621 Mile.
1 Kilometre (1000 m.)	=	0.62137 Mile.

Square Measure

Imperial		
1 Square Inch	=	6.4516 Square Centimetres.
1 Square Foot	=	9.2903 Square Decimetres.
1 Square Yard	=	0.836126 Square Metre
1 Rood	=	10.117 Ares.
1 Acre	=	0.40468 Hectare.
1 Square Mile	=	259.00 Hectares.

Metric		
1 Square Centimetre	=	0.15500 Square Inch.
1 Square Decimetre	=	15.500 Square Inches.
1 Square Metre	=	10.7639 Square Feet 1.1960 Square Yards.
1 Are	=	119.60 Square Yards.
1 Hectare	=	2.4711 Acres.

Equivalents of Imperial and Metric Weights and Measures—continued

IMPERIAL		METRIC	
Cubic Measure			
1 Cubic Inch .	= 16.387 Cubic Centimetres.	1 Cubic Centimetre .	= 0.0610 Cubic Ins.
1 Cubic Foot .	= 0.028317 Cubic Metre.	1 Cubic Decimetre (c.d.) . . .	= 61.024 Cubic Ins.
1 Cubic Yard .	= 0.764553 ,, ,,	1 Cubic Metre. .	= 35.3148 Cubic Feet. 1.307954 Cubic Yards.
Measure of Capacity			
1 Pint . .	= 0.568 Litre.	1 Centilitre (1/100 litre)	= 0.070 Gill.
1 Quart . .	= 1.136 Litres.	1 Decilitre (1/10 litre)	= 0.176 Pint.
1 Gallon . .	**= 4.5459631 Litres.**	**1 Litre . . .**	**= 1.75980 Pints.**
Weight			
Avoirdupois			*Avoirdupois*
1 Grain . .	= 0.0648 Gramme.	1 Milligramme (1/1000 grm.) .	= 0.015 Grain.
1 Dram . .	= 1.772 Grammes.	1 Centigramme (1/100 grm.) . .	= 0.154 ,,
1 Ounce . .	= 28.350 ,,	1 Gramme (1 grm.) .	= 15.432 ,,
1 Pound (7,000 Grains) . .	**= 0.45359243 Kilogramme.**	**1 Kilogramme (1,000 grm.) . .**	**= 2.2046223 Lb. or 15,432.3564 Grains.**
1 Hundredweight	= 50.80 Kilogrammes. 0.5080 Quintal.	1 Quintal (100 kilog.)	= 1.968 cwt.
1 Ton. . .	= 1.0160 Tonnes or **1016** Kilogrammes.	1 Tonne (1,000 kilog.)	= 0.9842 Ton.
1 Grain (Troy) .	= 0.0648 Gramme.	1 Gramme (1 grm.) .	= 0.03215 Oz. Troy. 15.432 Grains.
1 Troy Ounce .	= 31.1035 Grammes.		

DECIMAL EQUIVALENTS OF MILLIMETRES IN INCHES

mm	in	mm	in	mm	in	mm	in
0.01	0.00039	0.39	0.01535	0.77	0.03032	16	0.62992
0.02	0.00079	0.40	0.01575	0.78	0.03071	17	0.66929
0.03	0.00118	0.41	0.01614	0.79	0.03110	18	0.70866
0.04	0.00157	0.42	0.01654	0.80	0.03150	19	0.74803
0.05	0.00197	0.43	0.01693	0.81	0.03189	20	0.7874
0.06	0.00236	0.44	0.01732	0.82	0.03228	21	0.82677
0.07	0.00276	0.45	0.01772	0.83	0.03268	22	0.86614
0.08	0.00315	0.46	0.01811	0.84	0.03307	23	0.90551
0.09	0.00354	0.47	0.01850	0.85	0.03346	24	0.94488
0.10	0.00394	0.48	0.01890	0.86	0.03386	25	0.98425
0.11	0.00433	0.49	0.01929	0.87	0.03425	26	1.02362
0.12	0.00472	0.50	0.01969	0.88	0.03465	27	1.06299
0.13	0.00512	0.51	0.02008	0.89	0.03504	28	1.10236
0.14	0.00551	0.52	0.02047	0.90	0.03543	29	1.14173
0.15	0.00591	0.53	0.02087	0.91	0.03583	30	1.1811
0.16	0.00630	0.54	0.02126	0.92	0.03622	31	1.22047
0.17	0.00669	0.55	0.02165	0.93	0.03661	32	1.25984
0.18	0.00709	0.56	0.02205	0.94	0.03701	33	1.29921
0.19	0.00748	0.57	0.02244	0.95	0.03740	34	1.33858
0.20	0.00787	0.58	0.02283	0.96	0.03780	35	1.37795
0.21	0.00827	0.59	0.02323	0.97	0.03819	36	1.41732
0.22	0.00866	0.60	0.02362	0.98	0.03858	37	1.45669
0.23	0.00906	0.61	0.02402	0.99	0.03898	38	1.49606
0.24	0.00945	0.62	0.02441	1	0.03937	39	1.53543
0.25	0.00984	0.63	0.02480	2	0.07874	40	1.57484
0.26	0.01024	0.64	0.02520	3	0.11811	41	1.61417
0.27	0.01063	0.65	0.02559	4	0.15748	42	1.65354
0.28	0.01102	0.66	0.02598	5	0.19685	43	1.69291
0.29	0.01142	0.67	0.02638	6	0.23622	44	1.73228
0.30	0.01181	0.68	0.02677	7	0.27559	45	1.77165
0.31	0.01220	0.69	0.02717	8	0.31496	46	1.81102
0.32	0.01260	0.70	0.02756	9	0.35433	47	1.85039
0.33	0.01299	0.71	0.02795	10	0.3937	48	1.88976
0.34	0.01339	0.72	0.02835	11	0.43307	49	1.92913
0.35	0.01378	0.73	0.02874	12	0.47244	50	1.9685
0.36	0.01417	0.74	0.02913	13	0.51181	51	2.00787
0.37	0.01457	0.75	0.02953	14	0.55118	52	2.04724
0.38	0.01496	0.76	0.02992	15	0.59055	53	2.08661

DECIMAL EQUIVALENTS OF MILLIMETRES IN INCHES *(Continued)*

mm	in	mm	in	mm	in	mm	in
54	2.12598	92	3.62204	130	5.1181	168	6.6142
55	2.16535	93	3.66141	131	5.15747	169	6.6535
56	2.20472	94	3.70078	132	5.19684	170	6.6929
57	2.24409	95	3.74015	133	5.23621	171	6.7322
58	2.28346	96	3.77952	134	5.27558	172	6.7716
59	2.32283	97	3.81889	135	5.31495	173	6.8110
60	2.3622	98	3.85826	136	5.35432	174	6.85038
61	2.40157	99	3.89763	137	5.39369	175	6.88975
62	2.44094	100	3.937	138	5.43306	176	6.92912
63	2.48031	101	3.97637	139	5.47243	177	6.96849
64	2.51968	102	4.01574	140	5.5118	178	7.00786
65	2.55905	103	4.05511	141	5.55117	179	7.04723
66	2.59842	104	4.09448	142	5.59054	180	7.0866
67	2.63779	105	4.13385	143	5.62991	181	7.12597
68	2.67716	106	4.17322	144	5.66928	182	7.16534
69	2.71653	107	4.21259	145	5.70865	183	7.20471
70	2.7559	108	4.25196	146	5.74802	184	7.24408
71	2.79527	109	4.29133	147	5.78739	185	7.28345
72	2.83464	110	4.3307	148	5.82676	186	7.32282
73	2.87401	111	4.37007	149	5,86613	187	7.36219
74	2.91338	112	4.40944	150	5.9055	188	7.40156
75	2.95275	113	4.44881	151	5.94487	189	7.44093
76	2.99212	114	4.48818	152	5.98424	190	7.4803
77	3.03149	115	4.52755	153	6.02361	191	7.51967
78	3.07086	116	4.56692	154	6.06298	192	7.55904
79	3.11023	117	4.60629	155	6.10235	193	7.59841
80	3.1496	118	4.64566	156	6.14172	194	7.63778
81	3.18897	119	4.68503	157	6.18119	195	7.67715
82	3.22834	120	4.7244	158	6.22046	196	7.71652
83	3.2677	121	4.76377	159	6.25983	197	7.75589
84	3.30708	122	4.80314	160	6.2992	198	7.79526
85	3.34645	123	4.84251	161	6.33857	199	7.83463
86	3.38582	124	4.88188	162	6.37794	200	7.874
87	3.42519	125	4.92125	163	6.41731		
88	3.46456	126	4.96062	164	6.45668		
89	3.50395	127	5.00000	165	6.49605		
90	3.5433	128	5.03936	166	6.53542		
91	3.58267	129	5.07873	167	6.57479		

TABLE OF DECIMAL EQUIVALENTS

Fraction	Decimal	Fraction	Decimal
1/64	.015625	33/64	.515625
1/32	.03125	17/32	.53125
3/64	.046875	35/64	.546875
1/16	.0625	9/16	.5625
5/64	.078125	37/64	.578125
3/32	.09375	19/32	.59375
7/64	.109375	39/64	.609375
1/8	.1250	5/8	.6250
9/64	.140625	41/64	.640625
5/32	.15625	21/32	.65625
11/64	.171875	43/64	.671875
3/16	.1875	11/16	.6875
13/64	.203125	45/64	.703125
7/32	.21875	23/32	.71875
15/64	.234375	47/64	.734375
1/4	.2500	3/4	.7500
17/64	.265625	49/64	.765625
9/32	.28125	25/32	.78125
19/64	.296875	51/64	.796875
5/16	.3125	13/16	.8125
21/64	.328125	53/64	.828125
11/32	.34375	27/32	.84375
23/64	.359375	55/64	.859375
3/8	.375	7/8	.8750
25/64	.390625	57/64	.890625
13/32	.40625	29/32	.90625
27/64	.421875	59/64	.921875
7/16	.4375	15/16	.9375
29/64	.453125	61/64	.953125
15/32	.46875	31/32	.96875
31/64	.484375	63/64	.984375
1/2	.5000	1	1.0000

COMPARISON OF THERMOMETERS

The temperature scales most generally in use are the Celsius (Centigrade) scale, which is widely used on the Continent and to an increasing degree in the UK, and the Fahrenheit scale, which is still widely used in many English-speaking countries. The Kelvin scale is also widely used in scientific work.

On the Celsius scale, the freezing point of water is at 0° and the boiling point at 100°, at atmospheric pressure.

On the Fahrenheit scale, the freezing point of water is at 32° and the boiling point at 212°.

On the Kelvin scale, the freezing point of water is at 273·15° and boiling point is at 373·15°, the intervals on the scale being identical with those on the Celsius scale.

For converting a temperature on one scale to another, the following formulae apply:

$$\text{degrees F} = \frac{9 \times \text{degrees C}}{5} + 32$$

$$\text{degrees C} = \frac{5 \times (\text{degrees F} - 32)}{9}$$

$$\text{degrees K} = 273{\cdot}15 + \text{degrees C}$$

Absolute zero refers to that point on the thermometer scale, theoretically determined, where a lower temperature is inconceivable. This is located at − 273·15°C (= 0K) or − 459·7°F.

*TEMPERATURE CONVERSION TABLES

Albert Sauveur type of table. Values revised.

−459.4 to 6

C	C F	F	C	C F	F
−273	−459.4		−129	−200	−328
−268	−450		−123	−190	−310
−262	−440		−118	−180	−292
−257	−430				
−251	−420		−112	−170	−274
			−107	−160	−256
−246	−410		−101	−150	−238
−240	−400		− 96	−140	−220
−234	−390		− 90	−130	−202
−229	−380				
−223	−370		− 84	−120	−184
			− 79	−110	−166
−218	−360		− 73	−100	−148
−212	−350		− 68	− 90	−130
−207	−340		− 62	− 80	−112
−201	−330				
−196	−320		− 57	− 70	− 94
			− 51	− 60	− 76
−190	−310		− 46	− 50	− 58
−184	−300		− 40	− 40	− 40
−179	−290		− 34	− 30	− 22
−173	−280				
−169	−273	−459.4	− 29	− 20	− 4
			− 23	− 10	14
−168	−270	−454	− 17.8	0	32
−162	−260	−436	− 17.2	1	33.8
−157	−250	−418	− 16.7	2	35.6
−151	−240	−400	− 16.1	3	37.4
−146	−230	−382	− 15.6	4	39.2
−140	−220	−364	− 15.0	5	41.0
−134	−210	−346	− 14.4	6	42.8

Look up reading in middle column. If in degrees Centigrade, read Fahrenheit equivalent in right-hand column; if in Fahrenheit degrees, read Centigrade equivalent in left-hand column.

*Reproduced by courtesy of Edgar Allen & Co. Ltd.

CONTENTS

PREFACE

A programme for the adoption of the International System of units (SI system) by British industry was published by the British Standards Institution in 1967. The target date of this programme is 1975, by which time the greater part of industry should have changed over to the new system. The programme has been split into several sections. Following the initial work on the preparation of standards is the changeover of design and development stages and then the production planning stage, with a view to overall metrication of production by 1975.

Many British Standards have already been published, but other standards in more limited fields are still being issued. Some of the new standards are equivalent to the old ones, but, when this not so, a second choice is recommended which may be adopted until new designs are available to the new specifications. It is recognised by the Institution that recommended sizes will be revised if necessary, in the light of experience in the use of the new system.

The situation is continually changing, but the information in this pocket book was up-to-date in May 1971.

Acknowledgements are due to the following organisations who have provided tables and information: The Carborundum Co. Ltd.; James Booth & Co. Ltd.; The British Aluminium Co. Ltd.; Lee & Crabtree Ltd.; The Barber Colman Co. Ltd.; Crofts (Engineers) Ltd.; R. A. Skelton & Co. Ltd.; Edgar Allen & Co. Ltd.; MacReady's Metal Co. Ltd.; and the British Standards Institution.

First published in 1941 *by George Newnes Ltd.*
Second edition 1944
Third edition 1949
Fourth edition 1958
reprinted 1960
Fifth edition 1964
reprinted 1966, 1967, 1969, 1970
Sixth edition published by Newnes-Butterworth, 1971
reprinted 1975, 1976, 1978

ISBN 0 408 00059 7

Printed in Great Britain by
J. W. Arrowsmith, Bristol

NEWNES ENGINEER'S POCKET BOOK

Revised by
J. L. Nayler, M.A., C.Eng.,
F.R.Ae.S., F.A.I.A.A.

LONDON
NEWNES–BUTTERWORTHS

THE BUTTERWORTH GROUP

UNITED KINGDOM:

BUTTERWORTH & CO. (PUBLISHERS) LTD.
LONDON: 88 Kingsway, WC2B 6AB

AUSTRALIA:

BUTTERWORTHS PTY LTD.
SYDNEY: 586 Pacific Highway, Chatswood, NSW 2067
Also at MELBOURNE, BRISBANE, ADELAIDE and PERTH

CANADA:

BUTTERWORTH & CO. (CANADA) LTD.
TORONTO: 2265 Midland Avenue, Scarborough, Ontario MIP 4S1

NEW ZEALAND:

BUTTERWORTHS OF NEW ZEALAND LTD.
WELLINGTON: T & W Young Building, 77–85 Customhouse Quay 1, CPO 472

SOUTH AFRICA:

BUTTERWORTH & CO. (SOUTH AFRICA) (PTY) LTD.
DURBAN: 152–154 Gale Street

USA:

BUTTERWORTH (PUBLISHERS) INC.
BOSTON: 19 Cummings Park, Woburn, Mass. 01801

NEWNES ENGINEER'S POCKET BOOK

TEMPERATURE CONVERSION TABLES

(continued)

1450 to 1940

C	C F	F	C	C F	F
788	1450	2642	927	1700	3092
793	1460	2660	932	1710	3110
799	1470	2678	938	1720	3128
804	1480	2696	943	1730	3146
810	1490	2714	949	1740	3164
816	1500	2732	954	1750	3182
821	1510	2750	960	1760	3200
827	1520	2768	966	1770	3218
832	1530	2786	971	1780	3236
838	1540	2804	977	1790	3254
843	1550	2822	982	1800	3272
849	1560	2840	988	1810	3290
854	1570	2858	993	1820	3308
860	1580	2876	999	1830	3326
866	1590	2894	1004	1840	3344
871	1600	2912	1010	1850	3362
877	1610	2930	1016	1860	3380
882	1620	2948	1021	1870	3398
888	1630	2966	1027	1880	3416
893	1640	2984	1032	1890	3434
899	1650	3002	1038	1900	3452
904	1660	3020	1043	1910	3470
910	1670	3038	1049	1920	3488
916	1680	3056	1054	1930	3506
921	1690	3074	1060	1940	3524

Look up reading in middle column. If in degrees Centigrade, read Fahrenheit equivalent in right hand column; if in Fahrenheit degrees read Centigrade equivalent in left hand columm.

TEMPERATURE CONVERSION TABLES

(continued)

1950 to 2440

C	C F	F	C	C F	F
1066	1950	3542	1204	2200	3992
1071	1960	3560	1210	2210	4010
1077	1970	3578	1216	2220	4028
1082	1980	3596	1221	2230	4046
1088	1990	3614	1227	2240	4064
1093	2000	3632	1232	2250	4082
1099	2010	3650	1238	2260	4100
1104	2020	3668	1243	2270	4118
1110	2030	3686	1249	2280	4136
1116	2040	3704	1254	2290	4154
1121	2050	3722	1260	2300	4172
1127	2060	3740	1266	2310	4190
1132	2070	3758	1271	2320	4208
1138	2080	3776	1277	2330	4226
1143	2090	3794	1282	2340	4244
1149	2100	3812	1288	2350	4262
1154	2110	3830	1293	2360	4280
1160	2120	3848	1299	2370	4298
1166	2130	3866	1304	2380	4316
1171	2140	3884	1310	2390	4334
1177	2150	3902	1316	2400	4352
1182	2160	3920	1321	2410	4370
1188	2170	3938	1327	2420	4388
1193	2180	3956	1332	2430	4406
1199	2190	3974	1338	2440	4424

Look up reading in middle column. If in degrees Centigrade, read Fahrenheit equivalent in right hand column; if in Fahrenheit degrees, read Centigrade equivalent in left hand column.

TEMPERATURE CONVERSION TABLES

(continued)

2450 to 3000

C	C F	F	C	C F	F
1343	2450	4442	1499	2730	4946
1349	2460	4460	1504	2740	4964
1354	2470	4478	1510	2750	4982
1360	2480	4496	1516	2760	5000
1366	2490	4514	1521	2770	5018
1371	2500	4532	1527	2780	5036
1377	2510	4550	1532	2790	5054
1382	2520	4568	1538	2800	5072
1388	2530	4586	1543	2810	5090
1393	2540	4604	1549	2820	5108
1399	2550	4622	1554	2830	5126
1404	2560	4640	1560	2840	5144
1410	2570	4658	1566	2850	5162
1416	2580	4676	1571	2860	5180
1421	2590	4694	1577	2870	5198
1427	2600	4712	1582	2880	5216
1432	2610	4730	1588	2890	5234
1438	2620	4748	1593	2900	5252
1443	2630	4766	1599	2910	5270
1449	2640	4784	1604	2920	5288
1454	2650	4802	1610	2930	5306
1460	2660	4820	1616	2940	5324
1466	2670	4838	1621	2950	5342
1471	2680	4856	1627	2960	5360
1477	2690	4874	1632	2970	5378
1482	2700	4892	1638	2980	5396
1488	2710	4910	1643	2990	5414
1493	2720	4928	1649	3000	5432

Look up reading in middle column. If in degrees Centigrade, read Fahrenheit equivalent in right hand column; if in Fahrenheit degrees, read Centigrade equivalent in left hand column.

TABLE FOR CONVERTING MINUTES INTO DECIMALS OF A DEGREE

Min.	Dec. of Degree	Min.	Dec. of Degree	Min.	Dec. of Degree	Min.	Dec. of Degree	Min.	Dec. of Degree
¼	0.00416	12¼	0.20416	24¼	0.40416	36¼	0.60416	48¼	0.80416
½	0.00833	12½	0.20833	24½	0.40833	36½	0.60833	48½	0.80833
¾	0.01250	12¾	0.21250	24¾	0.41250	36¾	0.61250	48¾	0.81250
1	0.01666	13	0.21666	25	0.41666	37	0.61666	49	0.81666
1¼	0.02083	13¼	0.22083	25¼	0.42083	37¼	0.62083	49¼	0.82083
1½	0.02500	13½	0.22500	25½	0.42500	37½	0.62500	49½	0.82500
1¾	0.02916	13¾	0.22916	25¾	0.42916	37¾	0.62916	49¾	0.82916
2	0.03333	14	0.23333	26	0.43333	38	0.63333	50	0.83333
2¼	0.03750	14¼	0.23750	26¼	0.43750	38¼	0.63750	50¼	0.83750
2½	0.04166	14½	0.24166	26½	0.44166	38½	0.64166	50½	0.84166
2¾	0.04583	14¾	0.24583	26¾	0.44583	38¾	0.64583	50¾	0.84583
3	0.05000	15	0.25000	27	0.45000	39	0.65000	51	0.85000
3¼	0.05416	15¼	0.25416	27¼	0.45416	39¼	0.65416	51¼	0.85416
3½	0.05833	15½	0.25833	27½	0.45833	39½	0.65833	51½	0.85833
3¾	0.06250	15¾	0.26250	27¾	0.46250	39¾	0.66250	51¾	0.86250
4	0.06666	16	0.26666	28	0.46666	40	0.66666	52	0.86666

From Foote Bros., by permission.

TABLE FOR CONVERTING MINUTES INTO DECIMALS OF A DEGREE—(*Continued*)

Min.	Dec. of Degree	Min.	Dec. of Degree	Min.	Dec. of Degree	Min.	Dec. of Degree	Min.	Dec. of Degree
4¼	0.07083	16¼	0.27083	28¼	0.47083	40¼	0.67083	52¼	0.87083
4½	0.07500	16½	0.27500	28½	0.47500	40½	0.67500	52½	0.87500
4¾	0.07916	16¾	0.27916	28¾	0.47916	40¾	0.67916	52¾	0.87916
5	0.08333	17	0.28333	29	0.48333	41	0.68333	53	0.88333
5¼	0.08750	17¼	0.28750	29¼	0.48750	41¼	0.68750	53¼	0.88750
5½	0.09166	17½	0.29166	29½	0.49166	41½	0.69166	53½	0.89166
5¾	0.09583	17¾	0.29583	29¾	0.49583	41¾	0.69583	53¾	0.89583
6	0.10000	18	0.30000	30	0.50000	42	0.70000	54	0.90000
6¼	0.10416	18¼	0.30416	30¼	0.50416	42¼	0.70416	54¼	0.90416
6½	0.10833	18½	0.30833	30½	0.50833	42½	0.70833	54½	0.90833
6¾	0.11250	18¾	0.31250	30¾	0.51250	42¾	0.71250	54¾	0.91250
7	0.11666	19	0.31666	31	0.51666	43	0.71666	55	0.91666
7¼	0.12083	19¼	0.32083	31¼	0.52083	43¼	0.72083	55¼	0.92083
7½	0.12500	19½	0.32500	31½	0.52500	43½	0.72500	55½	0.92500
7¾	0.12916	19¾	0.32916	31¾	0.52916	43¾	0.72916	55¾	0.92916
8	0.13333	20	0.33333	32	0.53333	44	0.73333	56	0.93333

From Foote Bros., by permission.

TABLE FOR CONVERTING MINUTES INTO DECIMALS OF A DEGREE—(*Continued*)

Min.	Dec. of Degree	Min.	Dec. of Degree	Min.	Dec. of Degree	Min.	Dec. of Degree	Min.	Dec. of Degree
8¼	0.13750	20¼	0.33750	32¼	0.53750	44¼	0.73750	56¼	0.93750
8½	0.14166	20½	0.34166	32½	0.54166	44½	0.74166	56½	0.94166
8¾	0.14583	20¾	0.34583	32¾	0.54583	44¾	0.74583	56¾	0.94583
9	0.15000	21	0.35000	33	0.55000	45	0.75000	57	0.95000
9¼	0.15416	21¼	0.35416	33¼	0.55416	45¼	0.75416	57¼	0.95416
9½	0.15833	21½	0.35833	33½	0.55833	45½	0.75833	57½	0.95833
9¾	0.16250	21¾	0.36250	33¾	0.56250	45¾	0.76250	57¾	0.96250
10	0.16666	22	0.36666	34	0.56666	46	0.76666	58	0.96666
10¼	0.17083	22¼	0.37083	34¼	0.57083	46¼	0.77083	58¼	0.96083
10½	0.17500	22½	0.37500	34½	0.57500	46½	0.77500	58½	0.97500
10¾	0.17916	22¾	0.37916	34¾	0.57916	46¾	0.77916	58¾	0.97916
11	0.18333	23	0.38333	35	0.58333	47	0.78333	59	0.98333
11¼	0.18750	23¼	0.38750	35¼	0.58750	47¼	0.78750	59¼	0.98750
11½	0.19166	23½	0.39166	35½	0.59166	47½	0.79166	59½	0.99166
11¾	0.19583	23¾	0.39583	35¾	0.59583	47¾	0.79583	59¾	0.99583
12	0.20000	24	0.40000	36	0.60000	48	0.80000	60	1.00000

From Foote Bros., by permission.

EQUIVALENT PRESSURES

in Water	*in Mercury*	*lb/in²*	*N/m²*	*in Water*	*in Mercury*	*lb/in²*	*N/m²*
1	0·074	0·036	250	19	1·402	0·684	4750
2	0·148	0·072	500	20	1·476	0·720	5000
3	0·221	0·108	750	21	1·550	0·756	5250
4	0·295	0·144	1000	22	1·624	0·792	5500
5	0·369	0·180	1250	23	1·697	0·828	5750
6	0·443	0·216	1500	24	1·771	0·864	6000
7	0·517	0·252	1750	25	1·845	0·900	6250
8	0·590	0·288	2000	26	1·919	0·936	6500
9	0·664	0·324	2250	27	1·993	0·972	6750
10	0·738	0·360	2500	28	2·066	1·008	7000
11	0·812	0·396	2750	29	2·140	1·044	7250
12	0·886	0·432	3000	30	2·214	1·080	7500
13	0·959	0·468	3250	31	2·288	1·116	7750
14	1·033	0·504	3500	32	2·362	1·152	8000
15	1·107	0·540	3750	33	2·435	1·188	8250
16	1·180	0·576	4000	34	2·509	1·224	8500
17	1·255	0·612	4250	35	2·583	1·260	8750
18	1·328	0·648	4500	36	2·657	1·296	9000

Water

1 gallon = 4·54596 litres
1 ft³ = 62·321 lb = 0·028 tons = 6¼ gallons
head in feet × 0·4325 = pressure in pounds per square inch
tons water × 224 = gallons
1 gallon = 10 lbs
1 atmosphere = 14·7 lb/in² = 101·3 kN/m²
(One atmosphere is usually taken as being 15 lb/in²)

STRESS CONVERSION TABLES

KILOGRAMMES PER SQUARE MILLIMETRE TO TONS PER SQUARE INCH

kgf per sq. mm	0	1	2	3	4	5	6	7	8	9
	TONS PER SQ. INCH.									
0	0·00	0·64	1·27	1·90	2·54	3·17	3·81	4·44	5·08	5·71
10	6·35	6·98	7·62	8·25	8·89	9·52	10·16	10·79	11·43	12·06
20	12·70	13·33	13·97	14·60	15·24	15·87	16·51	17·14	17·78	18·41
30	19·05	19·68	20·32	20·95	21·59	22·22	22·86	23·49	24·13	24·76
40	25·40	26·03	26·67	27·30	27·94	28·57	29·21	29·84	30·48	31·11
50	31·75	32·38	33·02	33·65	34·29	34·92	35·56	36·19	36·83	37·46
60	38·10	38·73	39·37	40·00	40·64	41·27	41·91	42·54	43·18	43·81
70	44·45	45·08	45·72	46·35	46·99	47·62	48·26	48·89	49·53	50·16
80	50·80	51·43	52·07	52·70	53·34	53·97	54·60	55·24	55·88	56·51
90	57·15	57·78	58·42	59·05	59·69	60·33	60·96	61·59	62·23	62·86

1. Kilogrammes per Square Millimetre multiplied by 0·635=Tons per Sq. Inch.

STRESS CONVERSION TABLES—(*Continued*)

TONS PER SQUARE INCH TO KILOGRAMMES PER SQUARE MILLIMETRE

Tons per Sq. Inch	0	1	2	3	4	5	6	7	8	9
	KILOGRAMMES FORCE PER SQ. MILLIMETRE									
0	0·00	1·57	3·15	4·72	6·30	7·87	9·45	11·02	12·60	14·17
10	15·75	17·32	18·90	20·47	22·05	23·62	25·20	26·77	28·35	29·92
20	31·50	33·07	34·65	36·23	37·80	39·38	40·95	42·52	44·10	45·67
30	47·25	48·82	50·40	51·97	53·55	55·12	56·70	58·27	59·84	61·42
40	63·00	64·57	66·15	67·72	69·30	70·87	72·45	74·02	75·60	77·17
50	78·75	80·32	81·90	83·47	85·05	86·62	88·20	89·77	91·35	92·92
60	94·50	96·07	97·65	99·22	100·80	102·37	103·95	105·52	107·10	108·67
70	110·25	111·82	113·40	114·97	116·55	118·12	119·70	121·27	122·85	124·42
80	126·00	127·57	129·14	130·72	132·30	133·87	135·45	137·02	138·59	140·15
90	141·75	143·32	144·90	146·47	148·05	149·62	151·20	152·77	154·35	155·92

2. Tons per Sq. Inch multiplied by 1·575=Kilogrammes per Sq. Millimetre.

PREFERRED NUMBERS

For many purposes a series of preferred numbers has been agreed internationally and adopted for use in engineering. ISO Recommendations R3, R17, R497 and DR557 deal with this subject, which has been summarised in BS 2045. Their field of use has applications such as the grading of length, area etc; they are, for example, used for the diameters of the standard series of metric threads.

Preferred numbers are derived from geometric series having one of the ratios $\sqrt[5]{10}$, $\sqrt[10]{10}$, $\sqrt[20]{10}$, $\sqrt[40]{10}$, $\sqrt[80]{10}$ which are approximately 1·58, 1·26, 1·12, 1·06 and 1·03; they give successive increases in the respective series of 58 %, 26 %, 12 %, 6 % and 3 %. They are designated R5, R10, R20, R40 and R80, thus indicating the particular root of 10 on which the series is based. In addition other series of rounded-off numbers are proposed in BS 2045 and are given in the BS Handbook No. 18.

Typical series for R10 and R20 will be found on p. 119 for the basic thicknesses of sheet and diameters of wire.

WEIGHT AND MASS

In commerce, a kilogramme weight is written kg. Actual weights in kilogrammes are, however, often written as kgf, that is, a force of a kilogramme. Anywhere in space it is the mass of a weight in kilogrammes that counts, that is the weight W divided by the value of g on the earth. g, the acceleration due to the pull of the earth on the body, varies at different places on the earth's surface so that a kilogramme weight also varies in value, but the mass remains the same everywhere whether on the earth or in space. Owing to the flattening of the earth at the poles the value of gravity varies irregularly from 983 cm/s^2 at the poles to 978 cm/s^2 at the equator. A mean value of 981 cm/s^2 is commonly adopted.

ENERGY AND HORSEPOWER

The original unit of the rate of doing work was the horsepower, dating from the time when the horse was the most usual form of external power available for use by man.

ENERGY AND HORSEPOWER (*Continued*)

The unit of horsepower was based on the assumption that a horse could travel $2\frac{1}{2}$ miles per hour for 8 hours a day pulling a load of 150 lb vertically out of a shaft by means of a rope. This is the same as 33 000 lb raised vertically one foot per minute or, in other words, 33 000 foot-pounds per minute; alternatively, 550 foot-pounds (ft.lb) per second. With the coming of electric power with its easy transmission from place to place, the watt has come to be used concurrently, and the kilowatt has now been adopted as the SI unit of power. Its approximate value is given by

$$1 \text{ horsepower} = 0{\cdot}746 \text{ kilowatt.}$$

A kilowatt in SI terms is the force of 1000 newtons acting over a distance of one metre in one second.

The coherent units of energy and power are the joule (J) and watt (W), which are, in symbols,

$$J = N \times m = kg.m^2/s^2$$
$$W = J/s$$

The unit of power is derived from mechanics but the watt used in electricity is identical with it and is not defined independently. The ampere (A) is defined in terms of newtons and metres (see p. 20), followed by the volt (V) from,

$$V \times A = J/s = W$$

The unit of energy should be same throughout mechanics and physics, so that all branches of science share the same coherent unit of energy.

HEAT

Changes in temperature are measured in degrees Kelvin (K) where 1K = 1°C, that is the same temperature interval as in the Celsius (Centigrade) scale, but the zero is different. The Kelvin temperature is equal to the Celsius temperature plus 273·15*. The linear expansion coefficient is the reciprocal of the temperature, that is, unity divided by temperature in degrees K (1/K).

* Boiling point of water is 373·15K = 100°C
Freezing point of water is 273·15K = 0°C
also, 0K = −273·15°C

HEAT *(Continued)*

The quantity of heat is measured in joules (J) and the heat flow rate in watts (W). From these three base SI units a number of other quantities are defined as follows:

Heat capacity or entropy is the ratio of joules to kelvins (J/K) and specific heat capacity is the heat capacity per kilogramme (J/kgK). Specific latent heat or specific energy is the ratio of joules to kilogrammes (J/kg).

Density of heat flow rate is measured as watts per square metre (W/m^2). The coefficient of heat transfer is this ratio divided by K (W/m^2K). Thermal conductivity is measured as watts per metre per kelvin (W/mK).

TIME

1 sidereal second = 0·99727 second (mean solar)
1 second (mean solar) = 1·002738 sidereal second
Length of seconds pendulum latitude 45°
= 0·993555 metres.

VELOCITY

Velocity of wireless waves and light	= 299 800 kilometres per second
Velocity of earth	= 28 956 kilometres per second
Velocity of sound (air at 0°C)	= 331·82 metres per second
Velocity of sound (water)	= 1440·62 metres per second
Velocity of sound (brass)	= 3497·96 metres per second
Velocity of sound (cast steel)	= 4981·85 metres per second

ELECTRICAL EQUATIONS

Amperes × volts	= watts	A.V	= W
Amperes × ohms	= volts	A.Ω	= V*
Joules ÷ seconds	= watts	J/s	= W
Coulombs ÷ seconds	= amperes	C/s	= A

* This is Ohm's law for direct current.

ELECTRICAL EQUATIONS (Continued)

Coulombs ÷ volts	= farads	C/V	= F
Joules ÷ coulombs	= volts	J/C	= V
Webers ÷ second	= volts	Wb/s	= V
Webers ÷ square metre	= teslas	Wb/m^2	= T
Volt-seconds ÷ ampere	= henrys	Vs/A	= H

OHM'S LAW FOR D.C. AND A.C.

This is a law that gives the relations existing in any complete circuit between current, voltage and resistance; namely, current = voltage/resistance. If I represents current, V the applied voltage for direct current (d.c.) and R the resistance then

$$I = V/R$$

In the three following cases for alternating current, V is the root-mean-square (see below) of the applied voltage:

(a) Alternating current for a purely inductive circuit.

reactance $$X_L = 2\pi fL$$

where f is the frequency and L is the inductance in henrys and $I = V/X_L$, lagging by a quarter of a cycle.

(b) Alternating current for a purely capacitive circuit.

reactance $$X_C = 1/2\pi fC$$

where f is the frequency and C is the capacitance in farads and $I = V/X_C$, leading by a quarter of a cycle.

(c) Alternating current for a circuit consisting of R, X_L and X_C in series.

impedance $$Z = \sqrt{\{R^2 + (X_L - X_C)^2\}}$$

and $I = V/Z$ with a phase difference ϕ, where $\tan \phi = (X_L - X_C)/R$. If $\tan \phi$ is positive, the current lags behind the applied voltage by that angle.

ROOT-MEAN-SQUARE VALUE

To obtain an average measure of a series of differences, it is usual to take, instead of the arithmetical mean, the

ROOT-MEAN-SQUARE VALUE *(Continued)*

square root of the sum of all the squares of the differences divided by the number of the differences. The root-mean-square emphasises the effect of the larger differences. Between two sets of numbers $x_1\, x_2, \ldots x_n$, and $y_1\, y_2 \ldots y_n$, it is

$$\left\{\frac{1}{n}\sum_{i=1}^{n}(x_i - y_i)^2\right\}^{1/2}$$

ELECTRIC FLEX: COLOUR CODES

All new electrical equipment supplied by retailers must now be fitted with flexible cord that conforms to the new international colour code: i.e.

live—brown
neutral—blue
earth—green/yellow

The regulations do not apply to flex sold over the counter or to appliances fitted with two-core cords.

ELECTRICITY SUPPLY

Electric power is supplied and distributed in the form of a 3-phase supply. The windings of the 3-phase alternator generating the power can be connected either in star (Y) or delta (Δ), as shown below.

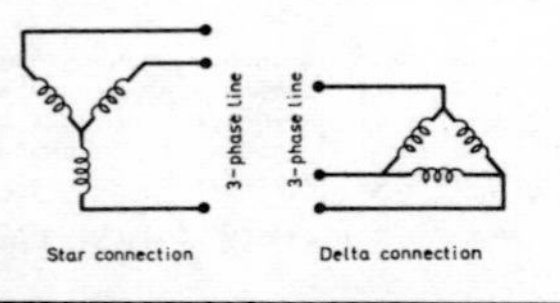

The star connection is often preferred because the neutral point at the centre of the star is often very useful. In this connection the line voltage equals $\sqrt{3}$ times the phase voltage. In the delta connection, the line and phase voltages are equal.

If the star connection is used and the neutral point is left unconnected, this is a "three-wire system". If the neutral point of the generator is connected to the neutral point of the load, it is then a "four-wire system."

Units of power in common use are the kilowatt (kW) $= 10^3$W, the megawatt (MW) $= 10^6$W, and the sub-multiples milliwatt (mW) and microwatt (μW). The relationships between the watt and other common units are as follows:

$$1\text{W} = 1\text{J/s} = 10^7\text{erg/s}$$
$$1\text{hp} = 550\text{ft lb/s} = 746\text{W}$$

The standard unit of electrical energy is the kilowatt-hour (kWh), which is equal to $3{\cdot}6 \times 10^6$ watt-seconds or joules.

In a single-phase alternating circuit, the mean value of the instantaneous power is $\frac{1}{2}VI\cos\phi$, where V is the voltage in volts, I is the current in amperes and ϕ is the phase angle. For three-phase alternating current, the instantaneous power is $\frac{3}{2}VI\cos\phi$. With the same notation the instantaneous voltage and current are given by the equations $v = V\sin\omega t$ and $i = I\sin(\omega t - \phi)$, where the frequency f is given by $\omega/2\pi$.

METRIC GAUGE BLOCKS

With the introduction of metrication, measurements of metric lengths will be required. BS 4311: 1968 specifies the requirements for gauge blocks of rectangular form in metric sizes up to 100 mm. For lengths greater than 100 mm the use of length bars is recommended as specified in BS 1790: *Length Bars and their Accessories.*

BS 4311 specifies five grades designated as AA, AA-Special, A, B and C with tolerances as regards flatness and parallelism. The grade most commonly used in the production of components, tools and gauges is grade B, and that for preliminary setting up with greater tolerances is grade C. Grade A is for inspection work and grade AA for calibrating other gauge blocks and for checking gauges in a comparator where the tolerance is one or two micrometres or less.

For checking actual distances the gauges are wrung together. Each gauge block has a rectangular section with a face width of $9^{+0}_{-0\cdot1}$ mm and a face length of 30 ± 0·5 mm for sizes up to and including 10 mm and 35 ± 0·5 mm for larger sizes. There are two size series based on 2 mm base and on 1 mm base. The range of the blocks is from 1·005 mm up to 100 mm, and various sets are recommended in this standard such as set No. M 47.

Range	Steps	Pieces
1·005	0·005	1
2·01 – 2·09	0·01	9
2·10 – 2·90	0·1	9
1 – 24	1·0	24
40, 60, 80, 100	20	4
		47

Usually the wringing together of five blocks or less will give any required length. As an example, using the M 47 set, the building up of 58·345 mm with five gauges is as follows: 1·005, 2·04, 2·30, 13 and 40. Except where single gauges can be used, only sizes above 4 mm can be built up if digits in tenths and hundredths are involved.

SINE BARS

The 5 in sine bar is being replaced by sine bars in sizes of 100 mm, 200 mm and 250 mm. The designating sizes represent the centre distances between the axes of the rollers. Three types of sine bar are illustrated diagrammatically in BS 3064, which also states the main desirable accuracies of the various parts as follows:

Upper and lower surfaces:
 0·0013 mm for 100 mm bars
 0·0025 mm for 200 mm and 250 mm bars

Side faces, (for all three):
 Flat within 0·005 mm
 Square to the upper surface of the bar within 0·0025 mm per 25 mm
 Square to the axis of the hinge roller 0·013 per 25 mm

End faces:
 Flat within 0·0025 mm
 Square to the upper surface of the bar within 0·0025 mm per 25 mm
 Parallel to the axis of the hinge roller within 0·013 mm per 25 mm

The natural sine tables (pp. 401–405) multiplied by 100, 200 and 250, respectively, give the required setting constants for the sine bars.

MACHINE-TOOL ADAPTATION FOR METRICATION

In 1969 the Ministry of Technology published a booklet describing some of the factors to be considered by machine-tool users in adapting machine tools for metric production, which is based on an article by the Director of the Machine Tools Industry Research Association. It deals with the problems of making newly designed metric components on existing inch machine tools and of making components to existing inch designs on new metric machine tools. Six figures in the form of outline drawings illustrate what can be done in the conversion of existing tools and the fitting of inch-metric feed-screw indicators.

MACHINE-TOOL ADAPTATION FOR METRICATION *(Continued)*

Feed Screws and Dials—The conversion of the feed mechanisms can be done either by replacing the feed screw and nut assemblies with units of metric pitch or by replacing the inch-reading indicator by one engraved in metric units. The former may be worth while if the machine is of fairly recent design with a considerable life before it. The latter, the conversion of inch-reading feed indicators to metric readings, can be done by replacing the indicator by a metric dial or adding a new dial alongside the old, standard units of this kind being available; by dividing one revolution of the handwheel into a whole number of millimetres, an approximate method; by fitting a dual-dial indicator, one fixed to the screw and giving the usual inch indication and the other, driven from the first, indicating in millimetres with the gearing chosen so that the inch-metric conversion is exact. The last, dual-dial units, are supplied with some machines.

Optical Position Indicators—Some makes of precision machine tools are being supplied with dual-reading optical indicators as standard equipment. Alternatively, all metric dimensions will need to be converted to inch units or the Imperial optical system replaced with a metric one.

Digital Read-out Equipment—(a) Digital read-out equipment, increasingly in use on machine tools for indicating the position of the tools or workpiece, can obtain a conversion facility giving the read-out in either inch or metric units. The fitting of such equipment may be preferable to changing the feed screws, etc.

(b) For coarse adjustment of tool or workpiece, such as depth indicators on drilling machines and taper indicator plates on cylindrical grinding machines, it is probably worth while fitting replacement plates to show either metric or inch/metric values.

(c) For indicating the feed rates of the cutting tool or workpiece, as on pillar and radial drilling machines, it is not usually necessary to modify the spindle feed arrangement, and the necessary conversion is merely the direct one of changing the information on the feed indicator plate from inch to metric units.

MACHINE-TOOL ADAPTATION FOR METRICATION *(Continued)*

It should be noted that when the speed rate is dependent on the pitch of the hand-feed screw, as in milling machines, the change to a screw of metric pitch alters the range of feed rates.

MACHINE-TOOL CONVERSION

(a) **Screw-cutting Lathes**—Modifications to the screw-cutting facilities of screw-cutting lathes is not advised. If a new metric lathe has to be purchased, the range and accuracy of its screw-cutting facilities must be carefully looked into. Providing a thread pitch indicator for a screw-cutting lathe to metric specification is a problem because, although standard on British lathes, the indicator is difficult to design for use with a metric feed screw: such an indicator is available as extra equipment on continental machines. Without an indicator, the manufacture of screw threads is a more lengthy and expensive operation.

(b) **Numerically Controlled Machine Tools**—In the case of numerically controlled machine tools, it is advisable to carry out conversions of the dimensional information at the programming stage. The conversions can be done manually or by computer according to the method of preparation of the tape.

(c) **Toolholders**—Many features of machine tool design have been standardised at international level, and are identical in Imperial and metric sizes, but in the change to metrication modifications of critical dimensions may take place. Thus a watch must be kept as regards existing tools to be used on new machines or their replacement. This applies to the thread on which chucks are to be mounted, to tooling mounted on arbors, to grinding wheels etc.

(d) **Measuring Equipment**—Metric micrometers, verniers, depth gauges, scales, etc. should be available. Dual-reading inch-metric scales, dial gauges and micrometers are available.

(e) **Standards-room Equipment**—Length-measuring instruments and comparators with dual-reading scales should be available. Angular-displacement measuring

MACHINE-TOOL ADAPTATION FOR METRICATION *(Continued)*

equipment can readily provide results in inch or metric units. Arithmetical conversion techniques can be used with existing Imperial equipment.

INCH-METRIC FEED-SCREW INDICATORS

Single-dial System—Equivalent inch and metric values can be marked on one dial, but a limitation arises on account of the exact internationally agreed conversion factor of 1 in = 25·4 mm. A single rotation of the hand-wheel of a 5 t.p.i. lead screw produces 5·08 mm axial movement and distances of more than this figure would be susceptible to error.

Dual-dial Systems—Several dual-dial systems are available including (a) belt systems, (b) simple systems with auxiliary shafts, (c) systems with single internal gearing, (d) systems with additional internal gearing, (e) connected-ring type with both inch and metric dials fixed to internal gear rings and (f) planetary-gear types with the sun (central) wheel fixed to the frame of the machine whilst the arm is fixed to the handwheel and rotates with the lead screw.

SCREW THREADS

Two complete systems for screw threads and fasteners will become standardised under the metrication programme. The ISO inch system is the unified screw thread and the associated range of unified fasteners, as manufactured in Britain and identical in all respects to the fasteners in use in the USA and Canada and wherever American practice is followed. The other internationally recognised screw thread is the ISO metric thread with its associated range of fasteners, now being introduced in the leading metric countries, which will completely eliminate the various metric standards that differ from it. The ISO standard dimensions and tolerances should be substituted for out-dated practices. The two systems have the same basic thread form; the selection of pitches in relation to particular diameters is very close to unified practice, metric coarse corresponding to unified coarse and metric fine corresponding approximately to unified fine; the tolerances in the two systems are identical. Standard fasteners will in future use coarse threads and only special applications will have fine threads (BS 4168, 4183 and 3692).

British screw threads, Whitworth, BSF and BA, become obsolete for fasteners in 1975, but the Whitworth thread will still be used for pipe threads. Although the basic thread forms of the ISO inch (unified) and ISO metric are similar, the threads are *not* interchangeable because the diameter/pitch combinations are not compatible. The general plan for screw threads is given in the following table; pipe threads are dealt with separately (see p. 75).

ISO METRIC SCREW THREADS

GENERAL PLAN SHOWING THREE CHOICES* UP TO 56mm DIAMETER

(All dimensions in millimetres)

Nominal Diameter			*Coarse Series*	*Fine Series*	*Constant-pitch Series*
First Choice	*Second Choice*	*Third Choice*			
1·6			0·35		
	1·8		0·35		
2			0·4		
	2·2		0·45		
2·5			0·45		0·35
3			0·5		0·35
	3·5		0·6		0·35
4			0·7		0·5
	4·5		0·75		0·5
5			0·8		0·5
		5·5			0·5
6			1		0·75
		7	1		0·75
8†			1·25	1	1, 0·75
		9	1·25		1, 0·75
10			1·5	1·25	1·25, 1, 0·75
		11	1·5		1, 0·75
12			1·75	1·25	1·5, 1·25, 1
	14		2	1·5	1·5, 1·25,‡ 1
		15			1·5, 1
16			2	1·5	1·5, 1

* Only first choice given in succeeding tables. The general plan continues to 300 mm.
† Below 8 mm, constant-pitch series is used for all fine threads. ‡ Only used for spark plugs for engines.

ISO METRIC SCREW THREADS (*Continued*)

Nominal Diameter			*Coarse Series*	*Fine Series*	*Constant-pitch Series*
First Choice	*Second Choice*	*Third Choice*			
		17			1·5, 1
	18		2·5	1·5	2, 1·5, 1
20			2·5	1·5	2, 1·5, 1
	22		2·5	1·5	2, 1·5, 1
24			3	2	2, 1·5, 1
		25			2, 1·5, 1
		26			1·5
	27		3	2	2, 1·5, 1
30			3·5	2	2, 1·5, 1
		32			2, 1·5
	33		3·5	2	2, 1·5
		35§			
36			4	3	3, 2, 1·5
		38			1·5
	39		4	3	3, 2, 1·5
		40			3, 2, 1·5
42			4·5	3	4, 3, 2, 1·5
	45		4·5	3	4, 3, 2, 1·5
48			5	3	4, 3, 2, 1·5
		50			3, 2, 1·5
	52		5	3	4, 3, 2, 1·5
		55			4, 3, 2, 1·5
56			5·5	4	4, 3, 2, 1·5

§ Used only for locking nuts on bearings.

BASIC FORM OF SCREW THREADS

The basic form of unified threads (BS 1580) and ISO metric threads (BS 3643) is identical with the form below, where p is the pitch of the thread and H is the depth.

$$p = \text{pitch} = \frac{1}{\text{no. of threads per in}}$$

a = depth of thread = $0{\cdot}866p$.
c = width of thread at top = $0{\cdot}335p$.
d = width of thread at bottom = $0{\cdot}310p$.
e = thickness of thread at pitch line = $0{\cdot}5p$.
f = depth of thread to pitch line = $0{\cdot}3183p$.

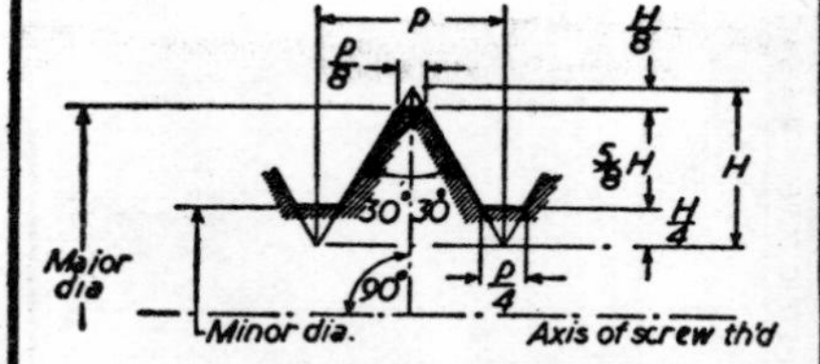

The basic form of Unified Threads.
For tables, see page 61.

BASIC DIMENSIONS FOR ISO INCH (UNIFIED) COARSE THREADS

Nom. Size	*Threads per in*	*Major Diameter*	*Effective Diameter*	*Minor Diameter External Threads*	*Minor Diameter Internal Threads*
$\frac{1}{4}$	20	0·250	0·2175	0·1887	0·1959
$\frac{5}{16}$	18	0·3125	0·2764	0·2443	0·2524
$\frac{3}{8}$	16	0·375	0·3344	0·2983	0·3073
$\frac{7}{16}$	14	0·4375	0·3911	0·3499	0·3602
$\frac{1}{2}$	13	0·500	0·450	0·4056	0·4167
$\frac{9}{16}$	12	0·5625	0·5084	0·4603	0·4723
$\frac{5}{8}$	11	0·625	0·5660	0·5135	0·5266
$\frac{3}{4}$	10	0·750	0·6850	0·6273	0·6417
$\frac{7}{8}$	9	0·875	0·8028	0·7387	0·7547
1	8	1·000	0·9188	0·8466	0·8647
$1\frac{1}{8}$	7	1·125	1·0322	0·9497	0·9704
$1\frac{1}{4}$	7	1·250	1·1572	1·0747	1·0954
$1\frac{3}{8}$	6	1·375	1·2667	1·1705	1·1946
$1\frac{1}{2}$	6	1·500	1·3917	1·2955	1·3196
$1\frac{3}{4}$	5	1·750	1·6201	1·5046	1·5335
2	$4\frac{1}{2}$	2·000	1·8557	1·7274	1·7594
$2\frac{1}{4}$	$4\frac{1}{2}$	2·250	2·1057	1·9774	2·0094
$2\frac{1}{2}$	4	2·500	2·3376	2·1933	2·2294
$2\frac{3}{4}$	4	2·750	2·5876	2·4433	2·4794
3	4	3·000	2·8376	2·6933	2·7294
$3\frac{1}{4}$	4	3·250	3·0876	2·9433	2·9794
$3\frac{1}{2}$	4	3·500	3·3376	3·1933	3·2294
$3\frac{3}{4}$	4	3·750	3·5876	3·4433	3·4794
4	4	4·000	3·8376	3·6933	3·7294

BASIC DIMENSIONS FOR ISO INCH (UNIFIED) FINE THREADS

Nom. Size	*Threads per in*	*Major Diameter*	*Effective Diameter*	*Minor Diameter External Threads*	*Minor Diameter Internal Threads*
$\frac{1}{4}$	28	0·250	0·2268	0·2062	0·2113
$\frac{5}{16}$	24	0·3125	0·2854	0·2614	0·2674
$\frac{3}{8}$	24	0·375	0·3479	0·3239	0·3299
$\frac{7}{16}$	20	0·4375	0·4050	0·3762	0·3834
$\frac{1}{2}$	20	0·500	0·4675	0·4387	0·4459
$\frac{9}{16}$	18	0·5625	0·5264	0·4943	0·5024
$\frac{5}{8}$	18	0·625	0·5889	0·5568	0·5649
$\frac{3}{4}$	16	0·750	0·7094	0·6733	0·6823
$\frac{7}{8}$	14	0·875	0·8286	0·7874	0·7977
1	12	1·0000	0·9459	0·8978	0·9098
$1\frac{1}{8}$	12	1·1250	1·0709	1·0228	1·0348
$1\frac{1}{4}$	12	1·250	1·1959	1·1478	1·1598
$1\frac{3}{8}$	12	1·3750	1·3209	1·2728	1·2848
$1\frac{1}{2}$	12	1·500	1·4459	1·3978	1·4098

ISO METRIC HEXAGON BLACK BOLTS AND SCREWS: PREFERRED SIZES

(All dimensions in millimetres)

Nominal Size and Thread Diameter	*Pitch of Thread: Coarse Pitch Series*	*Shank Diameter*		*Width Across Flats*		*Width Across Corners*		*Head Height*	
		max.	*min.*	*max.*	*min.*	*max.*	*min.*	*max.*	*min.*
M5	0·8	5·48	4·52	8·00	7·64	9·2	8·63	3·88	3·13
M6	1	6·48	5·52	10·00	9·64	11·5	10·89	4·38	3·63
M8	1·25	8·58	7·42	13·00	12·57	15·0	14·20	5·88	5·13
M10	1·5	10·58	9·42	17·00	16·57	19·6	18·72	7·45	6·55
M12	1·75	12·70	11·30	19·00	18·48	21·9	20·88	8·45	7·55
M16	2	16·70	15·30	24·00	23·16	27·7	26·17	10·45	9·55
M20	2·5	20·84	19·16	30·00	29·16	34·6	32·95	13·90	12·10
M24	3	24·84	23·16	36·00	35·00	41·6	39·55	15·90	14·10
M30	3·5	30·84	29·16	46·00	45·00	53·1	50·85	20·05	17·95
M36	4	37·00	35·00	55·00	53·80	63·5	60·79	24·05	21·95
M42	4·5	43·00	41·00	65·00	63·80	75·1	72·09	27·05	24·95
M48	5	49·00	47·00	75·00	73·80	86·6	83·39	31·05	28·95
M56	5·5	57·20	54·80	85·00	83·60	98·1	94·47	36·25	33·75
M64	6	65·20	62·80	95·00	93·60	109·7	105·77	41·25	38·75

Further details of black hexagon bolts and screws including non-preferred sizes are given in BS 4190:1967.

ISO METRIC HEXAGON BLACK NUTS AND HEXAGON THIN NUTS: PREFERRED SIZES

(All dimensions in millimetres)

Nominal Size and Thread Diameter	*Pitch of Thread: Coarse Pitch Series*	*Width Across Flats*		*Width Across Corners*		*Thickness*			
						Black Nuts		*Thin Nuts (faced both sides)*	
		max.	*min.*	*max.*	*min.*	*max.*	*min.*	*max.*	*min.*
M5	0·8	8·00	7·64	9·2	8·63	4·38	3·63	—	—
M6	1	10·00	9·64	11·5	10·89	5·38	4·63	—	—
M8	1·25	13·00	12·57	15·0	14·20	6·88	6·13	5·00	4·52
M10	1·5	17·00	16·57	19·6	18·72	8·45	7·55	6·00	5·52
M12	1·75	19·00	18·48	21·9	20·88	10·45	9·55	7·00	6·42
M16	2	24·00	23·16	27·7	26·17	13·55	12·45	9·00	8·42
M20	2·5	30·00	29·16	34·6	32·95	16·55	15·45	9·00	8·42
M24	3	36·00	35·00	41·6	39·55	19·65	18·35	10·00	9·42
M30	3·5	46·00	45·00	53·1	50·85	24·65	23·35	12·00	11·30
M36	4	55·00	53·80	63·5	60·79	29·65	28·35	14·00	13·30
M42	4·5	65·00	63·80	75·1	72·09	34·80	33·20	16·00	15·30
M48	5	75·00	73·80	86·6	83·39	38·80	37·20	18·00	17·30
M56	5·5	85·00	83·60	98·1	94·47	45·80	44·20	—	—
M64	6	95·00	93·60	109·7	105·77	51·95	50·05	—	—

Further details of hexagon black nuts and hexagon thin nuts including non-preferred sizes are given in BS 4190:1967.

ISO METRIC PRECISION HEXAGON BOLTS, SOME USFEUL DIMENSIONS: COARSE THREADS

(All dimensions in millimetres)

Nominal Size and Thread Diameter	*Pitch of Thread*	*Major Diameter*	*Effective Diameter (Bolt)*	*Minor Diameter (external)*	*Minor Diameter (internal)*	*Hexagons**	
						Across Flats max.	*Across Corners min.*
M1·6	0·35	1·60	1·37	1·17	1·22	3·20	3·70
M2	0·4	2·0	1·74	1·50	1·56	4·00	4·60
M2·5	0·45	2·50	2·21	1·95	2·01	5·00	5·80
M3	0·5	3·00	2·68	2·39	2·46	5·50	6·40
M4	0·7	4·00	3·55	3·14	3·24	7·00	8·10
M5	0·8	5·00	4·48	4·02	4·13	8·00	9·20
M6	1·0	6·00	5·36	4·77	4·92	10·00	11·50
M8	1·25	8·00	7·19	6·47	6·65	13·00	15·00
M10	1·5	10·00	9·03	8·16	8·38	17·00	19·60
M12	1·75	12·00	10·86	9·85	10·11	19·00	21·90
(M14)	2	14·00	12·70	11·55	11·84	22·00	25·4
M16	2	16·00	14·70	13·55	13·84	24·00	27·7
(M18)	2·5	18·00	16·38	14·93	15·29	27·00	31·2
M20	2·5	20·00	18·38	16·93	17·29	30·00	34·6
(M22)	2·5	22·00	20·38	18·93	19·29	32·00	36·9

ISO METRIC PRECISION HEXAGON BOLTS, SOME USEFUL DIMENSIONS: COARSE THREADS (Continued)

Nominal Size and Thread Diameter	*Pitch of Thread*	*Major Diameter*	*Effective Diameter (Bolt)*	*Minor Diameter (external)*	*Minor Diameter (internal)*	*Hexagons**	
						Across Flats max.	*Across Corners min.*
M24	3	24·00	22·05	20·32	20·75	36·00	41·6
(M27)	3	27·00	25·05	23·32	23·75	41·00	47·3
M30	3·5	30·00	27·73	25·71	26·21	46·00	53·1
(M33)	3·5	33·00	30·73	28·71	29·21	50·00	57·7
M36	4	36·00	33·40	31·09	31·67	55·00	63·5
(M39)	4	39·00	36·40	34·09	34·67	60·00	69·3
M42	4·5	42·00	39·08	36·48	37·13	65·00	75·1
(M45)	4·5	45·00	42·08	39·48	40·13	70·00	80·8
M48	5	48·00	44·75	41·87	42·59	75·00	86·6
(M52)	5	52·00	48·75	45·87	46·59	80·00	92·4

Sizes in brackets, e.g. (M14) are non-preferred.

* For bolt heads or nuts.

For further information see BS 3692. For nut threads and bolt threads, coarse thread series, medium quality, see R 965/11, February 1969.

ISO METRIC PRECISION HEXAGON BOLTS, SOME USEFUL DIMENSIONS: FINE THREADS

(All dimensions in millimetres)

Nominal Size and Thread Diameter First Choice	*Pitch of Thread*	*Major Diameter*	*Effective Diameter*	*Minor Diameter (external)*	*Minor Diameter (internal)*	*Hexagon**	
						Across Flats max.	*Across Corners max.*
M8	1	8·00	7·35	6·77	6·92	13·00	15·0
M10	1·25	10·00	9·19	8·47	8·65	17·00	19·6
M12	1·25	12·00	11·19	10·47	10·65	19·00	21·9
M16	1·5	16·00	15·03	14·16	14·38	24·00	27·7
M20	1·5	20·00	19·03	18·16	18·38	30·00	34·6
M24	2	24·00	22·70	21·55	21·84	36·00	41·6
M30	2	30·00	28·70	27·55	27·84	46·00	53·1
M36	3	36·00	34·05	32·32	32·75	55·00	63·5
M42	3	42·00	40·05	38·32	38·75		
M48	3	48·00	46·05	44·32	44·75		
M56	4	56·00	53·40	51·09	51·67		
M64	4	64·00	61·40	59·09	59·67		

* Bolts and nuts.

For further information see BS 3692. For nut and bolt threads fine-thread series, medium quality, see R 965/11, February 1969.

ISO METRIC SCREW THREADS: CONSTANT PITCH SERIES

(All dimensions in millimetres)

Nominal Size and Thread Diameter; First Choice	*Pitch of Thread*	*Major Diameter*	*Effective Diameter*	*Minor Diameter (external)*	*Minor Diameter (internal)*
Pitches 0·35, 0·5 and 0·75 mm					
M 2·5	0·35	2·50	2·27	2·07	2·12
3	0·35	3·00	2·77	2·57	2·62
4	0·5	4·00	3·68	3·39	3·46
6	0·75	6·00	5·51	5·08	5·19
8	0·75	8·00	7·51	7·08	7·19
10	0·75	10·00	9·51	9·08	9·19
Pitches 1 and 1·25 mm					
8	1	8·00	7·35	6·77	6·92
10	1·25	10·00	9·19	8·47	8·65
10	1	10·00	9·35	8·77	8·92
12	1·25	12·00	11·19	10·47	10·65
12	1	12·00	11·35	10·77	10·92
16	1	16·00	15·35	14·77	14·92

ISO METRIC SCREW THREADS CONSTANT PITCH SERIES *(Continued)*

(All dimensions in millimetres)

Pitches 1 *and* 1·25 mm

Nominal Size and Thread Diameter: First Choice	*Pitch of Thread*	*Major Diameter*	*Effective Diameter*	*Minor Diameter (external)*	*Minor Diameter (internal)*
20	1	20·00	19·35	18·77	18·92
24	1	24·00	23·35	22·77	22·92
30	1	30·00	29·35	28·77	28·92
14 *	1·25	14·00	13·19	12·47	12·65
Pitch 1·5 mm					
12	1·5	12·00	11·03	10·16	10·38
16	1·5	16·00	15·03	14·16	14·38
20	1·5	20·00	19·03	18·16	18·38
24	1·5	24·00	23·03	22·16	22·38
30	1·5	30·00	29·03	28·16	28·38
36	1·5	36·00	35·03	34·16	34·38
42	1·5	42·00	41·03	40·16	40·38
48	1·5	48·00	47·03	46·16	46·38
56	1·5	56·00	55·03	54·16	54·38

* This size is for spark plugs only. For further information see BS 3643.

ISO METRIC SCREW THREADS: CONSTANT PITCH SERIES (Continued)

(All dimensions in millimetres)

Nominal Size and Thread Diameter: First Choice	*Major Diameter*	*Effective Diameter*	*Minor Diameter (external)*	*Minor Diameter (internal)*
Pitch 2 mm				
20	20·00	18·70	17·55	17·84
24	24·00	22·70	21·55	21·84
30	30·00	28·70	27·55	27·84
36	36·00	34·70	33·55	33·84
42	42·00	40·70	39·55	39·84
48	48·00	46·70	45·55	45·84
56	56·00	54·70	53·55	53·84
Pitch 3 mm				
30	30·00	28·05	26·32	26·75
36	36·00	34·05	32·32	32·75
42	42·00	40·05	38·32	38·75
48	48·00	46·05	44·32	44·75
56	56·00	54·05	52·32	52·75

ISO METRIC SCREW THREADS: CONSTANT PITCH SERIES (Continued)

(All dimensions in millimetres)

Nominal Size and Thread Diameter; First Choice	*Major Diameter*	*Effective Diameter*	*Minor Diameter (external)*	*Minor Diameter (internal)*
Pitch 4 mm				
42	42·00	39·40	37·09	37·67
48	48·00	45·40	43·09	43·67
56	56·00	53·40	51·09	51·67

SLOTTED COUNTERSUNK 90° HEAD MACHINE SCREWS: METRIC SERIES (PREFERRED SIZES) (All dimensions in millimetres)

Nominal Size and Thread Diameter	Radius Below Head*	Thread Length min.	Thread Run Out	Slot Width	
				max.	min.
M1	0·1	†	0·50	0·45	0·31
M1·2	0·1	†	0·50	0·50	0·36
M1·6	0·1	15·0	0·70	0·60	0·46
M2·0	0·1	16·0	0·80	0·70	0·56
M2·5	0·1	18·0	0·90	0·80	0·66
M3	0·1	19·0	1·00	1·00	0·86
M4	0·2	22·0	1·40	1·20	1·06
M5	0·2	25·0	1·60	1·51	1·26
M6	0·25	28·0	2·00	1·91	1·66
M8	0·4	34·0	2·50	2·31	2·06
M10	0·4	40·0	3·00	2·81	2·56
M12	0·6	46·0	3·50	3·31	3·06
M16	0·6	58·0	4·00	4·37	4·07
M20	0·8	70·0	5·00	5·37	5·07

* These values are the maximum where the shank diameter is equal to the major diameter of the thread and the minimum where the shank diameter is equal to the effective diameter of the thread (see BS 4183).

† Threaded up to head only.

If d is the nominal size, the head diameters are max. $2d$ and min. $1{\cdot}75d$; the head heights are max. $0{\cdot}5d$ and min. $0{\cdot}45d$; slot depths max. $0{\cdot}3d$ and min. $0{\cdot}2d$. Slotted raised countersunk head machine screws have a height of raised portion $0{\cdot}25d$.

HEXAGON-SOCKET CAP HEAD SCREWS

(All dimensions in millimetres)

Nominal Size: First Choice	*Body Diameter; also Head Height*		*Head Diameter*		*Hexagon Socket Size*	*Key Engagement*
	max.	*min.*	*max.*	*min.*		*min.*
M3	3·00	2·86	5·50	5·20	2·50	1·30
M4	4·00	3·82	7·00	6·64	3·00	2·00
M5	5·00	4·82	8·50	8·14	4·00	2·70
M6	6·00	5·82	10·00	9·64	5·00	3·30
M8	8·00	7·78	13·00	12·57	6·00	4·30
M10	10·00	9·78	16·00	15·57	8·00	5·50
M12	12·00	11·73	18·00	17·57	10·00	6·60
M16	16·00	15·73	24·00	23·48	14·00	8·80
M20	20·00	19·67	30·00	29·48	17·00	10·70
M24	24·00	23·67	36·00	35·38	19·00	12·90

HEXAGON-SOCKET 90° COUNTERSUNK HEAD SCREWS

Nominal Size: First Choice	*Body Diameter*		*Head Diameter min.*	*Head Height*	*Hexagon Socket Size*	*Key Engagement min.*
	max.	*min.*				
M3	3·00	2·86	5·82	1·86	2·00	1·05
M4	4·00	3·82	7·78	2·48	2·50	1·49
M5	5·00	4·82	9·78	3·10	3·00	1·86
M6	6·00	5·82	11·73	3·72	4·00	2·16
M8	8·00	7·78	15·73	4·96	5·00	2·85
M10	10·00	9·78	19·67	6·20	6·00	3·60
M12	12·00	11·73	23·67	7·44	8·00	4·35
M16	16·00	15·73	29·67	8·80	10·00	4·89
M20	20·00	19·67	35·61	10·16	12·00	5·45

For further information, see BS 4168.

Hexagon wrenches: the range of hexagon wrench keys covers 13 sizes from 1·5 mm to 19 mm, the sizes being the same as the maximum dimensions across the flats: 1·5, 2·0, 2·5, 3·0, 4·0, 5·0, 6·0, 8·0, 10·0, 12·0, 14·0, 17·0, 19·0.

PIPE THREADS

Threads on pipes will continue to be BSP (British Standard Pipe) which have been adopted as the ISO pipe thread and are covered in ISO recommendation R7: "Pipe threads for gas list tubes and screwed fittings where pressure-tight joints are made on the threads ($\frac{1}{8}$ in to 6 in)".

The British Standard Whitworth thread form as in the figure is a symmetrical V-thread in which the angle between

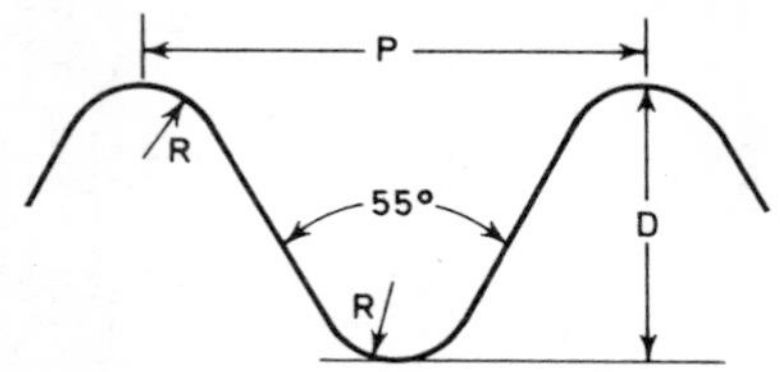

the flanks is 55°. One sixth of the sharp vee is rounded off equally at the crests and roots by circular arcs. The thread depth is 0·640327 times the nominal pitch and the radius of each circular arc is 0·137278 times the pitch. Some threads are tapered by 1 in 16, measured on the diameter, which is shown in exaggerated form in the figures on pages 94 and 95, relating to various terms that are defined here:

Gauge diameter: Basic major thread diameter.

Gauge length: Distance on external thread, parallel to the axis, from gauge plane to small end.

Washout thread: That part of thread not fully formed at the root.

Major cone: An enveloping cone touching the crests of an external thread or the roots of an internal thread.

Useful thread: All the thread excluding the washout thread.

Jointing threads: Jointing threads ensure that a pressure-tight seal is made by the mating of the threads.

British Standard taper pipe threads are designated BSP.Tr plus "EXT" or "INT" for external and internal threads.

Parallel internal threads are designated BSP.Pl.

WHITWORTH STANDARD PIPE THREADS

Selected values only of the Whitworth British Standard Pipe sizes have been adopted in the SI system. (ISO Recommendations R7, R228; BS 21 and BS 2517.) The basic dimensions are the same for jointing threads and where pressure-tight joints are not made on the threads ($\frac{1}{8}$ in to 6 in). Depth of thread = 0·64 divided by no. of threads per inch.

Nominal Bore in	*Threads per inch*	*Pitch* mm	*Depth of Thread*		*Major Gauge Diameter*		*Effective Diameter* mm	*Minor Diameter* mm	*Minimum Length of Thread on Pipe end* mm
			in	mm	in	mm			
$\frac{1}{8}$	28	0·91	0·0230	0·58	0·383	9·73	9·15	8·57	6·5
$\frac{1}{4}$	19	1·34	0·0335	0·86	0·518	13·16	12·30	11·45	9·7
$\frac{3}{8}$	19	1·34	0·0335	0·86	0·656	16·66	15·81	14·95	10·1
$\frac{1}{2}$	14	1·81	0·0455	1·16	0·825	20·96	19·79	18·63	13·2
$\frac{3}{4}$	14	1·81	0·0455	1·16	1·041	26·44	25·28	24·12	14·5
1	11	2·31	0·0580	1·48	1·309	33·25	31·77	30·29	16·8
$1\frac{1}{4}$	11	2·31	0·0580	1·48	1·650	41·91	40·43	38·95	19·1
$1\frac{1}{2}$	11	2·31	0·0580	1·48	1·882	47·80	46·32	44·85	19·1
2	11	2·31	0·0580	1·48	2·347	59·61	58·14	56·66	23·4
$2\frac{1}{2}$	11	2·31	0·0580	1·48	2·960	75·18	73·71	72·23	26·7
3	11	2·31	0·0580	1·48	3·460	87·88	86·41	84·93	29·8
$3\frac{1}{2}$	11	2·31	0·0580	1·48	3·950	100·33	98·85	97·37	31·4
4	11	2·31	0·0580	1·48	4·450	113·03	111·55	110·07	35·8
5	11	2·31	0·0580	1·48	5·450	138·43	136·95	135·47	40·1
6	11	2·31	0·0580	1·48	6·450	163·83	162·35	160·87	40·1

DIMENSIONS OF WING NUTS

Hot stamped or cast wing nuts types HS and DC (BS 856: 1969)

NOTE. Dotted outline illustrates alternative wing shape permissible. Exercise of this option would render dimension *F* redundant.

*Blank No.**	*Nominal Screw Thread Size*				*A*		*B*		*C*		*D*		*E*	
	ISO metric	*UNC UNF*	*BSW† BSF†*	*BA†*	mm	in	mm	in	mm	in	mm	in	mm	in
1	M3	No. 4 & No. 6	$\frac{1}{8}$ in	5 & 4	9	$\frac{11}{32}$	6·5	$\frac{1}{4}$	7	$\frac{9}{32}$	13·5	$\frac{17}{32}$	22	$\frac{7}{8}$
2	M4 & M5	No. 8 & No. 10	$\frac{3}{16}$ in	3 & 2	10	$\frac{13}{32}$	8	$\frac{5}{16}$	9	$\frac{11}{32}$	15	$\frac{19}{32}$	25·5	1
3	M6	No. 12 & $\frac{1}{4}$ in	$\frac{1}{4}$ in	1 & 0	13	$\frac{1}{2}$	9·5	$\frac{3}{8}$	11	$\frac{7}{16}$	18	$\frac{23}{32}$	30	$1\frac{3}{16}$
4	M8	$\frac{5}{16}$ in	$\frac{5}{16}$ in	—	16	$\frac{5}{8}$	12	$\frac{15}{32}$	13	$\frac{1}{2}$	23	$\frac{29}{32}$	38	$1\frac{1}{2}$
5	M10	$\frac{3}{8}$ in	$\frac{3}{8}$ in	—	17·5	$\frac{11}{16}$	14	$\frac{9}{16}$	14	$\frac{9}{16}$	25·5	1	44·5	$1\frac{3}{4}$
6	M12	$\frac{7}{16}$ in	$\frac{7}{16}$ in	—	19	$\frac{3}{4}$	16	$\frac{5}{8}$	15	$\frac{19}{32}$	28·5	$1\frac{1}{8}$	51	2
7	(M14)	$\frac{1}{2}$ in	$\frac{1}{2}$ in	—	22	$\frac{7}{8}$	17·5	$\frac{11}{16}$	17	$\frac{21}{32}$	32	$1\frac{1}{4}$	59	$2\frac{5}{16}$
8	M16	$\frac{5}{8}$ in	$\frac{5}{8}$ in	—	25·5	1	20·5	$\frac{13}{16}$	19	$\frac{3}{4}$	36·5	$1\frac{7}{16}$	63·5	$2\frac{1}{2}$
9	(M18)	$\frac{3}{4}$ in	$\frac{3}{4}$ in	—	32	$1\frac{1}{4}$	27	$\frac{11}{16}$	22	$\frac{7}{8}$	41	$1\frac{5}{8}$	78	$3\frac{1}{16}$

* The numbers given in Column 1 are the customary trade designations for the sizes of the nut blanks, which are common for Type HS and Type DC wing nuts. The values of the metric dimensions given are arithmetic conversions of the basic inch dimensions, rounded-off to the nearest 0·5 mm.

† BSW, BSF and BA threads have been rendered obsolescent and these threads should not be used for new designs.

NOTE. Sizes shown in brackets are non-preferred.

DIMENSIONS OF WING NUTS (Continued)

Blank No.*	Nominal Screw Thread Size				F		G		H		J	
	ISO metric	UNC UNF	BSW† BSF†	BA†	mm	in	mm	in	mm	in	mm	in
1	M3	No. 4 & No. 6	$\frac{1}{8}$ in	5 & 4	19	$\frac{3}{4}$	3·5	$\frac{9}{64}$	2·5	$\frac{3}{32}$	1·5	$\frac{1}{16}$
2	M4 & M5	No. 8 & No. 10	$\frac{3}{16}$ in	3 & 2	19	$\frac{3}{4}$	4	$\frac{5}{32}$	2·5	$\frac{3}{32}$	1·5	$\frac{1}{16}$
3	M6	No. 12 & $\frac{1}{4}$ in	$\frac{1}{4}$ in	1 & 0	19	$\frac{3}{4}$	5	$\frac{3}{16}$	2·5	$\frac{3}{32}$	1·5	$\frac{1}{16}$
4	M8	$\frac{5}{16}$ in	$\frac{5}{16}$ in	—	19	$\frac{3}{4}$	6·5	$\frac{1}{4}$	3	$\frac{1}{8}$	2·5	$\frac{3}{32}$
5	M10	$\frac{3}{8}$ in	$\frac{3}{8}$ in	—	19	$\frac{3}{4}$	7	$\frac{9}{32}$	5	$\frac{3}{16}$	3	$\frac{1}{8}$
6	M12	$\frac{7}{16}$ in	$\frac{7}{16}$ in	—	25·5	1	8	$\frac{5}{16}$	5	$\frac{3}{16}$	3	$\frac{1}{8}$
7	(M14)	$\frac{1}{2}$ in	$\frac{1}{2}$ in	—	25·5	1	9·5	$\frac{3}{8}$	5·5	$\frac{7}{32}$	4	$\frac{5}{32}$
8	M16	$\frac{5}{8}$ in	$\frac{5}{8}$ in	—	32	$1\frac{1}{4}$	10	$\frac{13}{32}$	6·5	$\frac{1}{4}$	5	$\frac{3}{16}$
9	(M18)	$\frac{3}{4}$ in	$\frac{3}{4}$ in	—	32	$1\frac{1}{4}$	12	$\frac{15}{32}$	7	$\frac{9}{32}$	5·5	$\frac{7}{32}$

The footnotes on the previous page apply to this table also.

DIMENSIONS OF WING NUTS (Continued)

Cold forged wing nuts type CF (BS 856: 1969)

Blank No.*	Nominal Screw Thread Size				A		B		C		D	
	ISO metric	UNC UNF	BSW† BSF†	BA†	mm	in	mm	in	mm	in	mm	in
1	M3	No. 4 & No. 6	$\frac{1}{8}$ in	5 & 4	8	$\frac{5}{16}$	5	$\frac{13}{64}$	3	$\frac{1}{8}$	9	$\frac{11}{32}$
2	M4 & M5	No. 8 & No. 10	$\frac{3}{16}$ in	3 & 2	11	$\frac{27}{64}$	6	$\frac{15}{64}$	4·5	$\frac{11}{64}$	11	$\frac{7}{16}$
3	M6	No. 12 & $\frac{1}{4}$ in	$\frac{1}{4}$ in	1 & 0	12	$\frac{31}{64}$	8	$\frac{5}{16}$	5	$\frac{13}{64}$	13	$\frac{1}{2}$
4	M8	$\frac{5}{16}$ in	$\frac{5}{16}$ in	—	14	$\frac{9}{16}$	9·5	$\frac{3}{8}$	6	$\frac{15}{64}$	16	$\frac{5}{8}$
5	M10	$\frac{3}{8}$ in	$\frac{3}{8}$ in	—	17	$\frac{43}{64}$	12	$\frac{15}{32}$	7	$\frac{17}{64}$	18	$\frac{23}{32}$
6	M12	$\frac{7}{16}$ in & $\frac{1}{2}$ in	$\frac{7}{16}$ in & $\frac{1}{2}$ in	—	22	$\frac{7}{8}$	16	$\frac{5}{8}$	9·5	$\frac{3}{8}$	23	$\frac{29}{32}$
8	M16 & M20	$\frac{5}{8}$ in & $\frac{3}{4}$ in	$\frac{5}{8}$ in & $\frac{3}{4}$ in	—	29·5	$1\frac{5}{32}$	22	$\frac{7}{8}$	13·5	$\frac{17}{32}$	35	$1\frac{3}{8}$

* The numbers given in Column 1 are the customary trade designations for the sizes of the nut blanks. The values of the metric dimensions given are arithmetic conversions of the basic inch dimensions, rounded-off to the nearest 0·5 mm.

† BSW, BSF and BA threads have been rendered obsolescent and these threads should not be used for new designs.

DIMENSIONS OF WING NUTS (Continued)

*Blank No.**	*Nominal Screw Thread Size*				*E*		*G*		*H*		*J*	
	ISO Metric	*UNC UNF*	*BSW† BSF†*	*BA†*	mm	in	mm	in	mm	in	mm	in
1	M3	No. 4 & No. 6	$\frac{1}{8}$ in	5 & 4	16·5	$\frac{21}{32}$	5	$\frac{13}{64}$	2·5	$\frac{3}{32}$	1	$\frac{3}{64}$
2	M4 & M5	No. 8 & No. 10	$\frac{3}{16}$ in	3 & 2	21·5	$\frac{27}{32}$	7	$\frac{17}{64}$	3	$\frac{1}{8}$	2	$\frac{5}{64}$
3	M6	No. 12 & $\frac{1}{4}$ in	$\frac{1}{4}$ in	1 & 0	27	$1\frac{1}{16}$	8	$\frac{5}{16}$	4	$\frac{5}{32}$	2·5	$\frac{3}{32}$
4	M8	$\frac{5}{16}$ in	$\frac{5}{16}$ in	—	31	$1\frac{7}{32}$	9·5	$\frac{3}{8}$	5	$\frac{3}{16}$	3	$\frac{7}{64}$
5	M10	$\frac{3}{8}$ in	$\frac{3}{8}$ in	—	36	$1\frac{13}{32}$	11	$\frac{7}{16}$	5·5	$\frac{7}{32}$	3	$\frac{1}{8}$
6	M12	$\frac{7}{16}$ in & $\frac{1}{2}$ in	$\frac{7}{16}$ in & $\frac{1}{2}$ in	—	47·5	$1\frac{7}{8}$	15	$\frac{19}{32}$	7·5	$\frac{19}{64}$	4·5	$\frac{11}{64}$
8	M16 & M20	$\frac{5}{8}$ in & $\frac{3}{4}$ in	$\frac{5}{8}$ in & $\frac{3}{4}$ in	—	68	$2\frac{11}{16}$	20·5	$\frac{13}{16}$	9·5	$\frac{3}{8}$	5	$\frac{3}{16}$

The footnotes on the previous page apply to this table also.

SCREWED STUDS FOR GENERAL PURPOSES: METRIC SERIES

The sizes given here are taken from BS 4439: 1969 which superseded BS 2693: 1956. Screwed studs are described as having an overall length consisting of a metal end (the end screwed into a component), a plain portion (the unthreaded middle length of the stud) and a nut end. The nominal lengths consist of the plain and nut end lengths and have tolerances as given below (in millimetres):

Nominal Length Range	12–16	20–50	55–80	85–120
Tolerance	± 0·35	± 0·50	± 0·60	± 0·70

Nominal Length Range	130–180	190–300	325–400	425–500
Tolerance	± 0·80	± 1·0	± 1·2	± 1·3

The thread lengths for the nut ends are specified as up to 125 mm, $2d + 6$ mm, over 125 mm up to 200 mm, $2d + 12$ mm; above 200 mm, $2d + 25$ mm, where d is the diameter of the plain portion. The metal ends shall have screw threads to the standard specified in BS 3643, Part 2, or alternatively may be oversize (OS) on the effective and minor diameters as tabulated below:

OVERSIZE (OS) THREADS FOR METAL ENDS OF SCREWED STUDS: PREFERRED NUMBERS

(All dimensions in millimetres)

Nominal Size and Thread Diameter	*Pitch (Coarse pitch Series)*	*Major Diameter*		*Pitch Effective Diameter*		*Minor Diameter*	
		max.	*min.*	*max.*	*min.*	*max.*	*min.*
M3	0·5	3·00	2·93	2·72	2·68	2·44	2·35
M4	0·7	4·00	3·91	3·60	3·55	3·20	3·09
M5	0·8	5·00	4·91	4·54	4·48	4·08	3·96
M6	1·0	6·00	5·89	5·42	5·35	4·84	4·70
M8	1·25	8·00	7·87	7·26	7·19	6·54	6·38
M10	1·5	10·00	9·85	9·11	9·03	8·24	8·05
M12	1·75	12·00	11·83	10·96	10·86	9·95	9·72
M16	2·0	16·00	15·82	14·80	14·70	13·65	13·40
M20	2·5	20·00	19·79	18·48	18·38	17·04	16·75
M24	3·0	24·00	23·76	22·18	22·05	20·44	20·10
M30	3·5	30·00	29·74	27·86	27·72	25·84	25·45
M36	4·0	36·00	35·70	33·54	33·40	31·23	30·80

METAL WASHERS

Metal washers are in three categories of size: normal, large and extra large. They are specified in BS 4320: 1968 for bright metal washers and black metal washers. Inside and outside diameters with thicknesses are tabulated for appropriate ranges of the nominal sizes of bolt or screw based on the M series* and are for round washers with round holes. Other shapes of washers, e.g. tapered and square, had not been considered at the time of issue of BS 4320. The material should be either steel made from cold rolled strip C54 in the hard condition or brass made from CZ 108 in the hard condition, flat and free from burrs or if chamfered (in the case of bright washers) with an optional chamfer angle of 30°. Tolerances are according to German Standard DIN 522.

BRIGHT WASHERS

The normal diameter series is suitable for hexagon headed bolts, screws and nuts specified in BS 3692, BS 4190 and BS 4183. For nominal sizes of bolt or screw, the inside and outside diameters, thicknesses for normal and light ranges with nominal, maximum and minimum sizes in millimetres are tabulated from M1·0 to M39 starting with 0·2 differences between the small M's and increasing to 3·0 for the largest M. Typically M6 has an inside diameter of 6·4 mm, outside diameter of 12·5 mm with tolerances of 0·39 and 0·4 respectively and thicknesses for normal and light of 1·6 and 0·8 respectively and smaller tolerances.

A larger-diameter bright washer series for M4 to M39 has larger outside diameters and similar tolerances, and is suitable for cases where the next larger size of hexagon is used for particular diameter rather than the normal nut or hexagon size given in ISO/R272. A light range of bright washers is also provided with thicknesses about 60% of those of normal range.

BLACK METAL WASHERS

Three similar tables are given for bolts M5 to M68, a large diameter series for bolts M8 to M39 and an extra large diameter series for bolts M5 to M39 having the outside diameters appreciably larger. Appendix B contains another table for sizes M72 to M150 giving inside and outside diameters, thicknesses and omitting tolerances. The black metal washers are listed in one thickness only.

* M refers to the shank diameter of bolt (mm).

SPRING WASHERS FOR GENERAL ENGINEERING AND AUTOMOBILE PURPOSES: METRIC SERIES

(All dimensions in millimetres)

Dimensions, tolerances and general requirements have been specified in BS 4464: 1969 for spring washers of single-coil square section (type A), single-coil rectangular section with normal (type B) and with deflected (type C) ends, and double-coil rectangular section (type D). The main dimensions of those with normal and deflected ends are the same. The materials referred to are steel spring (BS 970 and 1449), phospher-bronze spring (BS 2870 and 2873), copper-silicon spring (BS 2870 and 2873), and copper-beryllium spring (BS 2870 and 2873). The general dimensions of the preferred sizes are given below in tables.

SINGLE COIL SQUARE-SECTION SPRING WASHERS (TYPE A)

Nominal Size and Thread Diameter	*Inside Diameter*		*Outside Diameter*	*Thickness and Width*
	max.	*min.*		
M3	3·3	3·1	5·5	1
(M3·5)	3·8	3·6	6·0	1
M4	4·35	4·1	6·95	1·2
M5	5·35	5·1	8·55	1·5
M6	6·4	6·1	9·6	1·5
M8	8·55	8·2	12·75	2
M10	10·6	10·2	15·9	2·5
M12	12·6	12·2	17·9	2·5
(M14)	14·7	14·2	21·1	3
M16	16·9	16·3	24·3	3·5
(M18)	19·0	18·3	26·4	3·5
M20	21·1	20·3	30·5	4·5
(M22)	23·3	22·4	32·7	4·5
M24	25·3	24·4	35·7	5
(M27)	28·5	27·5	38·9	5
M30	31·5	30·5	43·9	6
(M33)	34·6	33·5	47·9	6
M36	37·6	36·5	52·1	7
(M39)	40·8	39·6	55·3	7
M42	43·8	42·6	60·3	8
(M45)	46·8	45·6	63·3	8
M48	50·0	48·8	66·5	8

SINGLE COIL RECTANGULAR-SECTION SPRING WASHERS (TYPE B)

(All dimensions in millimetres)

Nominal Size and Thread Diameter	*Inside Diameter*		*Outside Diameter*	*Width*	*Thickness*
	max.	*min.*	*max.*		
M1·6	1·9	1·7	3·5	0·7	0·4
M2	2·3	2·1	4·3	0·9	0·5
M2·5	2·8	2·6	5·0	1·0	0·6
M3	3·3	3·1	6·1	1·3	0·8
M4	4·35	4·1	7·55	1·5	0·9
M5	5·35	5·1	9·15	1·8	1·2
M6	6·4	6·1	11·7	2·5	1·6
M8	8·55	8·2	14·85	3	2
M10	10·6	10·2	18·0	3·5	2·2
M12	12·6	12·2	21·0	4	2·5
M16	16·9	16·3	27·3	5	3·5
M20	21·1	20·3	33·5	6	4
M24	25·3	24·4	39·8	7	5
M30	31·5	30·5	48·0	8	6
M36	37·6	36·5	58·1	10	6
M42	43·8	42·6	68·3	12	7
M48	50·0	48·8	74·5	12	7
M56	58·1	56·8	86·6	14	8
M64	66·3	64·9	93·8	14	8

Note: Type C has the same main dimensions as type B.

DOUBLE COIL RECTANGULAR-SECTION SPRING WASHERS (TYPE D)

(All dimensions in millimetres)

Nominal Size and Thread Diameter	*Inside Diameter*		*Outside Diameter max.*	*Width*	*Thickness*
	max.	*min.*			
M2	2·4	2·1	4·4	0·9	0·5
M2·5	2·9	2·6	5·5	1·2	0·7
M3·0	3·6	3·3	6·2	1·2	0·8
M4	4·6	4·3	8·0	1·6	0·8
M5	5·6	5·3	9·8	2	0·9
M6	6·6	6·3	12·9	3	1
M8	8·8	8·4	15·1	3	1·2
M10	10·8	10·4	18·2	3·5	1·2
M12	12·8	12·4	20·2	3·5	1·6
M16	17·0	16·5	27·4	5	2
M20	21·5	20·8	31·9	5	2
M24	26·0	25·0	39·4	6·5	3·25
M30	33·0	31·5	49·5	8	3·25
M36	40·0	38·0	60·5	10	3·25
M42	46·0	44·0	66·5	10	4·5
M48	52·0	50·0	72·5	10	4·5
M56	60·0	58·0	84·5	12	4·5
M64	70·0	67·0	94·5	12	4·5

The M number refers to the shank diameter.

CRINKLE WASHERS FOR GENERAL ENGINEERING PURPOSES: METRIC SERIES

(All dimensions in millimetres)

The form of crinkle washer whose dimensions are given in BS 4463: 1969 is the same as the crinkle washer illustrated in BS 3401 *Beryllium Copper Crinkle Washers* which has three equi-spaced radial corrugations. The materials mentioned are copper-beryllium with the requirements for CB 101 as specified in BS 2870, and stainless steel of material 301S21 specified in BS 1449, Part 4. The radii of the upper and lower corrugations should be equal. The inside and outside diameters of the preferred sizes are given below.

Nominal Size and Thread Diameter	*Inside Diameter*		*Outside Diameter*		*Height*	
	max.	*min.*	*max.*	*min.*	*max.*	*min.*
M1·6	1·8	1·7	3·7	3·52	0·51	0·36
M2	2·3	2·2	4·6	4·42	0·53	0·38
(M2·2)	2·5	2·4	5·2	5·02	0·53	0·38
M2·5	2·8	2·7	5·8	5·62	0·53	0·38
M3	3·32	3·2	6·4	6·18	0·61	0·46
(M3·5)	3·82	3·7	6·9	6·68	0·79	0·63
M4	4·42	4·3	8·1	7·88	0·84	0·69
M5	5·42	5·3	9·2	8·98	0·89	0·74
M6	6·55	6·4	11·5	11·23	1·14	0·99
M8	8·55	8·4	15·0	14·73	1·40	1·25
M10	10·68	10·5	19·6	19·27	1·70	1·55
M12	13·18	13·0	22·0	21·67	1·90	1·65
(M14)	15·18	15·0	25·5	25·17	2·06	1·80
M16	17·18	17·0	27·8	27·47	2·41	2·16
(M18)	19·21	19·0	31·3	30·91	2·41	2·16
M20	21·21	21·0	34·7	34·31	2·66	2·16

SLOTTED GRUB SCREWS: METRIC SERIES

Slotted grub screws can have cone or cup points. When the total length is less than or equal to the nominal diameter, the cone angle shall be 118° ± 2°, and when the total length is greater than the nominal diameter the cone angle shall be 90° ± 2°. For a cup point the angle of the cone is 90° to 100° with the diameter of the cup specified for each nominal size. Each thread is from the coarse-pitch series. (All dimensions in millimetres.)

Nominal Size and thread diameter (M)		1·6	2	2·5	3	4	5	6	8	10	12
Thread pitch		0·35	0·4	0·45	0·50	0·70	0·80	1·0	1·25	1·5	1·75
Cup point diameter	max.	—	—	—	1·40	2·00	2·50	3·00	5·00	6·00	8·0
	min.	—	—	—	1·00	1·60	2·10	2·60	4·52	5·52	7·42
Preferred lengths* range	min.	2·5	3	5	5	6	6	12	12	16	—
	max.	12	20	12	25	30	50	50	50	50	—

* The standard preferred lengths are 2, 2·5, 3, 4, 5, 6, 8, 10, 12, 14, 16, 20, 25, 30, 35, 40, 45, 50. Non-preferred lengths are 7, 9, 11, 18, 22, 28, 32, 38.

For further information see BS 4219: 1967.

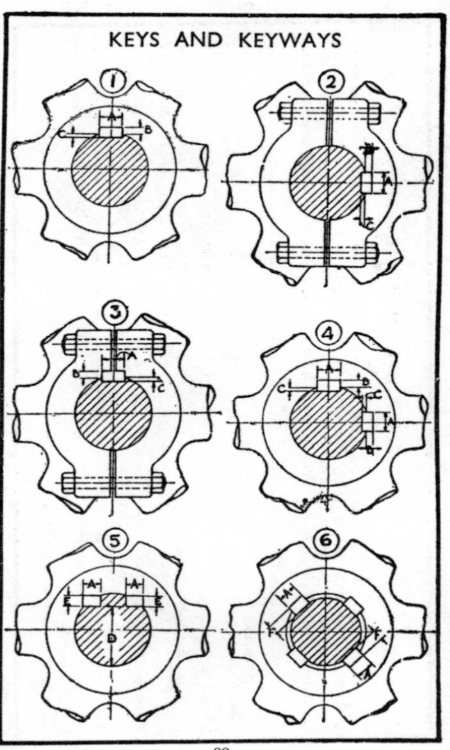
KEYS AND KEYWAYS
1
2
3
4
5
6
A
B
C
D
E
F

KEYS AND KEYWAYS—(*Continued*)

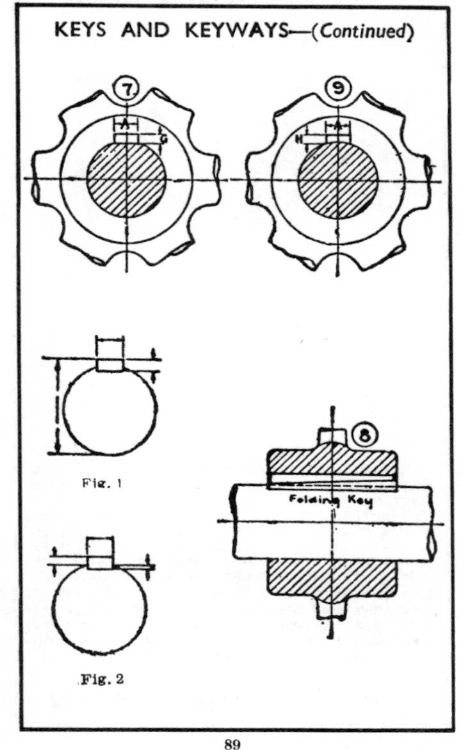

Fig. 1

Fig. 2

KEYS AND KEYWAYS—(*Continued*)

In defining the size of a key, two methods are adopted (See Figs. 1 and 2). If cutting keyways in the bores of wheels, as well as in shafts, is considered, the method shown by Fig. 2 is the most convenient for shop purposes. The method shown in Fig. 1 is extensively used in connection with very high-class work, but it does not necessarily follow that it is the most useful system.

Whichever system is used the diagram always represents the head or larger end of a tapered key.

In the case of pulleys, etc., required for shaft extensions, i.e., motor pulleys, the shallow end of the keyway will be at side of boss flush with rim.

Typical key proportions.

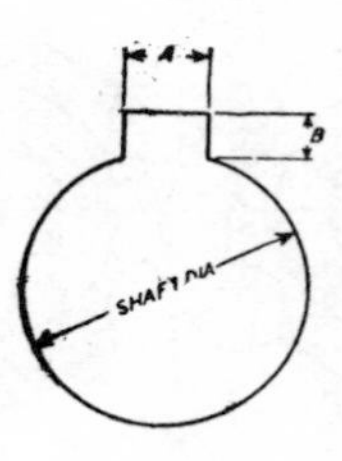

PARALLEL KEYS AND KEYWAYS

In each case the range of shaft diameter is from the quoted dimension to the subsequent figure.

Keyway	Shaft Diameter	6	8	10	12	17	22	30	38	44	50	58	65	75	85
	A	2	3	4	5	6	8	10	12	14	16	18	20	22	25
	B	2	3	4	5	6	7	8	8	9	10	11	12	14	14
Key	Chamfer min.	0·16	0·16	0·16	0·25	0·25	0·25	0·40	0·40	0·40	0·40	0·40	0·60	0·60	0·60
	Chamfer max.	0·25	0·25	0·25	0·40	0·40	0·40	0·60	0·60	0·60	0·60	0·60	0·80	0·80	0·80
	Length min.	6	6	8	10	14	18	22	28	36	45	50	56	63	70
	Length max.	20	36	45	56	70	90	110	140	160	180	200	220	250	280

Keyway	Shaft Diameter	95	110	130	150	170	200	230	260	290	330	380	440	(500)	—
	A	28	32	36	40	45	50	56	63	70	80	90	100	—	—
	B	16	18	20	22	25	28	32	32	36	40	45	50	—	—
Key	Chamfer min.	0·60	0·60	1·00	1·00	1·00	1·00	1·60	1·60	1·60	2·50	2·50	2·50	—	—
	Chamfer max.	0·80	0·80	1·20	1·20	1·20	1·20	2·00	2·00	2·00	3·00	3·00	3·00	—	—
	Length min.	80	90	100	—	—	—	—	—	—	—	—	—	—	—
	Length max.	320	360	400	—	—	—	—	—	—	—	—	—	—	—

For further information see BS 4235 (All dimensions in millimetres.)

STRAIGHT-SIDED SPLINES FOR CYLINDRICAL SHAFTS

The nominal dimensions are common to the shaft and hub. (All dimensions in millimetres)

Number of Splines	*Minor Diameter*	*Outside Diameter*		*Width of Splines*
		Light Series	*Medium Series*	
6	11	—	14	3
6	13	—	16	3·5
6	16	—	20	4
6	18	—	22	5
6	21	—	25	5
6	23	26	28	6
6	26	30	32	6
6	28	32	34	7
8	32	36	38	6
8	36	40	42	7
8	42	46	48	8
8	46	50	54	9
8	52	58	60	10
8	56	62	65	10
8	62	68	72	12
10	72	78	82	12
10	82	88	92	12
10	92	98	102	14
10	102	108	112	16
10	112	120	125	18

Note: ISO Recommendation R14 gives the proportions of straight-sided splines. The decision regarding involute splines has not been taken at the time of the current edition of the handbook. The present British practices for four, six and ten splines will be found on p. 269 *et seq*.

Tapers and Angles

Taper per Foot	Included			With Centre Line.			Taper per Inch	Taper per Inch from Centre Line.
	Deg.	Min	Sec	Deg.	Min	Sec.		
1/8	0	35	48	0	17	54	.010416	.005203
3/16	0	53	44	0	26	52	.015625	.007812
1/4	1	11	36	0	35	48	.020833	.010416
5/16	1	29	30	0	44	45	.026042	.013021
3/8	1	47	24	0	53	42	.031250	.015625
7/16	2	5	18	1	2	39	.036458	.018229
1/2	2	23	10	1	11	35	.041667	.020833
9/16	2	41	4	1	20	32	.046875	.023438
5/8	2	59	42	1	29	51	.052084	.026042
11/16	3	16	54	1	38	27	.057292	.028646
3/4	3	34	44	1	47	22	.062500	.031250
13/16	3	52	38	1	56	19	.067708	.033854
7/8	4	10	32	2	5	16	.072917	.036456
15/16	4	28	24	2	14	12	.078125	.039063
1	4	46	18	2	23	9	.083330	.041667
1¼	5	57	48	2	58	54	.104666	.052084
1½	7	9	10	3	34	35	.125000	.062500
1¾	8	20	26	4	10	13	.145833	.072917
2	9	31	36	4	45	48	.166666	.083332
2½	11	53	36	5	56	48	.208333	.104166
3	14	15	0	7	7	30	.250000	.125000
3½	16	35	40	8	17	50	.291666	.145833
4	18	55	28	9	27	44	.333333	.166666
4½	21	14	20	10	37	10	.375000	.187500
5	23	32	12	11	46	6	.416666	.208333
6	28	4	20	14	2	10	.500000	.250000

MORSE AND METRIC TAPERS

Angle of key 8°-19°
Taper 1¾ in 12

The Morse tapers 0 to 6 have been retained as standards but Morse taper 7, Reed, Jarno and other tapers have been discarded. Seven metric tapers, all with a taper of 1:20, have been adopted in addition; they have diameters of 4, 6, 80, 100, 120, 160 and 200 mm, the first two being smaller than M.T. "0", and the remaining five larger than M.T. "6". Details of the metric tapers are given in ISO Recommendation R296 (see also Lathe Centres p. 96).

MORSE TAPERS

No. of Taper	Diam. of Plug at Small End	Diam. at End of Socket	Standard Plug Depth	Whole Length of Shank	Depth of Hole	End of Socket to Keyway	Length of Keyway	Width of Keyway
	D	A	P	B	H	K	L	W
0	.252	.356	2	2 11/32	2 1/32	1 15/16	9/16	.160
1	.369	.475	2⅛	2 9/16	2 3/16	2 1/16	¾	.213
2	.572	.7	2 9/16	3⅛	2⅝	2½	⅞	.26
3	.778	.938	3 3/16	3⅞	3¼	3 1/16	1 3/16	.322
4	1.02	1.231	4 1/16	4⅞	4⅛	3⅞	1¼	.478
5	1.475	1.748	5 3/16	6⅛	5¼	4 15/16	1½	.635
6	2.116	2.494	7¼	8 9/16	7⅜	7	1¾	.76
7	2.75	3.27	10	11⅝	10⅛	9½	2⅝	1.135

MORSE TAPER SHANKS
(*Continued*)

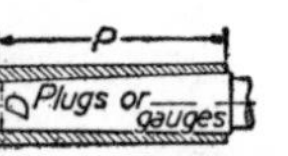

Length of Tongue	Diam. of of Tongue	Thick-ness of Tongue	Radius of Mill for Tongue	Radius of Tongue	Shank Depth	Taper per Foot	Taper per Inch	No. of Key
T	d	t	R	a	S			
1/4	.235	5/32	5/32	.04	2 7/32	.62460	.05205	0
3/8	.343	13/64	3/16	.05	2 7/16	.59858	.04988	1
7/16	17/32	1/4	1/4	.06	2 15/16	.59941	.04995	2
9/16	23/32	5/16	9/32	.08	3 11/16	.60235	.05019	3
5/8	31/32	15/32	5/16	.10	4 5/8	.62326	.05193	4
3/4	1 13/32	5/8	3/8	.12	5 7/8	.63151	.05262	5
1 1/8	2	3/4	1/2	.15	8 1/4	.62565	.05213	6
1 3/8	2 5/8	1 1/8	3/4	.18	11 1/4	.62400	.05200	7

LATHE CENTRES

LATHE CENTRES WITH SELF-HOLDING TAPER SHANKS (ISO RECOMMENDATION R298)

Type	*Metric* No. 4	*Metric* No. 6	*Morse* 0	*Morse* 1	*Morse* 2	*Morse* 3	*Morse* 4	*Morse* 5	*Morse* 6	*Metric* No. 80	*Metric* No. 100
Taper	0·05	0·05	0·05205	0·04988	0·04995	0·05020	0·05194	0·05263	0·05214	0·05	0·05
Diameter of Shank*	4	6	9·045	12·065	17·780	23·825	31·267	44·399	63·348	80	100
Maximum Diameter†	4·1	6·2	9·2	12·2	18·0	24·1	31·6	44·7	63·8	80·4	100·5
Insertion length maximum	23	32	50	53·5	64	81	102·5	129·5	182	196	232

* Near the large end of the taper and at the large end of the taper.

Note. The angle of the centre engaging the work piece is usually 60°; with very heavy workpieces wider angles may be used if necessary and 75° or 90° are recommended.

TWIST-DRILL GAUGE SIZES

The drill numbers and letters, 1 to 80 and A to Z become obsolete under the new system. The standard sizes to be adopted are given below in millimetres (with a few exceptions). The equivalents are usually less than one thousandth of an inch different therefrom and often the difference is much less.

Gauge number	*Metric*	*Gauge number*	*Metric*	*Gauge letter*	*Metric*
80	0·35 mm	40	2·50 mm	A	$\frac{15}{64}$ in
79	0·38 mm	39	2·55 mm	B	6·00 mm
78	0·40 mm	38	2·60 mm	C	6·10 mm
77	0·45 mm	37	2·65 mm	D	6·20 mm
76	0·50 mm	36	2·70 mm	E	$\frac{1}{4}$ in
75	0·52 mm	35	2·80 mm	F	6·50 mm
74	0·58 mm	34	2·80 mm	G	6·60 mm
73	0·60 mm	33	2·85 mm	H	$\frac{17}{64}$ in
72	0·65 mm	32	2·95 mm	I	6·90 mm
71	0·65 mm	31	3·00 mm	J	7·00 mm
70	0·70 mm	30	3·30 mm	K	$\frac{9}{32}$ in
69	0·75 mm	29	3·50 mm	L	7·40 mm
68	$\frac{1}{32}$ in	28	$\frac{9}{64}$ in	M	7·50 mm
67	0·82 mm	27	3·70 mm	N	7·70 mm
66	0·85 mm	26	3·70 mm	O	8·00 mm
65	0·90 mm	25	3·80 mm	P	8·20 mm
64	0·92 mm	24	3·90 mm	Q	8·40 mm
63	0·95 mm	23	3·90 mm	R	8·60 mm
62	0·98 mm	22	4·00 mm	S	8·80 mm
61	1·00 mm	21	4·00 mm	T	9·10 mm
60	1·00 mm	20	4·10 mm	U	9·30 mm
59	1·05 mm	19	4·20 mm	V	$\frac{3}{8}$ in
58	1·05 mm	18	4·30 mm	W	9·80 mm
57	1·10 mm	17	4·40 mm	X	10·10 mm
56	$\frac{3}{64}$ in	16	4·50 mm	Y	10·30 mm
55	1·30 mm	15	4·60 mm	Z	10·50 mm
54	1·40 mm	14	4·60 mm		
53	1·50 mm	13	4·70 mm		
52	1·60 mm	12	4·80 mm		
51	1·70 mm	11	4·90 mm		
50	1·80 mm	10	4·90 mm		
49	1·85 mm	9	5·00 mm		
48	1·95 mm	8	5·10 mm		
47	2·00 mm	7	5·10 mm		
46	2·05 mm	6	5·20 mm		
45	2·10 mm	5	5·20 mm		
44	2·20 mm	4	5·30 mm		
43	2·25 mm	3	5·40 mm		
42	$\frac{3}{32}$ in	2	5·60 mm		
41	2·45 mm	1	5·80 mm		

Note. A few well-established fractional inch sizes are at present retained in the system.

TWIST DRILLS: STANDARD METRIC SIZES

Type A—The steps are 0·20, 0·22, 0·25, 0·28, commencing from 0·20 up to 0·28 mm and similarly from 0·30, etc. up to 1·00.
Subsequent steps are every 0·05 mm from 1·00 to 3·00 mm
then every 0·10 mm from 3·00 to 14·00 mm
then every 0·25 mm from 14·00 to 16·00 mm.

Type B—The steps are 0·50, 0·80, 1·00, 1·20, commencing from 0·5 to 1·20 mm and similarly from 1·50, etc. up to 14·00.
Subsequent steps are every 0·50 mm from 14·00 to 20·00 mm
then every 1·00 mm from 20·00 to 25·00 mm.

Type C—The steps are by 0·1 mm commencing from 1·00 up to 14·00 mm.
Subsequent steps are every 0·25 mm from 14·00 to 25·00 mm.

Type D—The steps are 3·00, 3·20, 3·50, 3·80, commencing from 3·0 to 3·80 mm and similarly from 4·00, etc. up to 14·00.
Subsequent steps are every 0·25 mm from 14·00 to 32·00 mm
then every 0·50 mm from 32·00 to 51·00 mm
then every 1·00 mm from 51·00 to 100·00 mm.

Note—The above choices of sizes are based on systems of preferred numbers (see p. 46).

TWIST DRILLS

(Sizes in millimetres)

A. Parallel-shank jobber series—Drills with two helical flutes and a parallel shank of approximately the same diameter as the cutting end.

B. Stub drills—A form of type A drill with a shorter flute.

C. Parallel-shank long-series twist drills—A form of type A drill with a longer flute length.

D. Morse taper shank twist drills and core drills—The former have two and the latter three or four helical flutes, both with a standard morse taper shank (M.T.) for holding and driving. They are designed for enlarging holes and are not suitable for boring holes in solid material. The diameter ranges are from the quoted number to that immediately below.

	A		*B*		*C*		*D*				
Diameter Range	*Flute Length*	*Overall Length*	*Flute Length*	*Overall Length*	*Flute Length*	*Overall Length*	*Flute Length*	*Overall Length*	*MT No.*	*Oversize Length*	*MT No.*
0·19	2·5	19	1·5	19							
0·24	3	19	1·5	19							
0·30	4	19	2·0	19							
0·38	5	20	2·5	19							
0·48	6	22	3·0	20							
0·53	7	24	3·5	21							
0·60	8	26	4·0	22							
0·67	9	28	4·5	23							
0·75	10	30	5·0	24							
0·85	11	32	5·5	25							

TWIST DRILLS (Continued)

Diameter Range	A Flute Length	A Overall Length	B Flute Length	B Overall Length	C Flute Length	C Overall Length	D Flute Length	D Overall Length	D MT No.	Oversize Length	MT No.
0·95	12	34	6·0	26	33	56					
1·06	14	36	7·0	28	37	60					
1·18	16	38	8·0	30	41	65					
1·32	18	40	9·0	32	45	70					
1·50	20	43	10	34	50	76					
1·70	22	46	11	36	53	80					
1·90	24	49	12	38	56	85					
2·12	27	53	13	40	59	90					
2·36	30	57	14	43	62	95					
2·65	33	61	16	46	66	100	33	114	1		
3·00	36	65	18	49	69	106	36	117	↓		
3·35	39	70	20	52	73	112	39	120			
3·75	43	75	22	55	78	119	43	123			
4·25	47	80	24	58	82	126	47	128			
4·75	52	86	26	62	87	132	52	133			
5·30	57	93	28	66	91	139	57	138			
6·00	63	101	31	70	97	148	63	144			
6·70	69	109	34	74	102	156	69	150			
7·50	75	117	37	79	109	165	75	156			
8·50	81	125	40	84	115	175	81	162			
9·50	87	133	43	89	121	184	87	168	↑		
10·60	94	142	47	95	128	195	94	175	1		

TWIST DRILLS (Continued)

Diameter Range	A		B		C		D				
	Flute Length	Overall Length	Flute Length	Overall Length	Flute Length	Overall Length	Flute Length	Overall Length	MT No.	Oversize Length	MT No.
11·80	101	151	51	102	134	205	101	182	1	199	2
13·20	108	160	54	107	140	214	108	189	1	206	2
14·00	114	169	56	111	144	220	114	212	2		
15·00	120	178	58	115	149	227	120	218	↓		
16·00			60	119	154	235	125	223			
17·00			62	123	158	241	130	228			
18·00			64	127	162	247	135	233		256	3
19·00			66	131	166	254	140	238		261	↓
20·00			68	136	171	261	145	243	↑	266	
21·20			70	141	176	268	150	248		271	↑
22·40			72	146	180	275	155	253*	2	276	3
23·60			75	151	185	282	160	281	3		
25·00			78	156	190	290	165	286	↓		
26·50			81	162	195	298	170	291		319	4
28·00			84	168	201	307	175	296		324	↓
30·00			87	174	207	316	180	301	↑	329	↑
31·50			90	180			185	306†	3	334	4
33·50			93	186			190	339	4		
35·50			96	193			195	344	↓		
37·50			100	200			200	349	↑		
40·00							205	354	4	392	5

* Also 276 for diameter 23·02 and MT 3. † Also 334 for diameter 31·75 and MT 4.

TWIST DRILLS (Continued)

Diameter Range	A		B		C		D				
	Flute Length	Overall Length	Flute Length	Overall Length	Flute Length	Overall Length	Flute Length	Overall Length	MT No.	Oversize Length	MT No.
42·50							210	359	↓	397	5
45·00							215	364		402	↓
47·50							220	369	↑	407	↑
50·00							225	374	4	412	5
50·80							225	412	5		
53·00							230	417	↓		
56·00							235	422			
60·00							240	427			
63·00							245	432		499	6
67·00							250	437	↑	504	
71·00							255	442		509	
75·00							260	447	5	514	6
76·20							260	514	6		
80·00							265	519	↓		
85·00							270	524			
90·00							275	529	↑		
95·00							280	534			
100·00							285	539	6		

CUTTING SPEEDS FOR DRILLS

(All dimensions in millimetres)

The question of cutting speeds for drills has been considered by the ISO in relation to their testing both as regards the penetration per minute and the number and depth of the holes to be drilled. In particular, for flute twist drills, suggestions have been made, as a guide, that the peripheral speeds should be 9 to 12 metres per minute for carbon steel drills, and 21 to 24 metres per minute for high-speed steel drills. Details for various sizes of drills have been tabulated in BS 328 Part 1 from which the following data have been extracted to indicate the desirable penetration per minute in a steel test billet, for smaller drills. The tabulation is extended to a size of 76·2 mm with high-speed penetration of 50·8 mm.

Size		3·2	4·0	4·8	5·6	6·4	7·1	7·9	8·7	9·5	10·3	11·1
penetration per minute	carbon	78·7	81·3	83·6	83·1	81·3	77·5	73·2	69·6	66·0	62·2	58·4
	high-speed	215·9	223·5	231·1	236·2	241·3	238·8	236·2	233·7	231·1	227·3	223·5

Size		11·9	12·7	14·3	15·9	17·5	19·1	20·6	22·2	23·8	25·4
penetration per minute	carbon	54·6	51·3	46·7	43·2	40·6	38·1	36·6	35·6	34·3	33·0
	high-speed	219·7	215·9	208·3	200·7	193·0	185·4	177·8	172·7	165·1	158·8

The dimensions for various types of drills follow. If more details are required, reference should be made to BS 328 which is in accordance with ISO/R 235 for short series drills and draft ISO/DR 663 for long-series drills.

TAPPING-DRILL SIZES

A simple rule for ascertaining the required tapping-drill size is:

Subtract the pitch of one thread from the diameter of the tap and choose the nearest standard diameter.

Examples For M3 with pitch 0·5, the tapping-drill diameter is 2·5 mm.
For 10 mm diameter with pitch 0·75, the tapping-drill diameter is 9·2 mm.
For 12 mm diameter with pitch 1·25, the tapping-drill diameter is 10·8 mm.

WIRE THREAD INSERTS

Wire thread inserts are used to provide a wear-resisting fastening in light alloys and similar (soft) materials. There is a wide variation in their application and they may, for example, be called upon to meet design requirements or they may be used for salvage or repair purposes.

BS 4377: 1968 gives relevant particulars in three sections, namely:

Section 1 covers size limits for tapped holes to take wire thread inserts which provide ISO metric thread sizes from M2·5 (2·5 mm diameter) to M39 (39 mm diameter) inclusive.

Section 2 covers screwing taps for holes specified in Section 1, and also the marking of taps.

Section 3 describes a recommended gauging system for checking the tapped holes.

With these inserts the nominal sizes given refer to the size of male thread which mates with the internal thread of a fitted insert: accordingly, the actual diameter of a tap is larger than the nominal size marked on it, but the tap will also be marked "Insert BS 4377"; for example, "M16 × 2 Insert BS 4377".

The first-choice diameter coarse pitch series threads for these inserts are:

M2·5 × 0·45	M8 × 1·25	M20 × 2·5
M3 × 0·5	M10 × 1·5	M24 × 3
M4 × 0·7	M12 × 1·75	M30 × 3·5
M5 × 0·8	M16 × 2	M36 × 4
M6 × 1		

Note: A thread insert is usually made of wire of approximately rhombic cross-section formed into a spring-like helix for insertion into tapped holes and to retain another threaded member. Insertion is made with a driving tool engaging on a leading end tang which is broken off at a notch leaving the insert just above the top of the countersink in the tapped hole.

REAMERS

All dimensions in millimetres.
For further details than are given below see BS 122 part 2, BS 1916, BS 1660, ISO/R 236 and 237 and ISO/DR 664 and 665.
A Hand reamers.
B Parallel machine reamers.
C Machine (chucking) reamers with taper shanks.
D Machine (chucking) reamers with parallel shanks.
The diameter ranges are from the quoted number to that immediately below.

Cutting Diameter Ranges	*Driving Square for A*		*A*		*B*			*C*			*D*			
	Width Across Flats	*Length Size of Square*	*Overall Lengths*	*Cutting Edge Lengths*	*Overall Lengths*	*Cutting Edge Lengths*	*Morse Taper Number*	*Overall Lengths*	*Cutting Edge Lengths*	*Morse Taper No.*	*Overall Lengths*	*Cutting Edge Lengths*	*Shank Length*	*Shank Diameter*
1·32	1·12	4	41	20							42	9	22	1·25
1·50	1·25	4	44	21										
1·70	1·40	4	47	23							47	10	24	1·60
1·90	1·60	4	50	25										
2·12	1·80	4	54	27							54	12	26	2·00
2·36	2·00	4	58	29										
2·65	2·24	5	62	31							61	14	28	2·50
3·00	2·50	5	66	33										
3·35	2·80	5	71	35							70	16	31	3·15
3·75	3·15	6	76	38										
4·25	3·55	6	81	41							80	18	34	4·00
4·75	4·00	7	87	44										
5·30	4·50	7	93	47				119	21	1	91	21	37	5·00
6·00	5·00	8	100	50	130	50	1							
6·70	5·60	8	107	54	134	54	1	128	25	1	103	25	40	6·30
7·50	6·30	9	115	58	138	58	1							

REAMERS (Continued)

Cutting Diameter Ranges	Driving Square for A		A		B			C			D			
	Width Across Flats	Length Size of Square	Overall Lengths	Cutting Edge Lengths	Overall Lengths	Cutting Edge Lengths	Morse Taper Number	Overall Lengths	Cutting Edge Lengths	Morse Taper No.	Overall Lengths	Cutting Edge Lengths	Shank Length	Shank Diameter
8·50	7·10	10	124	62	142	62	1	140	29	1	119	29	44	8·00
9·50	8·00	11	133	66	146	66	1							
10·60	9·00	12	142	71	151	71	1	153	33	1	136	33	48	10·00
11·80	10·00	13	152	76	156	76	1							
13·20	11·20	14	163	81	161	81	1	169‖	38	1	156	38	52	12·50
15·00	12·50	16	175	87	187*	87	2							
17·00	14·00	18	188	93	193	93	2	203	44	2				
19·00	16·00	20	201	100	200	100	2							
21·20	18·00	22	215	107	207	107	2	225	51	2				
23·60	20·00	24	231	115	242†	115	3							
26·50	22·40	26	247	124	251	124	3	271¶	60	3				
30·00	25·00	28	265	133	260	133	3							
33·50	28·00	31	284	142	302‡	142	4	326**	69	4				
37·50	31·50	34	305	152	312	152	4							
42·50	35·50	38	326	163	323	163	4	363	80	4				
47·50	40·00	42	347	174	334	174	4							
53·00	45·00	46	367	184	381§	184	5							
60·00	50·00	51	387	194	391	194	5							
67·00	56·00	56	406	203	400	203	5							
75·00	63·00	62	424	212	409††	212	5							
85·00	—	—	—	—										

* Also 14·0; 181; 81; 2. † Also 23·02; 234; 107; 3. ‡ Also 31·75; 293; 133; 4.
§ Also 50·8; 371; 174; 5. ‖ Also 14·0; 184; 38; 2. ¶ Also 23·02; 244; 51; 3.
** Also 31·75; 296; 60; 4. †† Also 76·20; 479, 212, 6.

BALL BEARINGS AND CYLINDRICAL ROLLER BEARINGS: METRIC UNITS

(All dimensions are in millimetres)

Ball and rolling bearings are grouped under single-row radial, single row angular contact, magneto and double-row self-aligning ball bearings; cylindrical roller bearings; and thrust ball bearings. Each dimension series has a two-figure number indicating the width series 0 or 1 by the first number and the diameter series 2, 3, 4 or 8, 9, 0 by the second number. The tabulated data follow under the individual headings with capital letters for each type of bearing.

	Type	*ISO Dimension Series*	*Page*
Single-row radial ball bearings: miniature series	BRX, BRXX	10, 18, 19	109
Single-row radial ball bearings: extra-light series	BRE, BRX	10	110
Single-row radial ball bearings: light series	BRL	02	111
Single-row radial ball bearings: medium series	BRM	03	112
Single-row radial ball bearings: heavy series	BRH	04	113
Single-row angular contact ball bearings: extra light series	ACE	10	110
Single-row angular contact ball bearings: light series	ACL	02	111
Single-row angular contact ball bearings: medium series	ACM	03	112
Single-row angular contact ball bearings: heavy series	ACH	04	113
Single-row separable ball bearings: magneto type	BML	*	113

* This type has no dimension series number.

BALL BEARINGS AND CYLINDRICAL ROLLER BEARINGS: METRIC UNITS (Continued)

	Type	*ISO Dimension Series*	*Page*
Double-row self-aligning ball bearings: light series	BAL	02	111
Double-row self-aligning ball bearings: medium series	BAM	03	112
Double-row self-aligning ball bearings: heavy series	BAH	04	113
Cylindrical roller bearings: extra-light series	RRE, RRX	10, 19	110
Cylindrical roller bearings: light series	RRL	02	111
Cylindrical roller bearings: medium series	RRM	03	112
Cylindrical roller bearings: heavy series	RRH	04	113
Thrust ball bearings: extra-light series	TFX	11	114
Thrust ball bearings: light series	TFL	12	114
Thrust ball bearings: medium series	TFM	13	115
Thrust ball bearings: heavy series	TFH	14	115

In the tables, d is the internal diameter, D is the external diameter, B or H is the overall thickness or height.

For additional details to those given below reference should be made to BS 292: 1969.

All series 18 bearings designated BRXX (previously BRX); e.g. BRXX 0006
All series 19 bearings designated BRX; e.g. BRX 0010.

SINGLE-ROW RADIAL BALL BEARINGS

MINIATURE SERIES: ISO DIMENSION SERIES 18 AND 19

Series 18				*Series 19*			
Bearing Reference	*d*	*D*	*B*	*Bearing Reference*	*d*	*D*	*B*
0006	0·6	2·5	1·0	0010	1·0	4·0	1·6
0010	1·0	3·0	1·0	0015	1·5	5·0	2·0
0015	1·5	4·0	1·2	0020	2·0	6·0	2·3
0020	2·0	5·0	1·5	0025	2·5	7·0	2·5
0025	2·5	6·0	1·8	0030	3·0	8·0	3·0
0030	3·0	7·0	2·0	0040	4·0	11·0	4·0
0040	4·0	9·0	2·5	0050	5·0	13·0	4·0
0050	5·0	11·0	3·0				
0060	6·0	13·0	3·5				

EXTRA LIGHT SERIES: ISO DIMENSION SERIES 19

Single-row radial ball bearing (BRX) and cylindrical roller bearing (RRX) are specified from 100 to 500. The corresponding dimensions relating to each number after the capitals are identical for both, this number being the same as *d* in the table; e.g. the first two are BRX100 and RRX100.

d	*D*	*B*	*d*	*D*	*B*
100	140	20	260	360	46
110	150	20	280	380	46
120	165	22	300	420	56
130	180	24	320	440	56
140	190	24	340	460	56
150	210	28	360	480	56
160	220	28	380	520	65
170	230	28	400	540	65
180	250	33	420	560	65
190	260	33	440	600	74
200	280	38	460	620	74
220	300	38	480	650	78
240	320	38	500	670	78

RADIAL BEARINGS *(Continued)*

EXTRA LIGHT SERIES: ISO DIMENSION SERIES 10: METRIC SIZES

(All dimensions in millimetres)

Single-row radial ball bearings (BRE) are specified from 006 to 500, single-row angular contact ball bearings (ACE) from 010 to 240, and cylindrical roller bearings from (RRE) 020 to 500. The corresponding dimensions relating to each number after the capitals are identical for all, this number being the same as *d* in the table e.g. the first is BRE006.

d	*D*	*B*	*d*	*D*	*B*
6	17	6	110	170	28
7	19	6	120	180	28
8	22	7	130	200	33
9	24	7	140	210	33
10	26	8	150	225	35
12	28	8	160	240	38
15	32	9	170	260	42
17	35	10	180	280	46
20	42	12	190	290	46
25	47	12	200	310	51
30	55	13	220	340	56
35	62	14	240	360	56
40	68	15	260	400	65
45	75	16	280	420	65
50	80	16	300	460	74
55	90	18	320	480	74
60	95	18	340	520	82
65	100	18	360	540	82
70	110	20	380	560	82
75	115	20	400	600	90
80	125	22	420	620	90
85	130	22	440	650	94
90	140	24	460	680	100
95	145	24	480	700	100
100	150	24	500	720	100

RADIAL BEARINGS (*Continued*)

LIGHT SERIES: ISO DIMENSION SERIES 02: METRIC SIZES

Single-row radial ball bearings (BRL) are specified from 003 to 340, single-row angular contact ball bearings (ACL) from 010 to 240, double-row self-aligning ball bearings (BAL) from 005 to 110, cylindrical roller bearings (RRL) from 010 to 340. The corresponding dimensions relating to each number after the capitals are identical for all, this number being the same as *d* in the table; e.g. the first is BRL003, and *d* = 10 refers to BRL010, ACL010, BAL010, and RRL010. Thrust collar sizes are specified for sizes 017 to 260.

d	*D*	*B*	*d*	*D*	*B*
3·0	10·0	4·0	75	130	25
4·0*	13·0*	5·0*	80	140	26
5·0*	16·0*	5·0*	85	150	28
4	16	5	90	160	30
5	19	6	95	170	32
6	19	6	100	180	34
7	22	7	105	190	36
8	22	7	110	200	38
9	26	8	120	215	40
10	30	9	130	230	40
12	32	10	140	250	42
15	35	11	150	270	45
17	40	12	160	290	48
20	47	14	170	310	52
25	52	15	180	320	52
30	62	16	190	340	55
35	72	17	200	360	58
40	80	18	220	400	65
45	85	19	240	440	72
50	90	20	260	480	80
55	100	21	280	500	80
60	110	22	300	540	85
65	120	23	320	580	92
70	125	24	340	620	92

*Extra light for BRL only.

RADIAL BEARINGS *(Continued)*

MEDIUM SERIES: ISO DIMENSION SERIES 03: METRIC SIZES

Single-row radial ball bearings (BRM) are specified from 010 to 200, single-row angular contact ball bearings (ACM) from 010 to 200, double-row self-aligning ball bearings (BAM) from 010 to 140, cylindrical roller bearings (RRM) from 010 to 200. The corresponding dimensions relating to each number after the capitals are identical for all, this number being the same as *d* in the table; e.g., the first is BRM010, ACM010, BAM010 and RRM010. Thrust collar sizes are specified for *d*: 17 mm and upwards.

d	*D*	*B*	*d*	*D*	*B*
10	35	11	80	170	39
12	37	12	85	180	41
15	42	13	90	190	43
17	47	14	95	200	45
20	52	15	100	215	47
25	62	17	105	225	49
30	72	19	110	240	50
35	80	21	120	260	55
40	90	23	130	280	58
45	100	25	140	300	62
50	110	27	150	320	65
55	120	29	160	340	68
60	130	31	170	360	72
65	140	33	180	380	75
70	150	35	190	400	78
75	160	37	200	420	80

All dimensions in millimetres

RADIAL BEARINGS (*Continued*)

HEAVY SERIES: ISO DIMENSION SERIES 04: METRIC SIZES

Single-row ball bearings (BRH), single-row angular contact ball bearings (ACH), double-row self-aligning ball bearings (BAH), and cylindrical roller bearings (RRH) are all specified for the series 017 to 100. The corresponding dimensions relating to each number after the capitals are identical for all, this number being the same as *d* in the table; e.g. the first is BRH017, ACH017, BAH017 and RRH017. Thrust collar sizes are specified from 017 to 100.

d	*D*	*B*	*d*	*D*	*B*
17	62	17	60	150	35
20	72	19	65	160	37
25	80	21	70	180	42
30	90	23	75	190	45
35	100	25	80	200	48
40	110	27	85	210	52
45	120	29	90	225	54
50	130	31	95	240	55
55	140	33	100	250	58

SINGLE-ROW SEPARABLE BALL BEARINGS MAGNETO TYPE: METRIC SIZES

The sizes range from BML003 to BML020

d	*D*	*B*	*d*	*D*	*B*
3	16	5	12	32	7
4	16	5	13	30	7
5	16	5	14	35	8
6	21	7	15	35	8
7	22	7	16	38	10
8	24	7	17	44	11
9	28	8	18	40	9
10	28	8	19	40	9
11	32	7	20	47	12

THRUST BALL BEARINGS

EXTRA LIGHT SERIES: ISO DIMENSION SERIES 11

Single-direction thrust ball bearings of this series are all designated TFX and are specified for values of *d* from 10 to 240, the first being TFX 010. *H* is the overall thickness or height.

d	*D*	*H*	*d*	*D*	*H*	*d*	*D*	*H*
10	24	9	55	78	16	130	170	30
12	26	9	60	85	17	140	180	31
15	28	9	65	90	18	150	190	31
17	30	9	70	95	18	160	200	31
20	35	10	75	100	19	170	215	34
25	42	11	80	105	19	180	225	34
30	47	11	85	110	19	190	240	37
35	52	12	90	120	22	200	250	37
40	60	13	100	135	25	220	270	37
45	65	14	110	145	25	240	300	45
50	70	14	120	155	25			

THRUST BALL BEARINGS (*Continued*)

LIGHT SERIES: ISO DIMENSION SERIES 12

Single-direction thrust ball bearings of this series have the reference letters TFL and sizes are specified from 10 to 200, the first being TFL 010.

d	*D*	*H*	*d*	*D*	*H*	*d*	*D*	*H*
10	26	11	50	78	22	110	160	38
12	28	11	55	90	25	120	170	39
15	32	12	60	95	26	130	190	45
17	35	12	65	100	27	140	200	46
20	40	14	70	105	27	150	215	50
25	47	15	75	110	27	160	225	51
30	52	16	80	115	28	170	240	55
35	62	18	85	125	31	180	250	56
40	68	19	90	135	35	190	270	62
45	73	20	100	150	38	200	280	62

THRUST BALL BEARINGS *(Continued)*

MEDIUM SERIES: ISO DIMENSION SERIES 13

Single-direction thrust ball-bearings of this series are all designated TFM and are specified for values of *d* from 25 to 150, the first being TFM 025.

d	*D*	*H*	*d*	*D*	*H*
25	32	18	75	135	44
30	60	21	80	140	44
35	68	24	85	150	49
40	78	26	90	155	50
45	85	28	100	170	55
50	95	31	110	190	63
55	105	35	120	210	70
60	110	35	130	225	75
65	115	36	140	240	80
70	125	40	150	250	80

THRUST BALL BEARINGS *(Continued)*

HEAVY SERIES: ISO DIMENSION SERIES 14

Single-direction thrust ball-bearings of this series are all designated TFH and are specified for values of *d* from 25 to 100, the first being TFH 025.

d	*D*	*H*	*d*	*D*	*H*
25	60	24	65	140	56
30	70	28	70	150	60
35	80	32	75	160	65
40	90	36	80	170	68
45	100	39	85	180	72
50	110	43	90	190	77
55	120	48	100	210	85
60	130	51			

FERROUS BARS

BS 4229: Part 2: 1969 makes recommendations for metric sizes of ferrous bars, which are typically as follows:

1. Hot-rolled Non-alloy Steel Round Bars: Diameter in Millimetres

First choice 6·0 8·0 . . . 260 280 300
Second choice 6·5 7·0 9·0 . . . 270 290 320

2. Hot-rolled Non-alloy Steel Square Bars: Side in Millimetres

First choice 7·0 8·0 9·0 . . . 18·0 20·0 . . . 120 130 140 150
Second choice 19·0 125

3. Hot-rolled Non-alloy Steel Hexagon Bars: Width in Millimetres

First choice 10·0 11·0 . . . 85 90 95 100
Second choice 9·5 10·5 88 93 98

4. Non-alloy Bright Steel Round Bars: Diameter in Millimetres

First choice 14·0 15·0 16·0 . . . 75·0 80 85 . . . 180 200
Second choice 14·5 15·5 76·0

5. Non-alloy Bright Steel Square Bars: Side in Millimetres

First choice 13·0 14 15 16 18 . . . 35 40 45 . . . 90 100
Second choice 17 36 41

6. Non-alloy Bright Steel Hexagon Bars: Width in Millimetres

First choice 12·0 13 14 17 . . . 36 41 . . . 90 95 100
Second choice 15 16 37 38 40

FERROUS BARS *(Continued)*

7. Non-alloy Steel Round Bars for Concrete Reinforcement: Diameter in Millimetres

6·0 8·0 10·0 12·0 16·0 20·0 25·0 32·0 40·0

8. Non-alloy Rods for Wire Drawing: Diameter in Millimetres

5·5 6·0 6·5 7·0 · · · 11·5 12·0 12·5 13·0

9. Hot-rolled and Bright Alloy and Stainless-steel Round Bars: Diameter in Millimetres

First choice 5·0 6·0 · · · 28·0 30·0 · · · 240 260 280
Second choice 5·5 6·5 28·5 29·0 29·5 250 270

10. Hot-rolled and Bright Alloy and Stainless-steel Square Bars: Side in Millimetres

First choice 5·0 5·5 6·0 7·0 · · · 42 45 · · · 120 130 140 150
Second choice 6·5 43 44 125

11. Hot-rolled and Bright Alloy and Stainless-steel Hexagon Bars: Width in Millimetres

First choice 3·2* 5·5* 7·0 8·0 10·0 · · · 32 35 · · · 90 95 100
Second choice 4·0* 5·0* 6·0 93 98

12. Close-tolerance Drawn or Ground Tool Steel Round Bars: Diameter in Millimetres

First choice 1·0 1·1 1·2 1·4 · · · 4·2 4·5 · · · 9·5 10·0
Second choice 1·3 4·3 4·4 9·6 9·7 9·8 9·9

* Bright bars only.

DIMENSIONS OF STEEL TUBES AND SOCKETS

BS 1387 AND ISO RECOMMENDATIONS R50 AND R65

Nominal Bore		*Outside Diameter*	*Thickness, mm*			*Minimum Socket Length mm*		
mm	*in*	*approx. mm*	*Light*	*Medium*	*Heavy*	*Light*	*Medium*	*Heavy*
6	1/8	10·2	1·8	2·0	2·65	17	17	17
8	1/4	13·5	1·8	2·35	2·9	25	25	25
10	3/8	17·2	1·8	2·35	2·9	26	26	26
15	1/2	21·3	2·0	2·65	3·25	34	34	34
20	3/4	26·9	2·35	2·65	3·25	36	36	36
25	1	33·7	2·65	3·25	4·05	43	43	43
32	1 1/4	42·4	2·65	3·25	4·05	48	48	48
40	1 1/2	48·3	2·9	3·25	4·05	48	48	48
50	2	60·3	2·9	3·65	4·5	56	56	56
65	2 1/2	76·1	3·25	3·65	4·5	65	65	65
80	3	88·9	3·25	4·05	4·85	71	71	71
90	3 1/2	101·6	3·65	4·05	4·85	75	75	75
100	4	114·3	3·65	4·5	5·4	83	83	83
125	5	139·7	—	4·85	5·4	—	92	92
150	6	165·1	—	4·85	5·4	—	92	92

Note: Light tubes shall have brown colour bands, medium have blue, and heavy have red; the bands are 75 mm wide.

METAL SHEET, STRIP AND WIRE

ISO RECOMMENDATION R388

The designation of thicknesses of sheet and diameters of wire by arbitrary gauge numbers will become obsolete and will be replaced by metric ranges of sizes from 0·020 to 25 mm. The sizes should be selected from the series of preferred numbers (see p. 46) in the series R10, R20 or R40 with a preference in that order. Each series has the same basic size as that quoted below for the R10 series: the additional numbers for the R20 series are enumerated under those for the R10 series and the numbers, if required, for the R40 series can be obtained from the original recommendation. All sizes in millimetres.

R10	0·020	0·025	0·032	0·040	0·050	0·063	0·080	0·100	0·125
R20	0·022	0·028	0·036	0·045	0·056	0·071	0·090	0·112	0·140
R10	0·160	0·200	0·250	0·315	0·400	0·500	0·630	0·800	1·000
R20	0·180	0·224	0·280	0·355	0·450	0·560	0·710	0·900	1·12
R10	1·25	1·60	2·00	2·50	3·15	4·00	5·00	6·30	8·00
R20	1·40	1·80	2·24	2·80	3·55	4·50	5·60	7·10	9·00
R10	10·0	12·5	16·0	20·0	25·0				
R20	11·2	14·0	18·0	22·4					

WEIGHTS OF METRIC SIZE STEEL BARS (KILOGRAMMES PER METRE)

CIRCULAR SECTION

Dia. *mm*	*Weight* *kg*	*Dia.* *mm*	*Weight* *kg*	*Dia.* *mm*	*Weight* *kg*
4	0·099	38	8·903	80	39·460
5	0·154	39	9·378	82	41·460
6	0·222	40	9·865	84	43·500
7	0·302	41	10·360	85	44·550
8	0·395	42	10·880	86	45·600
9	0·499	43	11·400	88	47·750
10	0·617	44	11·940	90	49·940
11	0·746	45	12·480	92	52·180
12	0·888	46	13·050	94	54·480
13	1·042	47	13·620	95	55·640
14	1·208	48	14·200	96	56·820
15	1·387	49	14·800	98	59·210
16	1·578	50	15·410	100	61·650
17	1·782	51	16·040	105	67·970
18	1·998	52	16·670	110	74·600
19	2·226	53	17·320	115	81·540
20	2·466	54	17·980	120	88·780
21	2·719	55	18·650	125	96·330
22	2·984	56	19·340	130	104·200
23	3·261	57	20·030	135	112·400
24	3·551	58	20·740	140	120·800
25	3·853	59	21·460	145	129·600
26	4·168	60	22·200	150	138·700
27	4·495	62	23·700	160	167·840
28	4·834	64	25·250	170	178·200
29	5·185	65	26·050	180	199·760
30	5·549	66	26·860	190	222·560
31	5·925	68	28·510	200	246·600
32	6·313	70	30·210	220	298·400
33	6·714	72	31·960	240	355·120
34	7·127	74	33·760	260	416·800
35	7·553	75	34·680	280	483·200
36	7·990	76	35·610	300	554·800
37	8·440	78	37·510		

Table reproduced by permission of MacReady's Metal Co. Ltd.

WEIGHTS OF METRIC SIZE STEEL BARS (KILOGRAMMES PER METRE) (Continued)

SQUARE SECTION				HEXAGONAL SECTION			
Size *mm*	*Weight* *kg*	*Size* *mm*	*Weight* *kg*	*Size(a/f)* *mm*	*Weight* *kg*	*Size(a/f)* *mm*	*Weight* *kg*
4	0·126	32	8·038	4	0·109	32	6·961
5	0·196	33	8·549	5	0·170	33	7·403
6	0·283	34	9·075	6	0·244	34	7·859
7	0·385	35	9·616	7	0·333	35	8·328
8	0·502	36	10·170	8	0·435	36	8·811
9	0·636	37	10·750	9	0·551	37	9·307
10	0·785	38	11·340	10	0·680	38	9·817
11	0·950	39	11·940	11	0·823	39	10·340
12	1·130	40	12·560	12	0·979	40	10·880
13	1·327	41	13·200	13	1·149	41	11·430
14	1·539	42	13·850	14	1·332	42	11·990
15	1·766	43	14·520	15	1·530	43	12·570
16	2·010	44	15·200	16	1·740	44	13·160
17	2·269	45	15·900	17	1·965	45	13·770
18	2·543	46	16·610	18	2·203	46	14·390
19	2·834	47	17·340	19	2·454	47	15·020
20	3·140	48	18·090	20	2·719	48	15·660
21	3·462	49	18·850	21	2·998	50	17·000
22	3·799	50	19·630	22	3·290	52	18·340
23	4·153	52	21·230	23	3·596	55	20·520
24	4·522	54	22·890	24	3·616	60	24·420
25	4·906	55	23·750	25	4·249	65	28·660
26	5·307	56	24·620	26	4·596	70	32·230
27	5·723	58	26·410	27	4·956	75	38·150
28	6·154	60	28·260	28	5·330	80	43·410
29	6·602	62	30·180	29	5·717		
30	7·065	64	32·150	30	6·118		
31	7·544	65	33·170	31	6·533		

Stocks of bright steel bars of various brands became available in the United Kingdom during 1969 for rounds, squares and hexagons for a variety of steels including:

Rounds: 2 mm to 150 mm for EN 1A, EN 3B, EN 8, EN16T and EN 32
Squares: 4 mm to 13 mm for EN 1A
Hexagons: 4 mm to 15 mm for EN 1A

All the above in accordance with BS 970: 1955
Flats also available in sizes from 15 × 5 mm to 100 × 40 mm in mild steel.

Table reproduced by permission of MacReady's Metal Co. Ltd.

HOT-ROLLED STEEL ROUND BARS: METRIC SERIES: MARCH, 1969

ISO RECOMMENDATION R1035/1

Dimensions, sectional areas and weights are given in the above recommendation in millimetres, centimetres squared and kilogrammes per metre, respectively, for a range of diameters from 5 mm to 220 mm. A second table gives the equivalent inch values and masses in pounds per foot.

HOT-ROLLED STEEL SQUARE BARS: METRIC SERIES: MARCH, 1969

ISO RECOMMENDATION R1035/11

Widths, sectional areas and weights are given in the above recommendation in millimetres, centimetres squared and kilogrammes per metre, respectively, for a range of widths from 5 mm to 130 mm. A second table gives the equivalent inch values and masses in pounds per foot.

No specification for steels is given in either recommendation. This is perhaps not surprising since the number of steels developed throughout Europe is great with but small variations in some cases between steels that have different and yet well-known trade names. It is, however, possible that some types of mild steel might be standardised in the future by S.I. in spite of the above difficulty associated with various trade products.

SCREW THREAD TERMS

Angle of Thread (included angle)—The angle of thread is the angle included between the sides of the thread, measured in an axial plane.

Axis of Thread—The longitudinal central line through the screw from which all corresponding parts are equally distant.

Basic Size—The theoretical or nominal standard size from which all variations are made.

Clearance—The distance, measured perpendicular to the axis, between the design forms at the root of the internal thread and the crest of the external thread (major) and from crest to root (minor).

Convolution—One full turn of a screw.

Core Diameter—See Minor Diameter.

Crest—The top surface joining the two sides of a thread.

Depth of Engagement—The radial distance by which the thread-forms of two mating threads overlap each other.

Depth of Thread—The depth of thread, in profile, is the distance between the crest and the root of the thread measured normal to the axis.

Drunken Thread—A thread in which the advance of the helix is irregular in every convolution.

Effective (or pitch) diameter.—The diameter of the pitch cylinder of a parallel thread, or the pitch cone of a taper thread in a specified plane normal to the axis.

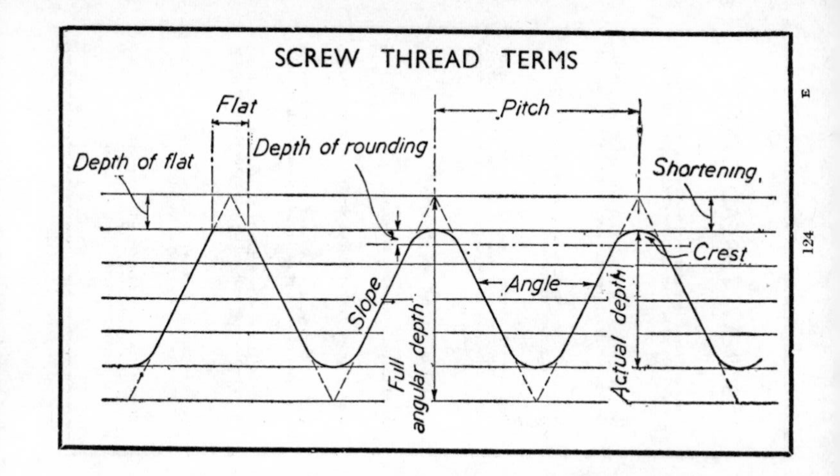
SCREW THREAD TERMS
Flat
Pitch
Depth of rounding
Depth of flat
Shortening,
Crest
Angle
Slope
Full angular depth
Actual depth

SCREW THREAD TERMS
(Continued)

Flank Angles—The angles between the flanks of a thread and a plane perpendicular to the axis, measured in an axial plane.

Flank of Thread—The surface of the thread which connects the crest with the root.

Full Diameter—See Major Diameter.

Helix Angle—The angle made by the helix of a thread at the pitch diameter with a plane perpendicular to the axis.

Lead—The distance a screw thread advances axially in one turn. On a single thread the lead and pitch are identical. On a double thread the lead is twice the pitch ; on a triple thread the lead is three times the pitch, etc.

Length of Engagement—The length of contact between two mating parts measured axially.

Limits—The extreme sizes which are prescribed for any dimension to provide variations in fit and workmanship.

Major Diameter—On a parallel screw thread the major diameter is the largest diameter of the thread on the screw or nut. The term " major diameter " replaces the term " outside diameter " as applied to the thread of a screw and also the term " full diameter " as applied to the thread of a nut.

Minor Diameter—On a parallel screw thread the minor diameter is the smallest diameter of the thread on the screw or nut. The term " minor diameter " and " root diameter " as applied to the thread on the screw or nut ; the root diameter. The term " inside diameter " is usually applied to the smallest diameter of a nut.

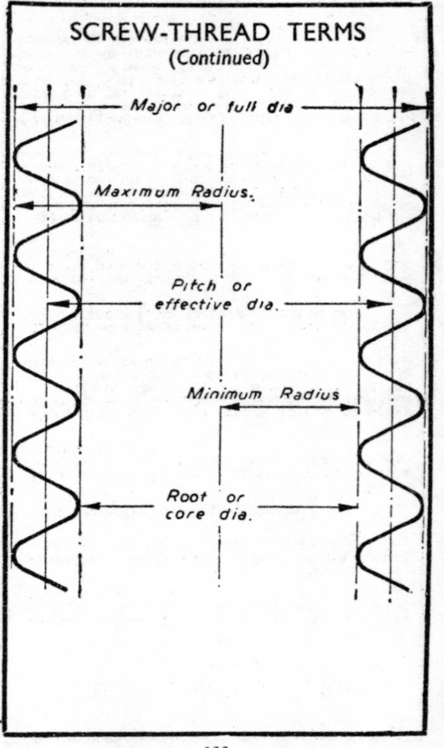
SCREW-THREAD TERMS
(Continued)
Major or full dia
Maximum Radius.
Pitch or
effective dia.
Minimum Radius
Root or
core dia.

SCREW THREAD TERMS

(Continued)

Number of Threads—The number of threads in a length of 1 inch.

Outside Diameter—See Major Diameter.

Pitch—The distance from a point on a screw thread to a corresponding point on the next convolution of the thread measured parallel to the axis. The pitch equals 1 divided by the number of threads per inch.

Pitch Diameter—See Effective Diameter.

Root—The bottom surface joining the sides of two adjacent threads.

Root Diameter—See Minor Diameter.

Screw Thread—A screw thread is a ridge of some desired profile generated in the form of a helix on the inside or outside of a cylinder or cone.

Spiral Angle—The spiral angle of a screw thread is found from the formula $\frac{P}{PD}$ = tan of spiral angle where P = Pitch and PD = Pitch diameter.

Thread, Single—A thread in which the lead is equal to the pitch.

Thread, Double—A thread in which the lead is equal to twice the pitch.

Thread Triple—A thread in which the lead is equal to three times the pitch.

Thread, Quadruple—A thread in which the lead is equal to four times the pitch.

(*Note—When ordering tools with multiple threads both pitch and lead should be stated.*)

Tolerance—The difference between the limits or maximum and minimum sizes specified for a given dimension of a part or gauge. A tolerance may be expressed as plus or minus or both, with reference to the basic size. A total tolerance is the sum of a plus and minus tolerance.

Thread Forms

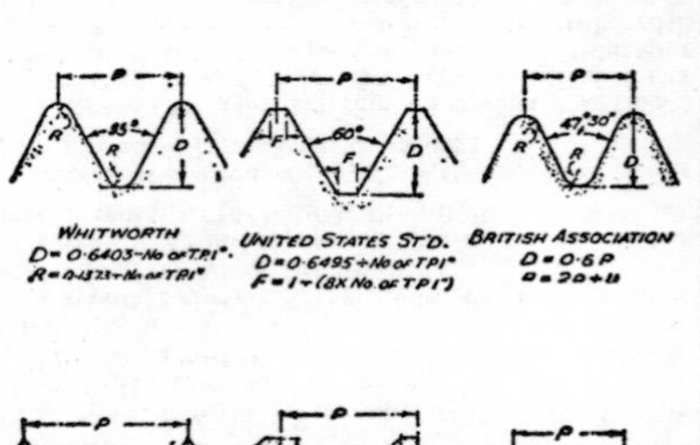

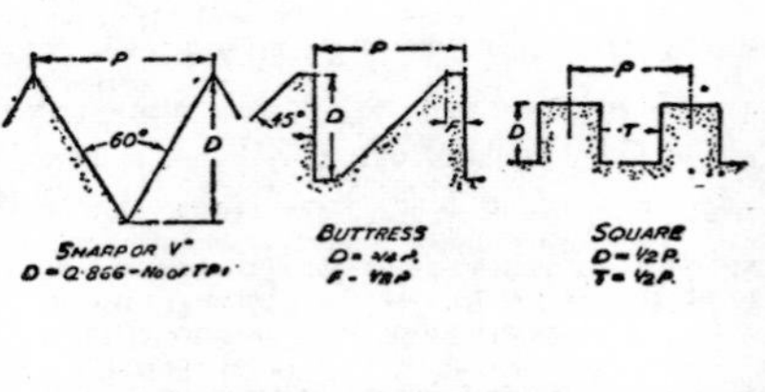

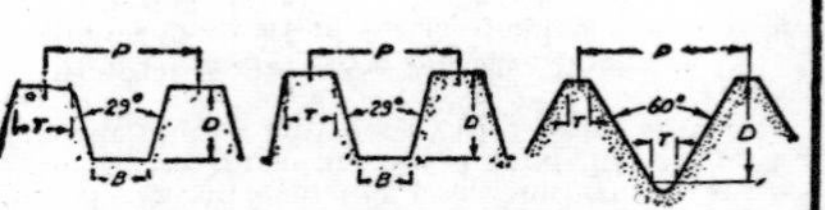

These diagrams indicate the thread forms and proportions of most of the standard screw threads.

Screw Threads

SINGLE-START THREAD

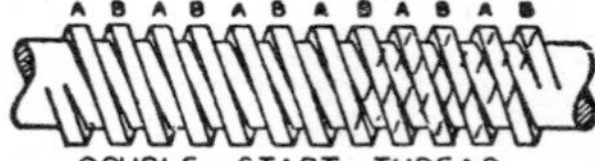

DOUBLE-START THREAD

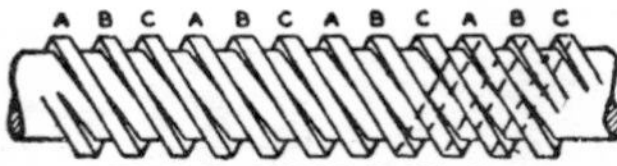

TRIPLE-START THREAD

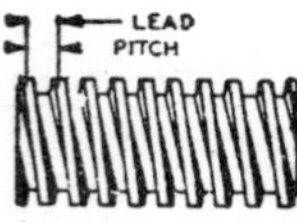

SINGLE THREAD

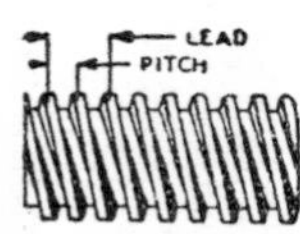

DOUBLE THREAD

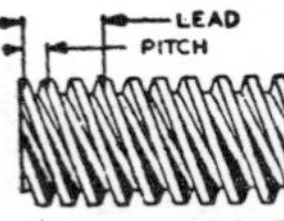

TRIPLE THREAD

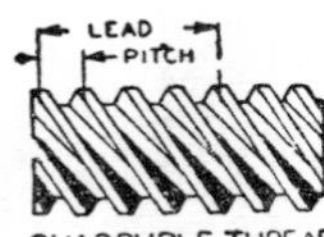

QUADRUPLE THREAD

STANDARD WORM THREAD

This is similar in form to the Acme thread. The angle made by the sides of the thread is 29 degrees. Depth of thread is greater, and so the width of the flat at top and bottom is less.

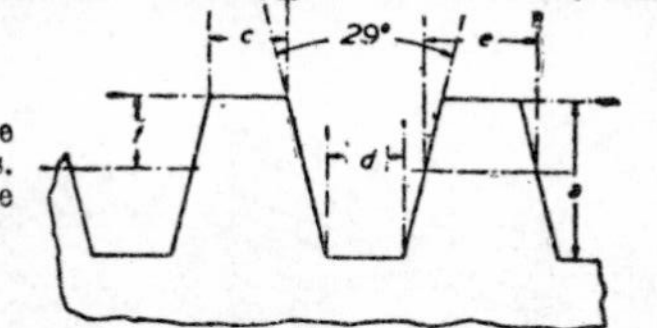

MEASUREMENT OF ACME THREADS

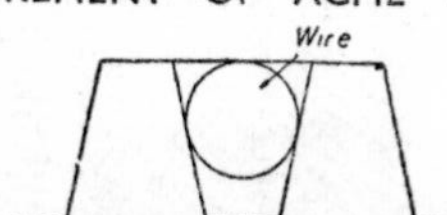

In measuring Acme threads the wire diameters are selected so that when the thread is of the correct shape and the pitch diameter is correct, the wire is flush with the tops of the thread in a tap and projects 0.010in. above the tops of the threads on a screw.

The formula for finding the diameter of the wire for any number of threads per in. is :

$$\text{Diameter of wire} = (p \times 0.6293 + 0.0052) \times 0.7743 \text{ where } p = \text{pitch} = \frac{1}{\text{no. of threads per in.}}$$

ACME STANDARD THREAD

The Acme Standard Thread is an adaptation of the most commonly used style of Worm Thread and is intended to take the place of the square thread.

It is a little shallower than the Worm Thread, but the same depth as the square thread and much stronger than the latter.

The various parts of the Acme Standard Thread are obtained as follows :

Width of Point of Tool for Screw or Tap Thread =

$$\frac{.3707}{\text{No. of Thds. per in.}} - .0052$$

Width of Screw or Nut Thread = $\frac{.3707}{\text{No. of Thds. per in.}}$

Diameter of Tap = Diameter of Screw + .020.

Diameter of Tap or Screw at Root =

Diameter of Screw

$$- \left\{ \frac{1}{\text{No. of Linear Thds. per in.}} + .020 \right\}$$

Depth of Thread = $\frac{1}{2 \times \text{No. of Thds. per in.}} + .010$

Acme Standard Screw Thread

Width of point of tool for screw or tap thread $= \dfrac{.3707}{\text{threads per in.}} - .0052$

Formula
$$\begin{cases} p = \text{pitch} = \dfrac{1}{\text{No. threads per inch}} \\ d = \text{depth} = \tfrac{1}{2}p + \cdot 010 \\ f = \text{flat on top of thread} = p \times .3707 \\ f' = \text{"on bottom"} = p \times .3707 - .0052 \end{cases}$$

Pitch	*No. of Thr'ds per In.*	*Depth of Thread*	*Width at Top of Thread*	*Width at Bottom of Thread*	*Space at Top of Thread*	*Thickness at Root of Thread*
2	½	1.010	.7414	.7362	1.2586	1.2637
1⅞	8/15	0.9475	.6950	.6897	1.1799	1.1850
1¾	4/7	0.8850	.6487	.6435	1.1012	1.1064
1⅝	8/13	0.8225	.6025	.5973	1.0226	1.0277
1½	⅔	0.7600	.5560	.5508	0.9439	0.9491
1 7/16	16/23	0.7287	.5329	.5277	0.9046	.09097
1⅜	8/11	0.6975	.5097	.5045	0.8652	0.8704
1 5/16	16/21	0.6662	.4865	.4813	0.8259	0.8311
1¼	⅘	0.635	.4633	.4581	0.7866	0.7918
1 3/16	16/19	0.6037	.4402	.4350	0.7472	0.7525
1⅛	8/9	0.5725	.4170	.4118	0.7079	0.7131
1 1/16	16/17	0.5412	.3938	.3886	0.6686	0.6739
1	1	0.510	.3707	.3655	0.6293	0.6345
15/16	1 1/15	0.4787	.3476	.3424	0.6686	0.5950
⅞	1 1/7	0.4475	.3243	.3191	0.5506	0.5558
13/16	1 3/13	0.4162	.3012	.2960	0.5112	0.5164

Acme Standard Screw Thread *cont.*

Pitch	*No. of Thr'ds per In.*	*Depth of Thread*	*Width at Top of Thread*	*Width at Bottom of Thread*	*Space at Top of Thread*	*Thickness at Root of Thread*
$\frac{3}{4}$	$1\frac{1}{3}$	0.385	.2780	.2728	0.4720	0.4772
$\frac{11}{16}$	$1^5/_{11}$	0.3537	.2548	.2496	0.4327	0.4379
$\frac{2}{3}$	$1\frac{1}{2}$	0.3433	.2471	.2419	0.4194	0.4246
$\frac{5}{8}$	$1\frac{3}{5}$	0.3225	.2316	.2264	0.3934	0.3986
$\frac{9}{16}$	$1^7/_9$	0.2912	.2085	.2033	0.3539	0.3591
$\frac{1}{2}$	2	0.260	.1853	.1801	0.3147	0.3199
$\frac{7}{16}$	$2^2/_7$	0.2287	.1622	.1570	0.2752	0.2804
$\frac{2}{5}$	$2\frac{1}{2}$	0.210	.1482	.1430	0.2518	0.2570
$\frac{3}{8}$	$2\frac{2}{3}$	0.1975	.1390	.1338	0.2359	0.2411
$\frac{1}{3}$	3	0.1766	.1235	.1183	0.2098	0.2150
$\frac{5}{16}$	$3\frac{1}{5}$	0.1662	.1158	.1106	0.1966	0.2018
$^2/_7$	$3\frac{1}{2}$	0.1528	.1059	.1007	0.1797	0.1849
$\frac{1}{4}$	4	0.1350	.0927	.0875	0.1573	0.1625
$^2/_9$	$4\frac{1}{2}$	0.1211	.0824	.0772	0.1398	0.1450
$\frac{1}{5}$	5	0.110	.0741	.0689	0.1259	0.1311
$\frac{3}{16}$	$5\frac{1}{3}$	0.1037	.0695	.0643	0.1179	0.1232
$\frac{1}{6}$	6	0.0933	.0617	.0565	0.1049	0.1101
$^1/_7$	7	0.0814	.0530	.0478	0.0899	0.0951
$\frac{1}{8}$	8	0.0725	.0463	.0411	0.0787	0.0839
$^1/_9$	9	0.0655	.0413	.0361	0.0699	0.0751
$^1/_{10}$	10	0.060	.0371	.0319	0.0629	0.0681
$\frac{1}{16}$	16	0.0412	.0232	.0180	0.0392	0.0444

TABLE OF ACME THREAD PARTS

Number of Threads per Inch	Depth of Thread	Width at Top of Thread	Width at Bottom of Thread	Space at Top of Thread	Thickness at Root of Thread
1	.5100	.3707	.3655	.6293	.6345
1⅓	.3850	.2780	.2728	.4720	.4772
2	.2600	.1853	.1801	.3147	.3199
3	.1767	.1235	.1183	.2098	.2150
4	.1350	.0927	.0875	.1573	.1625
5	.1100	.0741	.0689	.1259	.1311
6	.0933	.0618	.0566	.1049	.1101
7	.0814	.0529	.0478	.0899	.0951
8	.0725	.0463	.0411	.0787	.0839
9	.0655	.0413	.0361	.0699	.0751
10	.0600	.0371	.0319	.0629	.0681

Record
VICES ENGINEERS, MECHANICS, FITTERS
PIPE TOOLS CUTTERS. VICES. WELDING CLAMPS, WRENCHES
CRAMPS HEAVY DUTY G CRAMPS
BENDING MACHINES PORTABLE AND BENCH MOUNTED
TUBE AND CONDUIT BENDERS
BOLT CUTTERS 29 TYPES AND SIZES
* Send for catalogue E2 on the full range
C. & J, Hampton Ltd., Parkway Works, Sheffield S9 3BL

Gear Tooth Terms

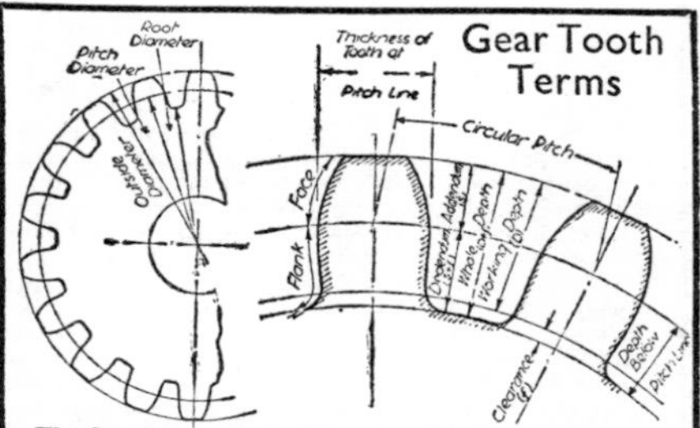

The *Involute* system of gear teeth is more generally used than any other system, the standard pressure angle being 14½ degrees. This means that a gear with this form of tooth will mesh with the teeth of a standard rack whose straight sides incline 14½ degrees from the vertical.

The *Stub-Tooth* system uses a shorter tooth than standard and the pressure angle is usually 20 degrees. This system is based on the combination of two diametral pitches. For example :—In a 4-5 pitch, the pitch diameter, circular pitch and thickness of tooth on pitch line are obtained from the 4 pitch, while the addendum and dedendum are obtained from the 5 pitch.

Circular Pitch—Distance from the centre of one tooth to the centre of the next measured on the pitch circle.

Diametral Pitch—Number of teeth divided by the pitch diameter or the number of teeth to each inch of diameter.

Pitch Diameter—Diameter of the pitch circle.

Pitch Line or Circle—A line which represents the touching of two cylinders, upon which the teeth are laid out, rolling upon one another, and is the line or circle upon which the *pitch* of teeth is measured.

Addendum (S)—Distance from pitch line to top of tooth.

Dedendum (S+f)—Distance from pitch line to root of tooth.

Clearance (f)—Amount by which the tooth space is cut deeper than the working depth.

Working Depth (D)—Depth in the tooth space to which the tooth of the mating gear extends.

Whole Depth (D+f)—Working depth plus clearance.

SPUR GEARING

The standard gear terms, forming the basis of calculations concerning tooth proportions, are shown below. It should be noted that *circular pitch* refers to the distance,

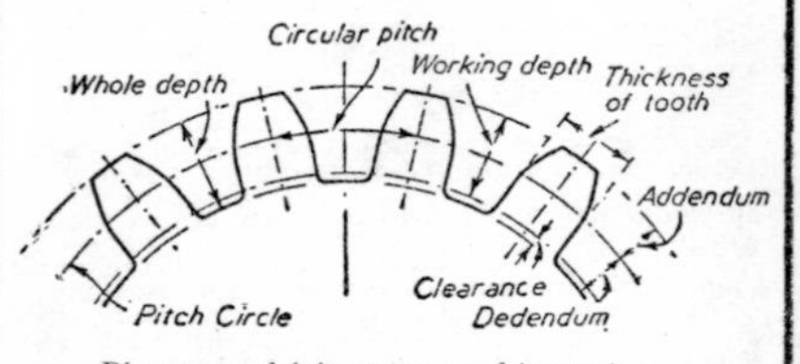

Diagram explaining terms used in gearing.

measured round the pitch line, from the centre of one tooth to the centre of the next tooth; whereas *diametral pitch* refers to the number of teeth in the pitch diameter. It is usual to refer to gears according to their diametral pitch.

Module is found by dividing the pitch diameter by the number of teeth. Standard normal modules should be in accordance with ISO/R54 "Modules and diametral pitches of cylindrical gears for general and heavy engineering" and the tolerances for various parts are given in BS 436, Part 2: 1970. The preferred modules are 1, 1·25, 1·5, 2, 2·5, 3, 4, 5, 6, 8, 10, 12, 16, 20, 25, 32, 40 and 50 with second choice modules at intermediate values. Part 2 includes a diagram similar to the above for the basic rack tooth profile for unit normal metric module based on a pressure angle of 20°, a pitch of 3·1416, an addendum of 1·00 and a dedendum of 1·25.

BEVEL GEARS
(Axes at Right Angles)

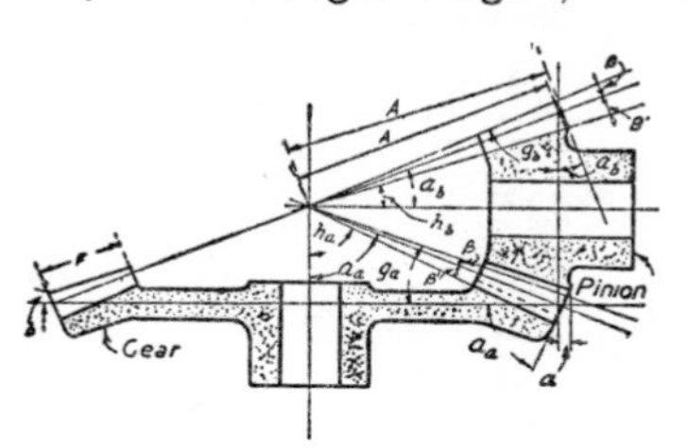

Diagram of symbols relating to right-angle bevel gears.

N_a = No. of teeth on gear.
N_b = No. of teeth on pinion.
P = diametral pitch.
P′ = circular pitch.
α_a = centre angle = angle of edge or pitch angle of gear.
α_b = centre angle = angle of edge or pitch angle of pinion.
β = angle of top.
β' = angle of bottom.
g_a = angle of face of gear.
g_b = angle of face of pinion.
h_a = cutting angle of gear.
h_b = cutting angle of pinion.
A = apex distance from pitch **circle.**
A′ = apex distance from large bottom of tooth.
D = outside diameter.
D′ = pitch diameter.

Bevel Gears (continued)

D'' = working depth of tooth.
s = addendum or module.
t = thickness of tooth at pitch line.
f = clearance at bottom of tooth.
$D''+f$ = whole depth of tooth.
2_a = diameter increment.
b = distance from top of tooth to plane of pitch circle.
F = width of face.

$\tan \alpha_a = \frac{N_a}{N_b}$; $\tan \alpha_b = \frac{N_b}{N_a}$; $\tan \beta = \frac{2 \sin \alpha}{N}$, or $\frac{s}{A}$.

$\tan \beta' = \frac{2.314 \sin \alpha}{N}$; or $\frac{s+f}{A}$.

$g_a = 90° - (\alpha_a + \beta)$; $g_b = 90° - (\alpha + \beta)$.

$h = \alpha - \beta'$.

$A = \frac{1}{2\,P} \sqrt{N_a^2 + N_b^2}$. $\quad A = \frac{N}{2\,P \sin \alpha}$.

$A' = \frac{A}{\cos \beta'}$, or $\frac{N}{2P \sin \alpha \cos \beta}$.

$A = \frac{\frac{1}{2}\,D}{\sin(\alpha + \beta)} \cos \beta$.

$P = \frac{N}{2\,A \sin \alpha}$, or $\frac{N + 2 \cos \alpha}{D}$, or $\frac{\pi}{P'}$.

$P' = \frac{\pi}{P}$.

$D' = \frac{N}{P}$, or $\frac{NP'}{\pi}$, or $\frac{DN}{N + 2 \cos \alpha}$, or $D - \frac{2 \cos \alpha}{P}$.

$D = D' + 2a$.

$2a = 2s \cos \alpha$.

$b = a \tan \alpha$ $\left\{ \begin{array}{l} a \text{ for gear} = b \text{ for pinion.} \\ b \text{ for gear} = a \text{ for pinion.} \end{array} \right.$

$s = \frac{1}{P}$, $\frac{P'}{\pi}$, or $0.3183\,P'$ or $A \tan \beta$.

$s + f = 0.3683\,P'$, or $A \tan \beta'$, or $\frac{1.157}{P}$.

$D'' = 2s$; $D'' + f = \frac{2.157}{P}$, or $0.6866\,P'$.

$t = \frac{P'}{2}$, or $\frac{\pi}{2P'}$.

$F = \frac{A}{3}$, or $\frac{5\,P'}{2}$ $\qquad f = \frac{t}{10}$.

BEVEL GEARS
(Axes at any angle)

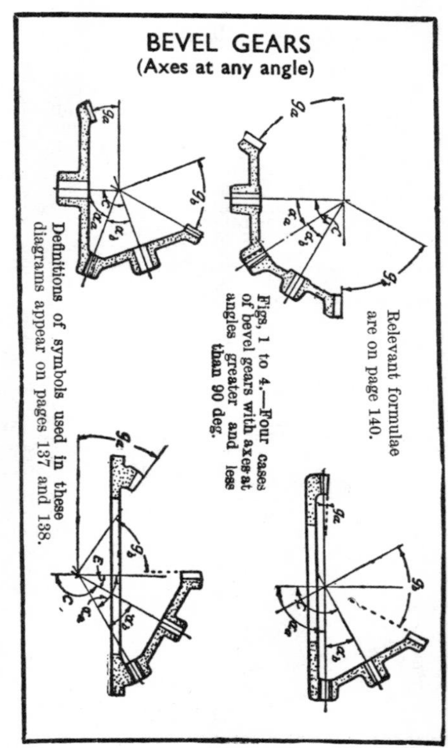

Relevant formulae are on page 140.

Figs. 1 to 4.—Four cases of bevel gears with axes at angles greater and less than 90 deg.

Definitions of symbols used in these diagrams appear on pages 137 and 138.

BEVEL GEARS

(Axes at any angle)

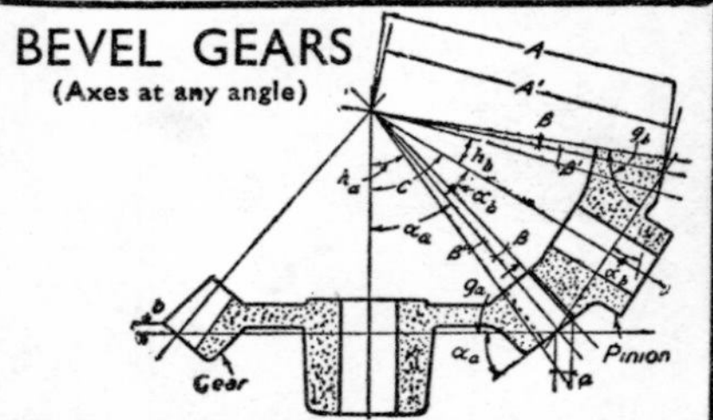

Symbols used in calculations relating to bevel gears with axes at any angle.

The formulae for tooth parts on page 136 apply in both cases. C = angle formed by axes of gears.

$$\tan\alpha_a=\frac{\sin C}{\frac{N_b}{N_a}+\cos C};\ \cot\alpha_a=\frac{N_b}{N_a\sin C}+\cot C.$$

$$\tan\alpha_b=\frac{\sin C}{\frac{N_a}{N_b}+\cos C};\ \cot\alpha_b\frac{N_a}{N_b\sin C}+\cot C.$$

Where the indicating letters *a* and *b* are not used the formulæ apply to gear and pinion.

In the case of Fig. 1 the formulæ given on pages 137 and 138 apply.

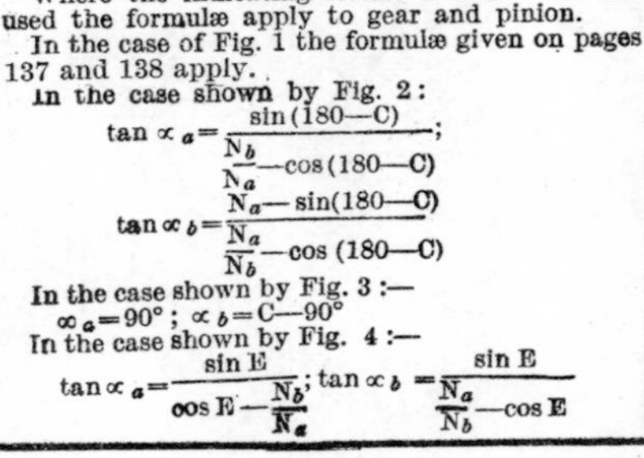

In the case shown by Fig. 2:

$$\tan\alpha_a=\frac{\sin(180-C)}{\frac{N_b}{N_a}-\cos(180-C)};$$

$$\tan\alpha_b=\frac{N_a-\sin(180-C)}{\frac{N_a}{N_b}-\cos(180-C)}$$

In the case shown by Fig. 3 :—

$$\alpha_a=90^\circ;\ \alpha_b=C-90^\circ$$

In the case shown by Fig. 4 :—

$$\tan\alpha_a=\frac{\sin E}{\cos E-\frac{N_b}{N_a}};\ \tan\alpha_b=\frac{\sin E}{\frac{N_a}{N_b}-\cos E}$$

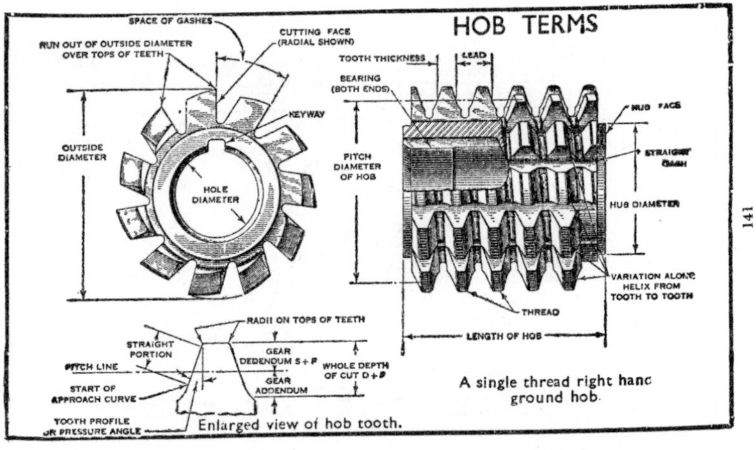

Enlarged view of hob tooth.

A single thread right hand
ground hob.

Circular Pitch

Circular Pitch is the Distance from the Centre of one Tooth to the Centre of the next Tooth, measuring along the Pitch Circle

To Get	*Having*	*Rule*	*Formula*
The Circular Pitch	The Diametral Pitch	Divide 3.1416 by the Diametral Pitch	$P' = \frac{3.1416}{P}$
The Circular Pitch	The Pitch Diameter and Number of Teeth	Divide Pitch Diameter by the product of 0.3183 and Number of Teeth	$P' = \frac{D'}{0.3183N}$
The Circular Pitch	The Outside Diameter and Number of Teeth	Divide Outside Diameter by the product of 0.3183 and Number of Teeth plus 2	$P' = \frac{D}{0.3183(N+2)}$
Pitch Diameter	Number of Teeth and the Circular Pitch	The continued product of the Number of Teeth, the Circular Pitch and 0.3183	$D' = NP'.3183$
Pitch Diameter	The Number of Teeth and the Outside Diameter	Divide the product of Number of Teeth and Outside Diameter by Number of Teeth plus 2	$D' = \frac{ND}{N+2}$
Pitch Diameter	The Outside Diameter and Circular Pitch	Subtract from the Outside Diameter the product of the Circular Pitch and 0.6366	$D' = D - (P'.6366)$
Pitch Diameter	Addendum and the Number of Teeth	Multiply the Number of Teeth by the Addendum	$D' = Ns$
Outside Diameter	Number of Teeth and the Circular Pitch	The continued product of the number of Teeth plus 2, the Circular Pitch and 0.3183	$D = (N+2)P'.3183$

Circular Pitch (*Continued*)

To Get	*Having*	*Rule*	*Formula*
Outside Diameter	The Pitch Diameter and Circular Pitch	Add to the Pitch Diameter the product of the Circular Pitch and 0.6366	$D = D' + (P'.6366)$
Outside Diameter	The Number of Teeth and the Addendum	Multiply Addendum by Number of Teeth plus 2	$D = s(N+2)$
Number of Teeth	The Pitch Diameter and Circular Pitch	Divide the product of Pitch Diameter and 3.1416 by the Circular Pitch	$N = \frac{D'3.1461}{P'}$
Thickness of Tooth	The Circular Pitch	One half the Circular Pitch	$t = \frac{P'}{2}$
Addendum	The Circular Pitch	Multiply the Circular Pitch by 0.3183 or $s = \frac{D'}{N}$	$s = P'.3183$
Dedendum	The Circular Pitch	Multiply the Circular Pitch by 0.3683	$s + f = P'.3683$
Working Depth	The Circular Pitch	Multiply the Circular Pitch by 0.6366	$D'' = P'.6366$
Whole Depth	The Circular Pitch	Multiply the Circular Pitch by 0.6866	$D' + f = P'.6866$
Clearance	The Circular Pitch	Multiply the Circular Pitch by 0.05	$f = P'.05$
Clearance	Thickness of Tooth	One tenth the Thickness of Tooth at Pitch Line	$f = \frac{t}{10}$

Diametral Pitch

Diametral Pitch is the Number of Teeth to each Inch of the Pitch Diameter.

To Get	*Having*	*Rule*	*Formula*
The Diametral Pitch	The Circular Pitch	Divide 3.1416 by the Circular Pitch	$P = \frac{3.1416}{P'}$
The Diametral Pitch	The Pitch Diameter and Number of Teeth	Divide Number of Teeth by Pitch Diameter	$P = \frac{N}{D}$
The Diametral Pitch	The Outside Diameter and Number of Teeth	Divide Number of Teeth plus 2 by Outside Diameter	$P = \frac{N+2}{D}$
Pitch Diameter	The Number of Teeth and Diametral Pitch	Divide Number of Teeth by the D.P.	$D' = \frac{N}{P}$
Pitch Diameter	The Number of Teeth and Outside Diameter	Divide the product of outside Diameter and Number of Teeth by Number of Teeth plus 2	$D' = \frac{DN}{N+2}$
Pitch Diameter	The Outside Diameter and Diametral Pitch	Subtract from the Outside Diameter the quotient of 2 divided by the D.P.	$D' = D - \frac{2}{P}$
Pitch Diameter	Addendum and Number of Teeth	Multiply Addendum by the Number of Teeth	$D' = sN$
Outside Diameter	The Number of Teeth and Diametral Pitch	Divide Number of Teeth plus 2 by the D.P.	$D = \frac{N+2}{P}$
Outside Diameter	The Pitch Diameter and Diametral Pitch	Add to the Pitch Diameter the quotient of 2 divided by the D.P.	$D = D' + \frac{2}{P}$

Diametral Pitch (*Continued*)

To Get	*Having*	*Rule*	*Formula*
Outside Diameter	The Pitch Diameter and Number of Teeth	Divide the Number of Teeth plus 2 by the Quotient of the Number of Teeth divided by the Pitch Diameter	$D = \frac{N+2}{\frac{N^1}{D^1}}$
Outside Diameter	The Number of Teeth and Addendum	Multiply the Number of Teeth plus 2 by Addendum	$D = (N+2)s$
Number of Teeth	The Pitch Diameter and Diametral Pitch	Multiply Pitch Diameter by the D.P.	$N = D'P$
Number of Teeth	The Outside Diameter and Diametral Pitch	Multiply Outside Diameter by the D.P. and subtract 2	$N = DP - 2$
Thickness of Tooth	The Diametral Pitch	Divide 1.5708 by the D.P.	$t = \frac{1.5708}{P}$
Addendum	The Diametral Pitch	Divide 1 by the D.P.	$s = \frac{1}{P}$ or $s = \frac{D}{N}$
Dedendum	The Diametral Pitch	Divide 1.157 by the D.P.	$s + f = \frac{1.157}{P}$
Working Depth	The Diametral Pitch	Divide 2 by the D.P.	$D'' = \frac{2}{P}$
Whole Depth	The Diametral Pitch	Divide 2.157 by the D.P.	$D'' + f = \frac{2.157}{P}$
Clearance	The Diametral Pitch	Divide 0.157 by the D.P.	$f = \frac{0.157}{P}$
Clearance	Thickness of Tooth	Divide Thickness of Tooth at Pitch Line by 10	$f = \frac{t}{10}$

Lubricants for Cutting Tools.

Material	*Drilling.*	*Reaming.*	*Milling.*	*Turning.*	*Tapping and Die Threading.*
Mild steel or machinery steel.	Soluble oil or good cutting compound.	Lard oil.	Soluble oil.	Soluble oil.	Soluble oil or lard oil.
Hard tool steel, alloy steel, forgings, etc.	Soda water, turpentine, soluble oil or lard oil.	Lard oil.	Soluble oil.	Soluble oil or lard oil.	Sulphur base or lard oil.
Brass or phosphor bronze.	Soluble oil or lard oil or dry.	Soluble oil or dry.	Soluble oil or dry.	Soluble oil.	Soluble oil or lard oil.
Copper.	Soluble oil.	Soluble oil.	Soluble oil or dry.	Soluble oil.	Soluble oil or lard oil.
Aluminium.	Soda water or lard oil and kerosene.	Lard oil.	Soluble oil or dry.	Soluble oil.	Soluble oil or lard oil.
Malleable iron castings, etc.	Soluble oil.	Soluble oil.	Soluble oil.	Soluble oil.	Soluble oil.
Monel metal.	Lard oil.	Lard oil.	Soluble oil.	Soluble oil.	Lard oil.
Cast iron.	Dry.	Dry.	Dry.	Dry.	Lard oil.

If possible, keep drill cooled with jet of compressed air.

THE SPARK TEST OF STEELS

A great advantage of this form of test is that it can be carried out on the steel at any point, e.g., as a billet, an ingot, a bar, a forging, or often a finished piece. The test is carried out on the steel as it stands, and the elaborate drilling of separate samples with the possibility of confusion is eliminated. At the present time, also, the test has great utility, because it enables pieces of undesired metal in a batch of different composition to be picked out quickly and cheaply, and set aside for scrap or salvage, whereas to have to analyse them chemically would constitute a prohibitive charge.

The principle on which the test is based is this. The effect of bringing a piece of steel into contact with the face or cutting edge of a grinding wheel is to force or wrench off tiny fragments of the steel. The wheel runs at a high speed, and the friction is so great that the temperature of these fragments is raised to such a height that they become white hot. This makes them brilliantly visible against a dark background, and their passage through the air as they are flung off has an almost comet-like trajectory, which is termed a " carrier line."

The basis of the test is that different metals give off sparks or particles of incandescent character each having a different trajectory and form. For example, wrought or ingot iron will give off a little bundle of individual lines called a " spark picture." A 0.2 per cent. carbon steel will give a line of brighter colour and will throw off a series of fine branches from this line known as " forks," or " primary bursts." These are due to the presence of carbon. It will thus be seen that wrought iron can readily be distinguished from carbon steel by means of the spark given off.

Raising the Temperature. The effect of raising the temperature of a metallic particle to white heat and hurling it through the air at great velocity is to cause any carbon existing in the fragment to combine with oxygen in the atmosphere to form carbon dioxide. The change from solid carbon to gaseous carbon dioxide results in an increase of volume. This increase of volume is withstood to the best of its ability by the particle, and the result is the setting up of an internal stress that ultimately leads to the complete disruption of the particle, thus

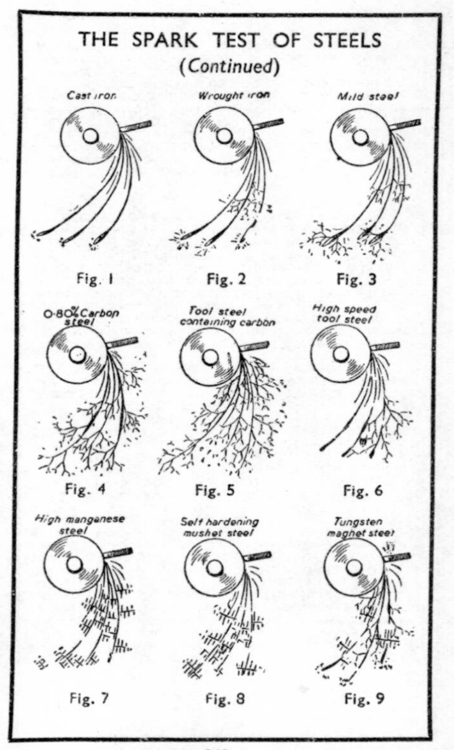
THE SPARK TEST OF STEELS
(Continued)
Cast iron
Wrought iron
Mild steel
Fig. 1
Fig. 2
Fig. 3
0·80% Carbon steel
Tool steel containing carbon
High speed tool steel
Fig. 4
Fig. 5
Fig. 6
High manganese steel
Self hardening mushet steel
Tungsten magnet steel
Fig. 7
Fig. 8
Fig. 9

causing the fork or burst responsible for the branching out of the line. This, at all events, is the theory. The greater the percentage of carbon in the steel, the more marked is the branching effect, and this has proved fairly conclusively that carbon is the element causing these forks or bursts.

Examples. A few examples will serve to illustrate these facts. Fig. 1 shows cast iron, which possesses a dull red, non-explosive spark that thickens towards the end. Fig. 2 shows wrought iron, whose spark is brighter, as indicated, and has a luminous extremity. If any traces of carbon are found in the iron, the extremity may reveal a burst or fork.

Fig. 3 shows mild steel. The thick, luminous iron spark is broken up by the branching due to carbon. Fig. 4 shows a 0.80 per cent. carbon steel spark. The tendencies have virtually vanished, and the carbon branching occurs nearer to the grinding wheel. Fig. 5 shows a high grade tool steel containing carbon. Fig. 6 is high-speed tool steel. An odd carbon spark or two are to be seen, but the rest are modified by the other alloying elements. The sparks are of an orange hue, and vary in brightness as they travel, giving the effect of an interrupted line, while they have a more luminous tip.

Fig. 7 is high manganese steel. In this case the spark is different from that of the carbon spark inasmuch as the explosive particle leaves the luminous line at right-angles, and the sub-division of explosions is also at 90 degrees, as against the 40-50 degrees of the carbon sparks at Fig. 3. Fig. 8 is self-hardening Mushet steel. Here an odd manganese spark is visible, and the relatively high tungsten percentage appears to give discontinuity to the spark. Finally, Fig. 9 is a tungsten magnet steel. Here can be perceived the respective sparks of manganese, tungsten and the like.

THE BRINELL HARDNESS TEST

and corresponding approximate Tensile Strength (Maximum Stress).

Size of Ball 10 m/m dia. Pressure 3,000 kgs.

Dia. of Ball Impression	Brinell Numeral	Tons per □ in.	Kilos. per □ mm.	Dia. of Ball Impression	Brinell Numeral	Tons per □ in.	Kilos. per □ mm.
2.3	713	170	267	3.75	262	58	91
2.4	652	156	245	3.8	255	56	88
2.5	600	135	208	3.85	248	55	86
2.6	555	125	192	3.9	241	54	85
2.65	532	122	179	3.95	235	52	82
2.7	512	120	176	4.0	228	51	80
2.75	495	116	170	4.1	217	49	77
2.8	477	112	165	4.2	207	47	74
2.85	460	107	160	4.3	196	44	69
2.9	444	102	155	4.4	187	42	66
2.95	430	98	150	4.5	179	40	63
3.0	418	94	145	4.6	170	39	61
3.05	402	90	139	4.7	163	37	58
3.1	389	87	134	4.8	156	36	57
3.15	375	84	130	4.9	149	34	54
3.2	364	81	126	5.0	143	33	52
3.25	351	77	122	5.1	137	31	49
3.3	340	74	118	5.2	131	30	47
3.35	332	72	114	5.3	126	29	46
3.4	321	70	111	5.4	121	28	44
3.45	311	68	107	5.5	116	27	43
3.5	302	66	104	5.6	112	26	41
3.55	293	64	101	5.7	107	25	39
3.6	286	63	99	5.8	103	24	38
3.65	277	61	96	5.9	99	22	35
3.7	269	60	93	6.0	95	21	33

BRINELL'S HARDNESS NUMBERS

DIAMETER OF STEEL BALL - 10m/m
PRESSURE - - - - 3,000 Kgr.

Diameter of Ball Impression.	Hardness Number.	Calculated Tonnage.	Diameter of Ball Impression.	Hardness Number.	Calculated Tonnage.	Diameter of Ball Impression.	Hardness Number.	Calculated Tonnage.
m/m			m/m			m/m		
2	946	206	3.70	269	59	5.35	124	28.5
2.05	898	196	3.75	262	57	5.40	121	28
2.10	857	187	3.80	255	55	5.45	118	27
2.15	817	178	3.85	248	54	5.50	116	26.5
2.20	782	171	3.90	241	52	5.55	114	26
2.25	744	162	3.95	235	51	5.60	112	25.5
2.30	713	155				5.65	109	25
2.35	683	149	4	228	50	5.70	107	24.5
2.40	652	142	4.05	223	49	5.75	105	24
2.45	627	136	4.10	217	47	5.80	103	23.5
2.50	600	131	4.15	212	46	5.85	101	23
2.55	578	126	4.20	207	45	5.90	99	22.75
2.60	555	121	4.25	202	44	5.95	97	22.5
2.65	532	116	4.30	196	43			
2.70	512	112	4.35	192	42	6	95	22
2.75	495	108	4.40	187	41	6.05	94	21.5
2.80	477	104	4.45	183	40	6.10	92	21
2.85	460	100	4.50	179	39.5	6.15	90	20.75
2.90	444	97	4.55	174	39	6.20	89	20.5
2.95	430	94	4.60	170	38.5	6.25	87	20
			4.65	166	38	6.30	86	19.75
3	418	91	4.70	163	37.5	6.35	84	19.25
3.05	402	88	4.75	159	36.5	6.40	82	19
3.10	387	84	4.80	156	36	6.45	81	18.75
3.15	375	82	4.85	153	35	6.50	80	18.5
3.20	364	79	4.90	149	34	6.55	79	18.25
3.25	351	76	4.95	146	33.5	6.60	77	17.75
3.30	340	74				6.65	76	17.5
3.35	332	72	5	143	33	6.70	74	17
3.40	321	70	5.05	140	32	6.75	73	16.75
3.45	311	68	5.10	137	31.5	6.80	71.5	16.5
3.50	302	66	5.15	134	31	6.85	70	16.25
3.55	293	64	5.20	131	30	6.90	69	16
3.60	286	62	5.25	128	29.5	6.95	68	15.75
3.65	277	60	5.30	126	29			

Reproduced by courtesy of Edgar Allen & Co., Ltd.

APPROXIMATE HARDNESS EQUIVALENTS

ROCKWELL 120° Cone C Scale			BRINELL	DIAMOND	SCLEROSCOPE	ROCK-WELL 1/16 Ball B Scale 100 kgs.	BRINELL 10 m/m Ball 3,000 kgs. or DIAMOND 30 kgs.	SCLEROSCOPE
150 kgs.	100 kgs.	60 kgs.	10 m/m Ball 3,000 kgs.	30 kgs.				
67	78	85	713	966	95	100	241	35
66	77	85	698	935	90	99	235	34
65	76	84	682	905	88	98	228	33
64	75	84	665	876	87	97	223	33
63	74	83	652	848	85	96	217	32
62	73	83	635	822	84	95	212	31
61	72	82	621	786	83	94	207	30
60	71	82	607	752	82	93	202	30
59	71	81	594	722	81	92	196	29
58	69	80	581	692	78	91	192	29
57	69	80	568	672	76	90	187	28
56	68	79	555	653	75	89	183	28
55	68	79	543	635	74	88	179	27
54	67	78	532	618	72	87	174	27
53	66	78	518	593	71	86	170	26
52	65	77	506	579	69	85	166	26
51	64	76	495	564	68	84	163	25
50	64	76	482	550	67	83	159	25
49	63	75	470	536	66	82	156	24
48	62	75	460	523	64	81	153	24
47	61	74	447	499	61	80	150	23
46	61	74	436	477	60	79	147	23
46	60	73	430	460	59	78	144	22
44	59	73	418	442	57	77	141	22

APPROXIMATE HARDNESS EQUIVALENTS (*Continued*)

ROCKWELL 120° Cone C Scale			BRINELL	DIAMOND	SCLEROSCOPE	ROCKWELL $^1/_{16}$ Ball B Scale 100 kgs.	BRINELL 10 m/m Ball 3,000 kgs. or DIAMOND 30 kgs.	SCLEROSCOPE
150 kgs.	100 kgs.	60 kgs.	10 m/m Ball 3,000 kgs.	30 kgs.				
43	58	72	402	430	55	76	139	—
42	58	72	394	419	54	75	137	—
41	57	71	387	401	53	74	135	—
40	56	71	375	388	52	73	132	—
39	55	70	364	380	50	72	130	—
38	54	69	351	369	49	71	127	—
37	53	69	340	354	47	70	125	—
36	52	68	332	340	46	69	123	—
35	52	68	321	327	45	68	121	—
34	51	67	311	315	44	67	119	—
33	50	67	302	306	43	66	117	—
32	50	67	293	295	42	65	116	—
31	49	66	286	289	41	64	114	—
30	49	66	277	281	40	63	112	—
29	48	65	272	275	39	62	110	—
28	47	65	269	272	38	61	108	—
27	46	64	262	263	37	60	107	—
26	45	64	255	256	36	59	106	—
25	45	63	248	248	36	58	104	—
24	44	63	241	241	35	57	103	—
23	43	62	235	235	34	56	101	—
22	42	62	228	228	33	55	100	—
21	41	61	223	223	33	54	98	—
20	41	61	217	217	32	53	97	—

RELATIVE CONVERSION TABLE OF HARDNESS VALUES

	Rockwell	Sclero-scope	Brinell	Rockwell	Sclero-scope	Brinell
1-16″ Ball—100 Kg.	84–B	22	156	C–36	48	337
	86–B	22	159	C–37	50	347
	87–B	22	163	C–38	51	357
	87–B	23	166	C–39	52	367
	88–B	24	170	C–40	53	377
	89–B	25	174	C–41	54	387
	90–B	26	179	C–42	56	398
	91–B	27	183	C–43	57	408
	92–B	28	187	C–44	58	419
	93–B	29	192	C–45	59	430
	94–B	30	196	C–46	61	442
	95–B	30	202	C–47	62	453
	96–B	30	207	C–48	63	464
	97–B	33	217	C–49	65	476
	98–B	34	223	C–50	66	488
	99–B	35	228	C–51	67	500
	100–B	35	230	C–52	69	512
				C–53	70	524
120°-Cone—150 Kg.	C–21	35	230	C–54	71	536
	C–22	35	235	C–55	73	548
	C–23	36	241	C–56	74	561
	C–24	36	247	C–57	76	574
	C–25	37	253	C–58	77	587
	C–26	38	259	C–59	78	600
	C–27	39	265	C–60	80	613
	C–28	40	272	C–61	81	627
	C–29	41	279	C–63	85	652
	C–30	42	286	C–65	88	683
	C–31	43	294	C–66	90	697
	C–32	44	301	C–67	92	715
	C–33	45	309	C–68	94	729
	C–34	46	318	C–69	96	748
	C–35	47	327	C–70	98	769

Reproduced by courtesy of Edgar Allen & Co., Ltd.

CO-ORDINATES FOR JIG BORING

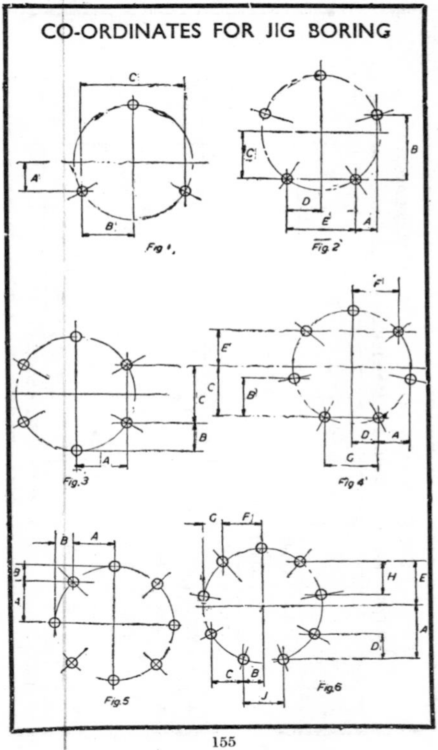

Fig 1.

Fig. 2

Fig. 3

Fig 4

Fig. 5

Fig. 6

CO-ORDINATES FOR JIG-BORING (*Continued*)

MULTIPLY VALUE SHOWN BY DIAMETER OF CIRCLE BEING CALCULATED

Fig. No.	No. of Holes	A	B	C	D	E	F	G	H	J	K
1	3	.250	43302	.86603							
2	5	.18164	.55902	.40451	.29389	.58779					
3	6	.43302	.250	.500							
4	7	.27052	.33920	.45049	.21694	.31175	.39090	.43388			
5	8	.35355	.1465								
6	9	.46985	.17101	.2620	.21985	.38302	.32139	.17101	.29620	.34202	
7	10	.29389	.09549	.18164	.2500	.15451					
8	11	.47975	.14087	.23701	.15232	.11704	.25627	.42063	.27032	.18449	.21319
9	12	.22415	.12941	.48297	.12941	.25882					

CO-ORDINATES FOR JIG BORING
(Continued)

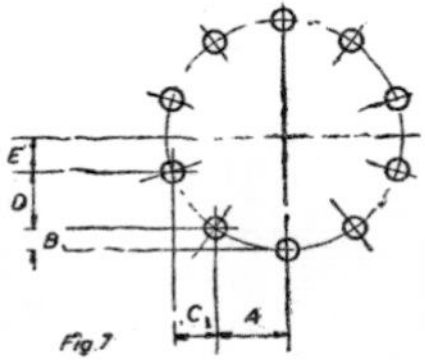

Fig. 7

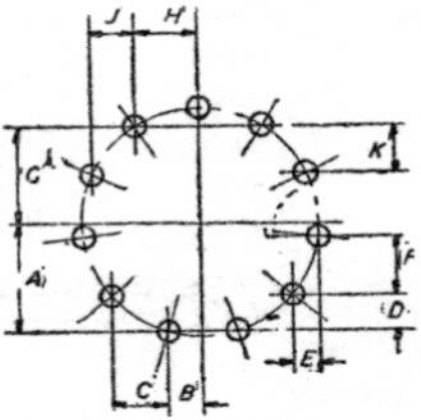

Fig. 8

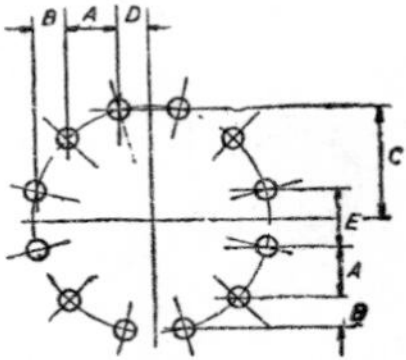

Fig. 9

DIAGRAMMATIC VIEW OF A COMPRESSION SPRING SHOWING PRINCIPAL DIMENSIONS

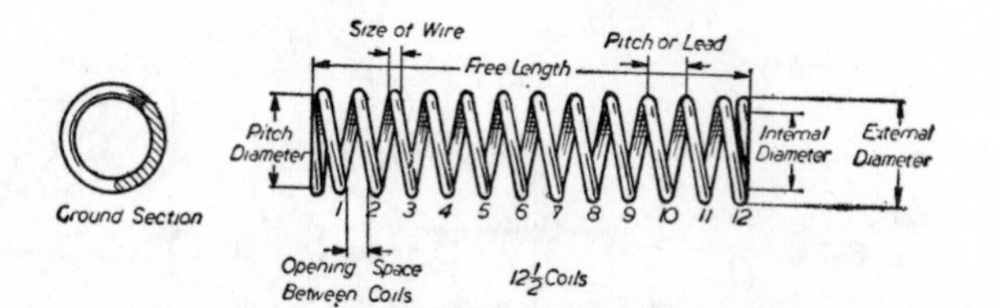

THE compression spring comprises an open coil spiral spring which is capable of resisting a compressive stress. Springs of this type find extensive industrial and other application. They are manufactured in numerous forms and from various types of wire, according to their purpose. Though it is essential in certain instances to employ square, rectangular or special sectioned or shaped wire, most compression springs are manufactured from round wire. **A typical compression spring is shown above.**

DIAGRAMMATIC VIEW OF AN EXTENSION SPRING SHOWING PRINCIPAL DIMENSIONS

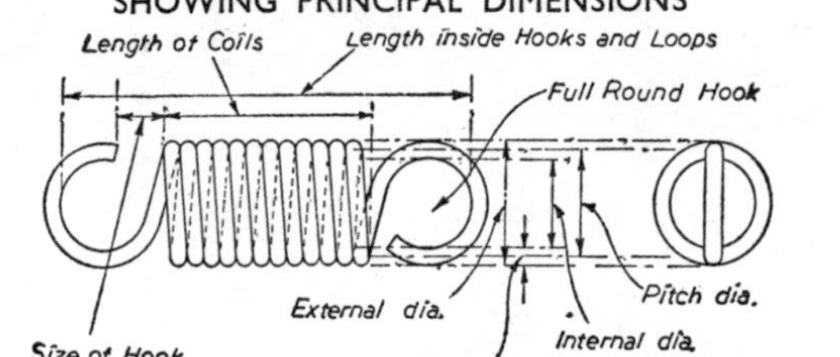

AN extension spring may be described as a closely-coiled spiral or helical spring, offering resistance to a pulling force. They are manufactured from round and square wire. The coils are generally close-wound and touching one another. Their difference from the compression spring is that whereas the coils of the latter are separated and only forced together by the compression stress, the coils of the former are in contact, and wound so firmly together that a considerable effort is necessary to separate them. This load, built up by coiling, is termed initial tension and to some degree can be controlled. The illustration shows an extension spring and its dimensional features.

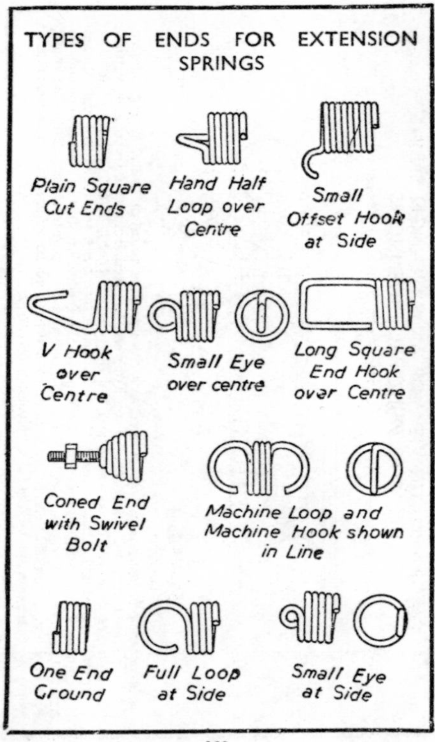
TYPES OF ENDS FOR EXTENSION SPRINGS
Plain Square Cut Ends
Hand Half Loop over Centre
Small Offset Hook at Side
V Hook over Centre
Small Eye over centre
Long Square End Hook over Centre
Coned End with Swivel Bolt
Machine Loop and Machine Hook shown in Line
One End Ground
Full Loop at Side
Small Eye at Side

TYPES OF ENDS FOR EXTENSION SPRINGS—(Continued)

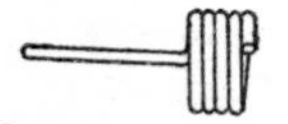

Straight End Annealed to allow forming

Extended Eye from either Centre or Side

Double Twisted Full Loop over Centre

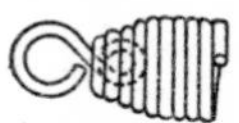

Coned End with Short Swivel Eye

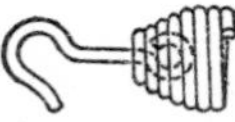

Coned Rod with Swivel Hook

Machine Loop & Machine Hook shown at Right Angles

Full Loop on Side and Small Eye from Centre

Machine Half Hook over Centre

TYPES OF ENDS FOR EXTENSION SPRINGS—(Continued)

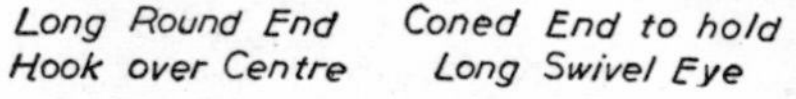

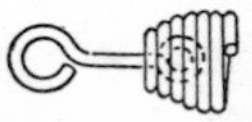

VALVE SPRINGS

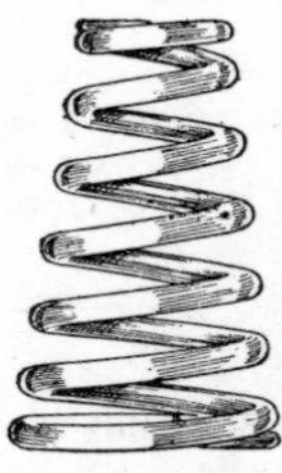

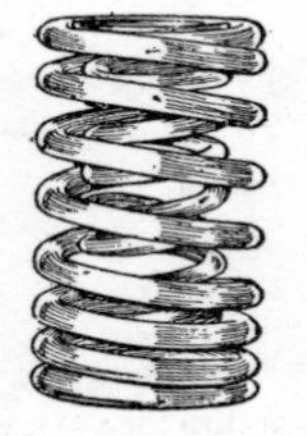

TYPES OF SPRING ENDS

Open Ends, R.H. Coiled

Squared Ends, Unground, R.H. Coiled

Squared & Ground Ends. L.H Coiled

Open Ends Ground, L.H. Coiled

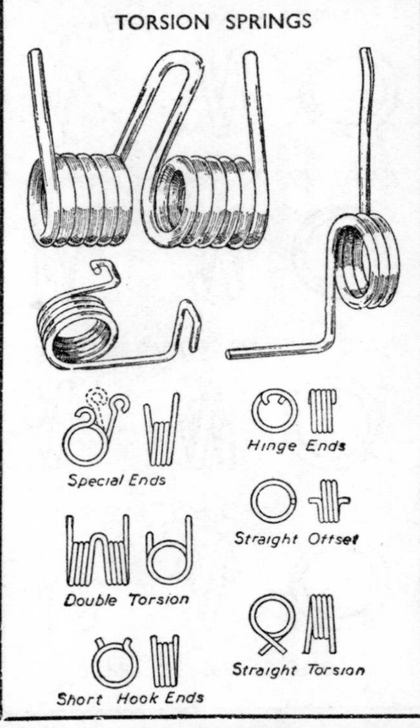
TORSION SPRINGS
Special Ends
Hinge Ends
Straight Offset
Double Torsion
Straight Torsion
Short Hook Ends

TORSION SPRINGS
(*Continued*)

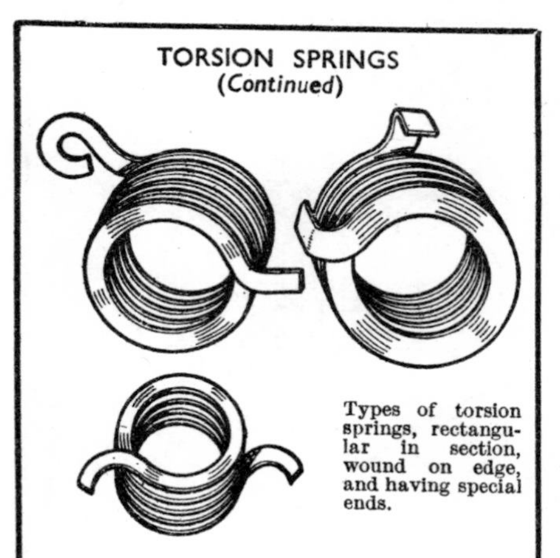

Types of torsion springs, rectangular in section, wound on edge, and having special ends.

Load in torsion springs is governed by the location of the ends. The ends vary, as a result of winding or coiling, in accordance with the variation in the number of coils and diameter of wire to spring diameters. Springs of this type manufactured from wire of 0·11–0·25 cm diameter, measuring about 0·95 cm internal diameter, and having fewer than four coils, could have their ends maintained in free position within 10-20 deg. If, however, the number of coils were increased to 10-15, there might be a variation of 30-50 deg. The length of straight ends is governed in variation by the recoil of the material, and this variation is usually between 0·04 cm to 0·24 cm. One end can be kept to rather narrower tolerances than the other because of cut-off mechanism. This will be the long end, should different lengths be desired. For accurate length, it will be essential to trim.

INITIAL TENSION IN SPRINGS

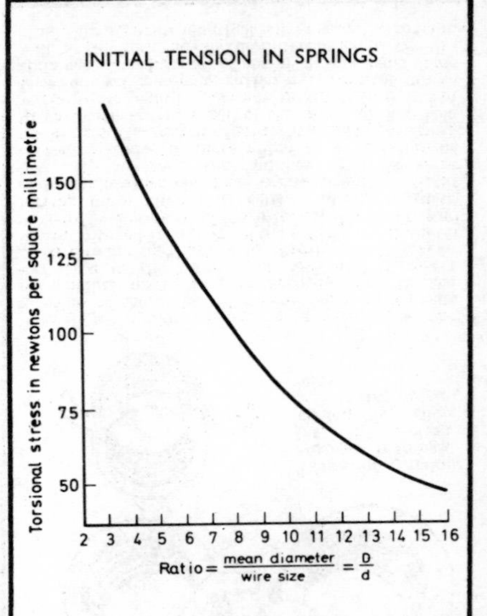

Initial tension in extension springs of oil-tempered wire or hard drawn spring wire as-coiled in an automatic coiler.

TYPES OF FLAT SPRINGS AND STAMPINGS

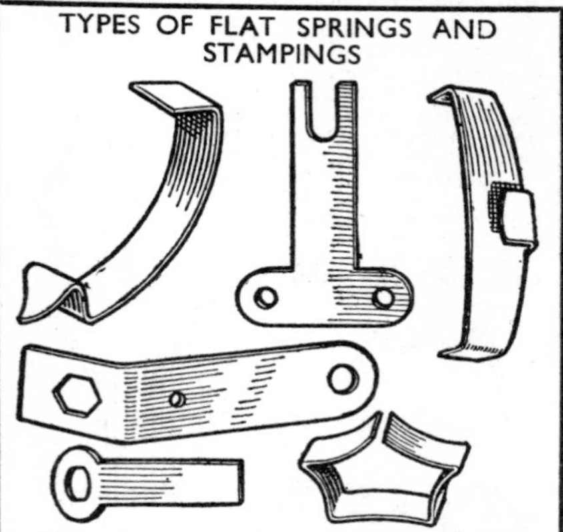

Flat springs are mostly manufactured from steels fairly high in carbon percentage, but additional materials quite often employed include the austenitic nickel-chromium stainless steels, the non-ferrous alloys, brass, monel metal, phosphor bronze, etc. The steel is usually annealed, and has to be given a suitable heat-treatment after it has been formed, because in the pre-tempered state it will not easily bend. On the other hand, so long as the section required is not thick, it is possible to use pre-tempered steel. If any doubt as to section thinness is felt, however, the annealed steel should be used. Otherwise, there will be heavy wear on the blanking dies, and upkeep cost will be heavy.

Flat springs are widely employed in various types of mechanisms. Numerous features can be worked into their design, and by using them, parts necessary for actions in the use of wire springs can be eliminated. As a rule, their applicability is controlled by the rate of deflection, which regulates length of travel or movement.

TYPES OF WIRE FORMS

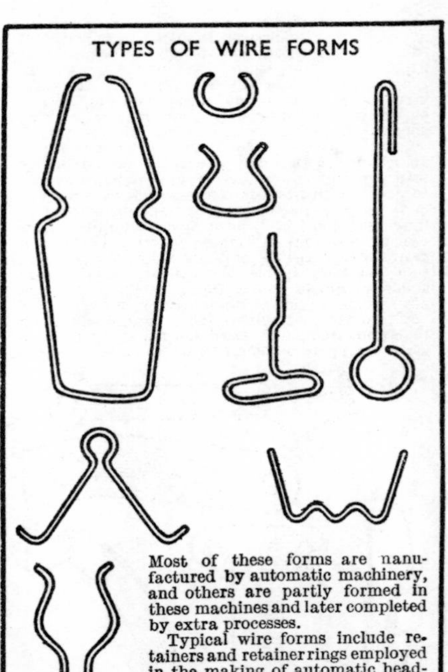

Most of these forms are nanufactured by automatic machinery, and others are partly formed in these machines and later completed by extra processes.

Typical wire forms include retainers and retainer rings employed in the making of automatic headlamps, for securing the lenses; piston pin lock rings; spring clips, etc. These forms have numerous applications. Most are employed as holding actions.

SPRING DATA

The Basic Law in Spring Manufacture.—The basic law in spring manufacture and design is that first formulated by Hooke, who proved that stress divided by strain is a constant within the elastic limit of the material employed. Thus, if a steel rod is twisted 10 deg. by the application of a load, then a 20 deg. twist would call for double that load. This, of course, ignores hardness or other physical properties of the steel. The basic formula for torsional stress can be obtained from most text-books on mechanics, and is: $S = \frac{MV}{J}$ for material in torsion. In this formula, M is the moment of the force $\frac{D}{2} \times P$; V is the distance from fibre under consideration to centre of gravity —generally the external fibre. For wire, therefore, $\frac{d}{2}$. J is the polar moment of inertia; $\frac{3.1416d^4}{32}$ for round sections.

Consideration of the stress to which a spring is subjected is highly necessary to intelligent spring design.

Calculations Affecting Design.—We now come to those calculations affecting the design of round wire springs. When calculating stress, the above values are substituted and the expression is simplified, so that the formula now becomes

$$S = \frac{P \times D \times 2.55}{(\text{diameter of wire})} \text{ or } d^3.$$

Here, P equals load in lb. on the spring. D equals mean or pitch diameter, i.e., the external diameter of the spring less the wire dimension in ins. d is the wire diameter in ins. S is the stress in lb. per sq. in. in torsion.

The above is torsional stress only, and does not take into account total stress to which the wire is subjected as the ratio of wire dimension to mean spring diameter changes.

SPRING DATA—(*Continued*)

A formula to meet this case and due to Mr. A. M. Wahl is as follows :

$$S\ max. = \frac{\frac{16\ PR}{3.1416^3}\left(\frac{(4C-1)}{(4C-4)}\right)+\frac{0.615}{C}}{Y}$$

Here R is mean radius in ins. ; d is wire diameter in ins. ; P is load in lb. ; C is $\frac{2R}{d}$. By obtaining the numerical ratio of mean diameter to wire diameter it is possible to obtain a factor from a curve based on this formula, which is Y in the formula. Then total stress equals $S \times Y$.

All calculations of important high-duty springs with a mean diameter ratio to wire size below eight should be made by this formula, which is applicable to both round and square wire.

Ascertaining Size of Spring.—The next requirement is a means of ascertaining the size of the spring. This is obtained by means of the following formula : $S = \frac{2.55PD}{d^3}$. In this, S is the stress in lb. per sq. in. in torsion, P is the load in lb. on the springs, D the mean or pitch diameter, and d the diameter of the wire in inches. Thus, if the stress is once known, the wire diameter cubed can be obtained, and by working out the formula the value of d, i.e., the wire size, is obtained.

It is then required to calculate the rate of the spring, which is done, once the wire size is known, by means of the formula : $F = \frac{8PD^3N}{Gd^4}$. In this formula, F is the deflection in inches, P the load in lb., D the mean diameter of the spring, N the number of active coils, G the modulus of the material in torsion, i.e., 11,500,000 for steel, d is the wire diameter in inches. If solid height is the decisive factor, certain substitutions will have to be made in this formula. As a rule, it is inadvisable to compress any spring in its working range to such an extent that the coils are in close contact. There should be a little space separating each coil when taking solid height into the reckoning.

Hardness of Material.—An important point to be observed is that the hardness of the material of which the spring is made makes no difference to the calculation of free length. In other words, springs made from a material having high elastic limit will be more extensively deflected in advance of attaining their elastic limit and therefore sustain a heavier load, because their rate per unit of deflection does not alter whether the material be hard or soft. As free length affects stress in design, it is possible to coil with a good deal of space between the coils, according to the solid stress and the elastic limit of the material. This space is regulated by the extent of set when solid compressed allowed for in coiling. The maximum load should not be over 62,000lb. stress per sq. in., however, if satisfactory service is desired. Since the torsional modulus is an essential part of the deflection formula, and varies with the type of material of which the spring is made, it will be useful to give a brief table of values, based on the principal spring-making materials. This will be found in the table below.

	Steel Music Wire < 0.015′ dia.	Steel Music Wire > 0.015′ dia.	Carbon Steel Wire
	12,000,000	11,500,000	11,500,000
lbs. per sq. in.	Chrome Vanadium Steel Wire	Hot Rolled Steel Rods > ½in. dia.	Stainless Steel Hard-drawn Wire
	11,500,000	10,500,000	10,500,000
	Monel Metal	Phosphor Bronze	Brass 66-34
	9,200,000	6,250,000	5,000,000

SPRING DATA—(*Continued*)

It will be observed that the hot rolled rods have a lower modulus value than most of the other types of steel materials. This is because the effect of hot rolling is to form a soft or decarburised surface layer on the material. It must be remembered that the extent to which decarburisation of the surface has proceeded, i.e., the extent to which carbon content has been reduced, has an effect on the modulus value, so that the figure given in the table must not be regarded as universally applicable. It is merely an average value. If the rods are ground on the surface after rolling, in order to remove this soft skin, the modulus value will rise to a figure equivalent to that of the carbon and low alloy steels as given in the table.

Square Wires.—We may now turn to the square wires. For calculating the stress, the formula used is $S = \frac{2.4PD}{d^3}$. Deflection is calculated by the aid of the formula $F = \frac{P5.58D^3N.}{Gd^4.}$

With the aid of these two formulæ, it is possible to work out also wire size. In calculating solid height, the point to be remembered is that if any square spring is coiled to a small mean diameter, the square wire, when coiled, takes on a keystone form, the inner side of the wire becoming broader than the outer. Thus, the wire's cross section has a trapezoidal form. Obviously, then, it is not possible to calculate correctly the solid height of the spring from the wire size, and a new formula has to be used, for which we are indebted to Stewart (S.A.E. Journal, Aug., 1925). This is as follows:

$$d_1 = 0.48d\left\{\frac{O.D.}{P.D.}+1\right\}$$

(In those instances in which the ratio of wire diameter to mean spring diameter is larger than 1 : 12, the keystone effect becomes negligible and the formula valueless.) It is possible to compute the solid height of the spring once the value of

d_1 is obtained. In this formula, O.D. is the external diameter of the spring, P.D. the pitch diameter in inches, d the original wire size, and d_1 the width of wire inside after being coiled

The table on page 174 gives formula for round and square wire, as arranged by Messrs. Barnes, Gibson, Raymond, of Detroit, and also gives suggested total stresses.

In connection with the stress table given on page 375, it must be remembered that these stresses are not absolute. The narrower the range of stress, the greater will be the safety margin of the spring. If feasible, it is better to design springs in such a manner that the stress range is never more than $\frac{1}{2}$-$\frac{2}{3}$ the maximum working stress. Thus, a spring stressed at 25,000-60,000 from oil-tempered wire will normally give many millions of load applications, so long as corrosion does not affect it, in which case its life will inevitably be uncertain.

Influence of Temperature.—Pages 175 and 176 show the influence of temperature on stress, and have been compiled by the well-known firm of spring manufacturers previously mentioned. They indicate the combinations of stress and temperature necessary to create load losses of 2 per cent. and 10 per cent. for different spring materials. The greater the allowable loss of load, the higher is the temperature at which any given material may be employed. For the carbon and chrome vanadium steels, stress is the most vital factor at temperatures lower than 205 deg. C. At higher temperatures than this, it is not safe to employ these steels. It will be seen that it is not safe to employ any of these materials at working temperatures of over 315 deg. C. The stresses represented were computed by means of the Wahl formula, and are therefore total shear stresses. Where the working temperatures are likely to exceed those indicated in the diagrams, different materials will have to be employed, and are available for temperatures up to 425 deg. C. It is advisable to consult the spring maker before specifying any particular spring material for use under these conditions.

FORMULA FOR ROUND AND SQUARE WIRE

	Load Carried		Deflection per Coil		Deflection in ins.	Stress		Wire Dia.		$\frac{P}{F}$ lbs. (scale) per in. Deflection
	(1)	(2)	(1)	(2)		(1)	(2)	(1)	(2)	
Round	$\frac{\pi S d^3}{8D}$	$\frac{Gfd^4}{8D^3}$	$\frac{\pi S D^2}{Gd}$	$\frac{8PD^3}{Gd^4}$	$\frac{8PD^3N}{Gd^4}$	$\frac{8PD}{\pi d^3}$	$\frac{fGd}{\pi d^2}$	$\frac{\pi D^2 S}{Gf}$	$\sqrt[3]{\frac{8PD}{\pi S}}$	$\frac{Gd^4}{8D^3N}$
Square	$\frac{0.416Sd^3}{D}$	$\frac{Gfd^4}{5.58D^3}$	$\frac{2.325D^2}{Gd}$	$\frac{5.58D^3P}{Gd^4}$	$\frac{5.58D^3NP}{Gd^4}$	$\frac{2.4PD}{d^3}$	$\frac{fGd}{2.32D^2}$	$\frac{2.32SD^2}{Gf}$	$\sqrt[3]{\frac{2.4PD}{S}}$	$\frac{Gd^4}{5.58D^3N}$

TOTAL STRESSES

	Steel Music Wire	Steel Wire Oil Tempered	Hard Drawn Spring Steel Wire	Stainless Steel Wire	Monel Metal	Phosphor Bronze	Brass
Max. Working Stress	lbs. per sq. in.	lbs. per sq.in.	lbs. per sq. in.	lbs. per sq. in	lbs.per sq. in.	lbs.per sq.in.	lbs.per sq.in.
	70,000	60,000	50,000	50,000	35,000	35,000	25,000
Max. Solid Stress	120,000	100,000	80,000	80,000	70,000	70,000	50,000

VALUES OF X AND B FOR VARIOUS RATIOS OF $\frac{b}{c}$

$\frac{b}{c}$	1.0	1.5	1.75	2.0	2.5	3.0	4.0	6.0	8.0	10.0	Infinite
X	0.208	0.231	0.239	0.246	0.258	0.267	0.282	0.299	0.307	0.313	0.333
B	0.141	0.196	0.214	0.229	0.249	0.263	0.281	0.299	0.307	0.313	0.333

ALLOWABLE STRESSES IN VARIOUS SPRING MATERIALS AT HIGH TEMPERATURE

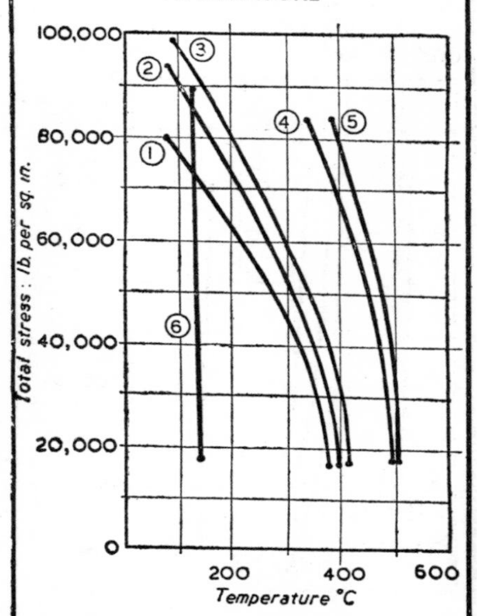

1. Monel metal. 2. Special valve spring wire. 3. Chromium, vanadium, steel. 4. Stainless steel (18/8). 5. Stainless steel (12/14% chromium). 6. Music wire. Loss in load is not more than 2%.

ALLOWABLE STRESSES IN VARIOUS SPRING MATERIALS AT HIGH TEMPERATURE—(*Continued*)

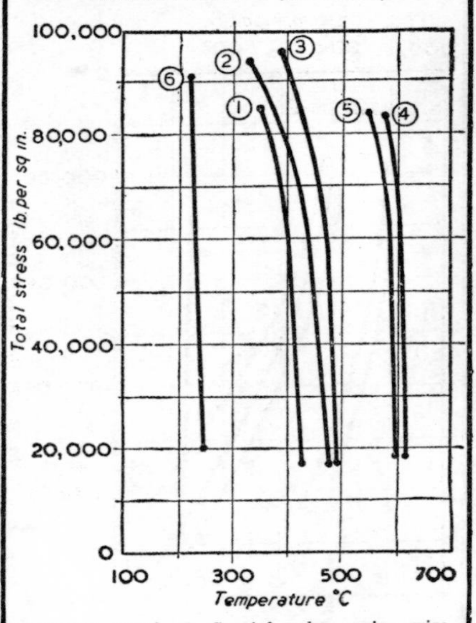

1. Monel metal. 2. Special valve spring wire. 3. Chromium, vanadium, steel. 4. Stainless steel (18/8). 5. Stainless steel (12/14% chromium). 6. Music wire. Loss in load is not more than 10%.

We next come to rectangular section wire, the calculations for which do not differ greatly from those for round wire. Deflection is calculated from the formula $F = \frac{2\pi NPR^3}{Bbc^3G}$.

In this formula b is width, and c is thickness. The other symbols are as in previous formulæ. Stress is calculated from the formula $S = \frac{PR}{Xbc^3}$.

Values of X and B for various ratios of $\frac{b}{c}$ will be found on page 375.

The Wahl constant does not hold for rectangular sections, though it holds good for square and round. New constants can be developed from the original data, if desired, by the spring manufacturers.

Conical Springs.—Evenly tapered or conical springs can be calculated from the formula already provided, so long as the maximum mean diameter or radius is substituted in the stress formulæ. In the deflection formulæ it is necessary to employ the average mean diameter of the spring. At the same time, the reader must bear in mind that in both instances the formulæ will hold good only when the deflection is small enough so that one coil does not become inactive wire as a result of touching another.

When a spring of this type has deflected sufficiently to eliminate one coil, deflection per coil for the spring can once more be computed, employing the average mean diameter of the still active portion of the spring. A kindred method would solve rectangular wire springs. The barrel-shaped springs, which have their maximum diameter in the middle, or those of hour-glass shape, with the minimum diameter in the middle, can be solved by making them represent two conical springs or by obtaining the precise mean diameter and substituting in the formulæ given.

SPRING DATA—(*Continued*)

We may now turn to the effect of suddenly applied load. On both extension and compression springs, it is often essential that the designer should calculate not only the influence of loads applied slowly, but also that of those suddenly applied, as well as those applied by bodies possessing kinetic energy. These are complex calculations, because one term of the formula is often unknown, i.e., the velocity. Nevertheless the force exerted on the spring multiplied by the space exactly equals the amount of energy the spring absorbs, so that it is not difficult to work out the answer. This ignored hysteresis of the spring, of course. The undermentioned formulæ are then applicable:

(1) for loads applied very slowly: $F=\frac{P}{W}$

(2) for loads applied suddenly: $F=\frac{2P}{W}$

(3) for loads allowed to fall from a specified height: $F^2=\frac{2P(s+F)}{W}$.

Here W is the rate per inch of spring deflection, P is the load on the spring, and s is the height in inches the load is allowed to fall.

Stress and Deflection.—In a few instances, springs are made from wire of special forms or sections. For these certain formulæ have been worked out by manufacturers for stress and deflection, as under.

(1) For elliptical wire: $S=\frac{16PR}{\pi e^2 h}$

$F=\frac{8(\pi)^3 PR^3 I_p N}{A^4 G}$. Here, e is the minor axis of ellipse, h the major axis of ellipse, I_p is $\frac{\pi}{64}(eh^3+e^3h)$, A is $\frac{eh}{4}$.

(2) For triangular wire of equilateral section: $S=\frac{20PR}{g^3}$. $F=\frac{92.4PR^3N(\pi)}{g^4G}$. Here, g is the base of the triangle.

SPRING DATA—(*Continued*)

(3) For octagonal wire: $S=\frac{PR}{0.223Ai}$.

$F=\frac{2PR^3N(\pi)}{0.130Ai^2G}$. Here, i is the diameter of an inscribed circle in the octagon, N is the number of coils, R the pitch radius, I_p is the polar moment of inertia, A is the area.

(4) For hexagonal wire: $S=\frac{PR}{0.217Ai}$.

$F=\frac{2PR^3N(\pi)}{0.133Ai^2G}$

A new set of formulæ comes into play

$n=531\sqrt{R/W}\,\frac{\Omega}{\mho}\frac{826,500d}{ND^2}$. In this formula, n is the complete number of free vibrations per minute of the spring as vibrating in itself. R is the rate of the spring. W is the weight of the active mass in the spring. d is the wire diameter. D is the pitch or mean diameter of the spring, and N is the number of active coils. The other, stated by L. F. G. Simonds in Paper No. 241 of the Advisory Committee for Aeronautics of Great Britain, is:

$n=\frac{1}{LP}\sqrt{\frac{WN}{8\pi Po}}$. Here, L is the total length of wire in the spring, W the sectional area, P the principal radius of curvature, N the modulus of rigidity 11,500,000, Po is the density, i.e. $\frac{1}{365}$ for steel. It is also possible to express this as frequency per minute $=\frac{62.4dBN}{R^2x}$. Here, X is the effective length of the spring in inches, R the mean radius of the coil in inches, B the pitch of coil in inches. In calculating X, it should be remembered that it is the length excluding any closed portion of the spring at each end.

TOTAL STRESSES FOR TORSIONAL SPRINGS

	Wire size below $\frac{1}{8}$in. dia.		Wire sizes $\frac{1}{8}$in.-$\frac{1}{4}$in.	
	Max. Working Stress, in lbs. per sq. in.	Max. Total Stress, in lbs. per sq. in.	Max. Working Stress, in lbs. per sq. in.	Max. Total Stress, in lbs. per sq. in.
Music Wire	170,000	200,000	150,000	180,000
Oil-tempered Wire ..	150,000	180,000	125,000	150,000
Hard-drawn Wire ..	130,000	150,000	110,000	130,000
18/8 Stainless Steel ..	130,000	150,000	100,000	120,000
Monel Metal .. .. Phosphor Bronze ..	} 50,000	70,000	40,000	60,000
Brass	40,000	60,000	30,000	50,000
Stresses for Flat Springs in Sections Under $\frac{1}{8}$in. Thick				
Steel	150,000	20,000		
Non-ferrous Alloys ..	40,000	60,000		

ALUMINIUM ALLOYS

The following pages relate to British alloys of aluminium which may in due course be replaced by new metric standards. For the present the references are therefore to British units. Should there be a choice which enables a change to be made to other alloys, the following references will be of value as regards wrought aluminium and aluminium alloys:

BS 4300/ 6: 1969	NS31	Sheet and strip
BS 4300/ 7: 1969	NS41	Sheet and strip
BS 4300/ 8: 1969	NS51	Plate, sheet and strip
BS 4300/ 9: 1969	NG41	Wire
BS 4300/10: 1969	NT51	Drawn tube
BS 4300/11: 1969	NF51	Forging stock and forgings
BS 4300/12: 1969	NE51	Bar, extruded round tube and sections
BS 4300/13: 1969	NG52	Welding wire

NS31 : Aluminium magnesium manganese alloy
NS41 : Aluminium magnesium alloy
NS51 : Aluminium magnesium manganese chromium alloy
NG41 : Aluminium magnesium alloy
NT51, NE51, NF51 : Aluminium magnesium manganese chromium alloys
NG52 : Aluminium magnesium manganese chromium titanium alloy.

The limits of other elements present in the above alloys are specified in the above BS 4300 metric standards, together with data on desirable sizes and tolerances.

PROPERTIES OF ALUMINIUM

Property	Value	Authority
PHYSICAL CONSTANTS		
Atomic Weight (Oxygen = 16)	26.97	Int. Atomic Wt. Comm., 1929
Spec. Ht., 20°–400° C. (av.) cals.	0.24	Int. Crit. Tables Formula.
Spec. Thermal Conductivity in cals. per cm. cube per degree C. per sec., at 0° C. ..	0.502	Bailey, Proc. Roy. Soc. A., 134, 57-76, 1931.
Approx. Relative Heat Conductivity (Silver = 100%)	51.8	
Melting Point—		
(99.97% pure) °C. ..	659.8	Edwards, J., American Electrochem. Soc., 1925.
(99.66% pure) °C. ..	658.7	
Boiling Point, °C. ..	1800	Greenwood, Proc. Roya Soc. 82, 1909.
Latent Ht. of Fusion, cals. per gm.	92.4	Awbery & Griffiths, Proc. Phys. Soc., Lond, Vol. 38, pt. 5, Aug. 15, 1926.
Total Ht. referred to 20° C., calories per gm. 400° C...	88	
,, ,, 600° C...	146	
,, ,, 700° C...	267	
Vapour Pressure at 658.7° C., mm. of Mercury	1.0×10-43	Richards, Jour. Franklin Inst., Vol. 187, 1919.
Heat of Combustion to Al_2O_3 per gm. mol., cals.	383,900	A.S.S.T. Handbook.
Coeff. of Linear Expansion /°C. H.D. Wire, 0°–30° C. ..	23×10-6	B.S. Spec. 215
Rolled metal (normal purity), Average value—		
20°-100° C. ..	24 ×10-6	Based on Hidnert, U.S. Bureau of Standards Paper No. 497.
20°-300° C. ..	26.7×10-6	
20°-600° C. ..	28.6×10-6	

PROPERTIES OF ALUMINIUM
(*Continued*)

Property	Value	Authority
Physical Constants (cont.)		
Specific Gravity :		
H.D. Wire (electrical conductors).. ..	2.703	B.S.S. No. 215, 1934.
Rolled sheet (normal purity)	2.71	British Aluminium Co. Ltd.
Molten (99.75% pure)		
658.7° C.	2.382	Edwards & Moorman, Chem. Met. Eng., Vol. 24, pp. 61-64, 1921.
1100° C.	2.262	
Wt. of 1 cubic ft. of Aluminium (normal purity), lb.	169.18	Calculated from Specific Gravity.
MECHANICAL CONSTANTS		
Modulus of Elasticity, lb./sq. in.	9.9×10^6	B.S.S. No. 215, 1934.
Torsion Modulus, lb./sq. in.	3.87×10^6	Koch & Dannecker, Ann. d. Phys., 1915.
Poisson's Ratio	.36	Bureau of Standards, Circ. No. 76, 1919.
Tensile Strength of Sheet:		
Annealed, tons/sq.in.	5–6½	B.S.S. No. 2L17, 1922.
Half Hard, tons/sq.in.	7–8½	B.S.S. No. 2L16, 1922.
Hard, tons/sq.in. ..	9 (min.)	B.S.S. No. 2L4, 1922.
Percentage Elongation in 2 in. :		
Pure Castings, Sand ..	20–30	British Alumin. Co., Ltd.
,, ,, Chill ..	30–40	,, ,, ,,
Pure Sheet, Annealed	12–40	,, ,, ,,
,, ,, Half Hard	5–12	,, ,, ,,
,, ,, Hard ..	2–8	,, ,, ,,
H.D. Wire	4–7	,, ,, ,,
Scleroscope Hardness (mag. ham.) :		
Annealed or Cast ..	5 to 5½	,, ,, ,,
Cold rolled (.128"–.020")	15 to 22	,, ,, ,,

PROPERTIES OF ALUMINIUM
(Continued)

Property	Value	Authority
Mechanical Constants (cont.)		
Brinell Hardness, 1 mm./5 kgm. :		
Cast	20 to 28	British Alumin. Co., Ltd.
Annealed Sheet ..	19 to 23	,, ,, ,,
Hard Sheet	38 to 45	,, ,, ,,
ELECTRICAL CONSTANTS		
Max. Specific Res. for H.D. Wire at 20° C. — microhms/cm. cube	2.8735	B.A. Co. Guar. Max'm.
Max. Specific Res. for H.D. Wire at 20° C. — microhms/in. cube	1.1313	,, ,, ,,
Standard Specific Res. for H.D. Wire at 20° C. — microhms/cm. cube	2.845	B.S.S. No. 215, 1934.
Standard Specific Res. for H.D. Wire at 20° C. — microhms/in. cube	1.1199	,, ,,
Coefficient of increase of res. with temp. for H.D. wire at 15.6°C. (60°F.) — °C.	0.00407	,, ,,
Coefficient of increase of res. with temp. for H.D. wire at 15.6°C. (60°F.) — °F.	0.00226	,, ,,
Electrochem. equivalent' grs. per coulomb ..	0.00009316	Calculated from stand value. for silver.
Electrolytic solution potential against a normal hydrogen electrode (in normal aluminium sulphate) volts	1.3	Bureau of Stands. Circ 346.
Thermo - electromotive force against pure platinum for 99.97% Al. at 100°C., millivolts	+.416	,, ,, ,,
Magnetic Susceptibility at 18°C.	.63 × 10-6	I.C.T., Vol. VI, p. 354.

USEFUL FORMULAE FOR ALUMINIUM

WEIGHTS

Sheet, per sq. ft. $= 14.1\ t$ lb.
Circles, each $= 0.0769\ D^2 t$ lb.
Tubes, per ft. $= 3.682\ t(D-t)$ lb.
Extruded sections, per ft.
$= 1.180\ a.$ lb.

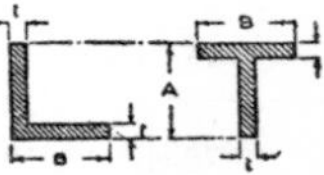

Extruded angles or 'T' bars, per ft.
$= 1.180\ t(P-t)$ lb. where $P = A + B$
Extruded channels or 'I' bars per ft.
$= 1.180\ t(P-2t)$ lb. where $P = 2A + B$

Round Rod or Wire—
Normal purity, per ft.
$= 0.923\ D^2$ lb.
Electrical purity, per ft.
$= 0.920\ D^2$ lb.
Busbars, per ft.
$= 1.172\ a.$ lb.

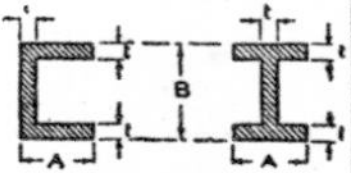

Where a = cross-section, sq. in.; D = overall diam., in.: t = thickness, in.

TEMPERATURE SCALES

To convert Fahr. to Cent.—
Add 40, multiply by 5/9, subtract 40
To convert Cent. to Fahr.—
Add 40, multiply by 9/5, subtract 40

CURRENT-CARRYING CAPACITY OF ALUMINIUM CONDUCTORS

Flat Bars— $C = ka^{0.45}\ S^{0.5}$
Where C = current in amperes
a = cross-sectional area in sq. inches
S = perimeter in ins.
k = 385 for a 40 deg. C. temperature rise, and 438 for a 50 deg. C. temperature rise.

Round Rod— $C = kD^{1.4}$
Where D = diameter in inches.
k = 659 for a 40 deg. C. temperature rise, and 749 for a 50 deg. C. temperature rise.

Bare Cable— $C = ka^{0.45}\ D^{0.5}$
Where a = effective cross-section of aluminium in sq. in. (neglecting the steel core in steel-cored cable)
D = overall diameter in inches.
k = 733 for a 40 deg. C. temperature rise, and 832 for a 50 deg. C. temperature rise.

ROUND ALUMINIUM ROD

(Nos. BA.24MS and BA.35)

For Screwing and Machining

Dia. In.	Weight per ft. run–lb. BA/24 MS	BA/35	Brass	Steel	Dia. In.	Weight per ft. run—lb. BA/24 MS	BA/35	Brass	Steel
6	33.1	34.0	101.7	96.1	3	8.28	8.49	25.4	24.0
5 7/8	31.7	32.6	97.5	92.1	2 7/8	7.60	7.80	23.4	22.1
5 3/4	30.4	31.2	93.4	88.2	2 3/4	6.96	7.13	21.4	20.2
5 5/8	29.1	29.8	89.4	84.4	2 5/8	6.34	6.50	19.5	18.4
5 1/2	27.8	28.5	85.5	80.7	2 1/2	5.75	5.90	17.7	16.7
5 3/8	26.6	27.3	81.6	77.1	2 3/8	5.19	5.32	15.9	15.1
5 1/4	25.4	26.0	77.9	73.6	2 1/4	4.66	4.78	14.3	13.5
5 1/8	24.2	24.8	74.2	70.1	2 1/8	4.15	4.26	12.8	12.1
5	23.0	23.6	70.7	66.7	2	3.68	3.77	11.3	10.7
4 7/8	21.9	22.4	67.2	63.4	1 7/8	3.23	3.32	9.93	9.38
4 3/4	20.8	21.3	63.8	60.2	1 3/4	2.82	2.89	8.65	8.17
4 5/8	19.7	20.2	60.4	57.1	1 5/8	2.42	2.49	7.46	7.05
4 1/2	18.6	19.1	57.2	54.0	1 1/2	2.07	2.12	6.36	6.00
4 3/8	17.6	18.1	54.1	51.1	1 3/8	1.74	1.78	5.34	5.05
4 1/4	16.6	17.0	51.0	48.2	1 1/4	1.44	1.47	4.42	4.17
4 1/8	15.6	16.1	48.1	45.4	1 1/8	1.16	1.19	3.58	3.38
4	14.7	15.1	45.2	42.7	1	0.920	0.943	2.83	2.67
3 7/8	13.8	14.2	42.4	40.1	7/8	0.704	0.722	2.16	2.04
3 3/4	12.9	13.3	39.7	37.5	3/4	0.517	0.531	1.59	1.50
3 5/8	12.1	12.3	37.1	35.1	5/8	0.359	0.368	1.10	1.04
3 1/2	11.3	11.6	34.6	32.7	1/2	0.230	0.236	0.706	0.667
3 3/8	10.5	10.7	32.2	30.4	3/8	0.129	0.133	0.397	0.375
3 1/4	9.71	9.96	29.8	28.2	1/4	0.0575	0.0590	0.177	0.167
3 1/8	8.98	9.21	27.6	26.1	1/8	0.0144	0.0147	0.0442	0.0417

The figures in this table have been calculated with the following specific gravities:

No. BA.24MS Aluminium Alloy	2.70
No. BA:35 Aluminium Alloy	2.77
Brass (66/34)	8.30
Steel	7.84

HEXAGON ALUMINIUM ROD

(Nos. BA.24MS and BA.35)

For Screwing and Machining to Standard Whitworth Threads

Bolt Size			Weight per ft. run—lb.			
Dia. In.	Width across Flats in.		BA/24 MS	BA/35	Brass	Steel
2	—	3.15	10.06	10.32	30.9	29.2
1¾	—	2.76	7.72	7.92	23.7	22.4
1⅝	—	2.58	6.75	6.92	20.7	19.6
1½	—	2.41	5.89	6.04	18.1	17.1
1⅜	—	2.21	4.95	5.08	15.2	14.4
1¼	—	2.05	4.26	4.37	13.1	12.4
1⅛	—	1.86	3.51	3.60	10.8	10.2
1	—	1.67	2.83	2.90	8.69	8.21
⅞	—	1.48	2.22	2.28	6.82	6.45
13/16	—	1.39	1.96	2.01	6.02	5.69
¾	—	1.30	1.71	1.76	5.27	4.97
11/16	—	1.20	1.46	1.50	4.49	4.24
⅝	—	1.10	1.23	1.26	3.77	3.56
9/16	—	1.01	1.03	1.06	3.18	3.00
—	1	—	1.01	1.04	3.12	2.94
½	—	0.919	0.856	0.878	2.63	2.49
—	⅞	—	0.776	0.796	2.39	2.25
7/16	—	0.820	0.682	0.699	2.10	1.98
—	¾	—	0.570	0.585	1.75	1.66
⅜	—	0.709	0.510	0.523	1.57	1.48
—	⅝	—	0.396	0.406	1.22	1.15
5/16	—	0.601	0.366	0.376	1.13	1.06
—	9/16	—	0.321	0.329	0.986	0.931
¼	—	0.525	0.280	0.287	0.859	0.811
—	½	—	0.254	0.260	0.779	0.736
3/16	7/16	—	0.194	0.199	0.596	0.563
—	⅜	—	0.143	0.146	0.438	0.414
—	5/16	—	0.099	0.102	0.304	0.287
—	¼	—	0.0634	0.0650	0.195	0.184
—	3/16	—	0.0357	0.0366	0.110	0.103
—	⅛	—	0.0158	0.0163	0.049	0.046

The figures in this table have been calculated with the following specific gravities : No. BA.24MS Aluminium Alloy, 2.70 ; No. BA.35 Aluminium Alloy, 2.77 ; Brass (66/34), 8.30 ; Steel 7.84.

EXTRUDED ALUMINIUM ALLOYS

Properties

Alloy	Thickness	Tensile Strength Tons/sq. in. Min.	Tensile Strength Tons/sq. in. Av.	0.1% Proof Stress Tons/sq. in. Min.	0.1% Proof Stress Tons/sq. in. Av.	Elongation on 2" % * Min.	Elongation on 2" % * Av.	Brinell Hardness Av.	Specific Gravity Av.	Weight per ft. per sq. in. lb. Av.
Pure Aluminium										
99.0% Min. Purity	All sections	4.75	5.5	—	2.25	20.0	30.0	20–25	2.710	1.180
General Purpose Alloys										
BA/24	All sections	10.0	11.0	5.5	6.5	18.0	25.0	52	2.701	1.171
BA/25	Up to ¾"	13.5	15.0	8.0	10.0	18.0	25.0	75	2.693	1.168
BA/60A	Up to ½"	6.5	7.5	2.0	3.5	30.0	36.0	30	2.738	1.187
Alloys for Machining										
BA/24 MS	Up to 2"	15.0	16.0	13.0	14.0	8.0	12.0	90	2.701	1.171
BA/35	Up to 0.3"	23.0	25.0	19.0	21.0	4.0	6.0	110	2.769	1.201
BA/35	Over 0.3" to 2½"	16.0	18.5	7.0	9.5	12.0	20.0	—	2.769	1.201

* *Elongation.*—Where the size of the section permits, test pieces are in accordance with British Standards Specification 18, test piece C being the standard for round bars. In other cases rectangular test pieces are used. Elongation is measured on a gauge length of 2" for all standard test pieces and other test pieces with a cross sectional area exceeding 0.1 sq. in. For test pieces with smaller cross sectional area the quoted elongation value is only guaranteed for test pieces having a gauge length equal to $4\sqrt{\text{area}}$.

ALUMINIUM ALLOY SHEET
Properties of BA.60A, Sheet and Strip

Temper and Gauge	Tensile Strength Tons/sq.in.	0.1% Proof Stress Tons/sq. in.		Elongation on 2" %		Bend Test	Brinell Hardness	
		Min.	Av.	Min.	Av.		Min.	Av.
Soft								
.128"–.064" (10–16 s.w.g.)				35	45			
.063"–.036" (16–20 s.w.g.)	6.0–7.5	—	2.7	30	43	Flat Bend	27	30
.035"–.022" (20–24 s.w.g.)				27	41			
.021"–.012" (24–30 s.w.g.)				22	31			
Quarter Hard								
.128"–.064" (10–16 s.w.g.)				12	17			
.063"–.036" (16–20 s.w.g.)	7.5–9.5	6.9	7.6	7	15	Flat Bend	35	40
.035"–.022" (20–24 s.w.g.)				7	15			
.021"–.012" (24–30 s.w.g.)				4	9			
Half Hard						180°		
.128"–.064" (10–16 s.w.g.)				7	10	Bend		
.063"–.036" (16–20 s.w.g.)	9.0–11.0	8.2	8.9	4	8	r=½t	40	44
.035"–.022" (20–24 s.w.g.)				4	6			
.021"–.012" (24–30 s.w.g.)				3	4			
Three-quarter Hard						180°		
.128"–.064" (10–16 s.w.g.)				5.5	8.5	Bend		
.063"–.036" (16–20 s.w.g.)	10.5–12.	9.4	10.0	2.5	4	r=t	42	47
.035"–.022" (20–24 s.w.g.)				2	4			
.021"–.012" (24–30 s.w.g.)				2	3			
						180°		
Hard						Bend		
.128"–.064" (10–16 s.w.g.)				3	7.5	r=2t		
.063"–.036" (16–20 s.w.g.)	11.75 min.	9.75	—	2	4	r=3t	44	—
.035"–.022" (20–24 s.w.g.)				2	3	r=3t		
.021"–.012" (24–30 s.w.g.)				2	3	r=4t		

ALUMINIUM ALLOY SHEET

Properties of BA.20, Sheet and Strip
(Now B.S. L.46)

Temper and Gauge	Tensile Strength Tons/sq. in. Range	0.1% Proof Stress Tons/sq. in.	Elongation on 2" %	Bend Test	Brinell Hardness
Soft	Min. Av.	Min. Av.	Min. Av.		Min. Av.
.128".–064" (10–16 s.w.g.)			15 23	Flat Bend	
.063"–.036" (16–20 s.w.g.)	11.0 11.25	3.5 4.5	15 20		42 45
.035"–.022" (20–24 s.w.g.)			15 18		
under .022"			12 16		
Half Hard					
.128"–.064" (10–16 s.w.g.)			5 7	180° Bend	
.063"–.036" (16–20 s.w.g.)	14.0 14.5	12.0 13.5	3 4.5	r=t	63 68
.035"–.022" (20–24 s.w.g.)			3 4		
under .022"			2 2.5		
Hard					
.128"–.064" (10–16 s.w.g.)			4 5.5	180° Bend	
.063"–.036" (16–20 s.w.g.)	16.0 16.5	14.0 15.5	3 4	r=2t	70 75
.035"–.022" (20–24 s.w.g.)			2 3.5		
under .022"			— 2.0		

All D.T.D. Specifications covering B.A. 20 have been cancelled. L.46 now covers BA.20, soft temper, and calls for 11 tons/sq. in. ultimate tensile strength and 18% elongation for sheet thicker than 0.104 in.

Specification Number	Corresponding temper	Tensile Strength Tons/sq. in.	0.1% Proof Stress Tons/sq. in.	Elongation on 2" % *	Bend Test
		Min.	Min.	Min.	180°
D.T.D.278	Soft	11.0	—	18.0	Flat
D.T.D.266	Half Hard	14.0	12.0	5.0	r=2t
D.T.D.249	Hard	16.0	14.0	5.0	r=4t

*These values are only specified for sheets thicker than 0.104" (12 s.w.g.).

ALUMINIUM PIPE SIZES

BRITISH STANDARD SIZES AND THREADS

Nominal Bore.	Approx. External Dia.	Thickness of Walls. Gas.	Thickness of Walls. Water.	Thickness of Walls. Steam.	Approximate Weight/ft. Gas.	Approximate Weight/ft. Water.	Approximate Weight/ft. Steam.	Dia. Top of Thread.	Depth of Thread.	Core Diameter.	No. of Threads per inch.	Length of Thread. Pipe End.	Length of Thread. In Coupler Min.
in.	in.	s.w.g.	s.w.g.	s.w.g.	lb.	lb.	lb.	in.	in.	in.	in.	in.	in.
1/8	13/32	14	13	12	.096	.1055	.116	.383	.0230	.337	28	.375	.75
1/4	17/32	14	13	12	.133	.147	.162	.518	.0335	.451	19	.437	.87
3/8	11/16	13	12	11	.202	.221	.241	.656	.0335	.589	19	.500	1.00
1/2	27/32	12	11	10	.281	.309	.336	.825	.0455	.734	14	.625	1.25
5/8	15/16	11	10	9	.352	.383	.420	.902	.0455	.811	14	.625	1.25
3/4	1 1/16	11	10	9	.402	.437	.485	1.041	.0455	.950	14	.750	1.50
7/8	1 7/32	11	10	9	.467	.512	.467	1.189	.0455	1.098	14	.750	1.50
1	1 11/32	10	9	8	.569	.631	.690	1.309	.0580	1.193	11	.875	1.75
1¼	1 11/16	9	8	7	.816	.895	.975	1.650	.0580	1.534	11	1.000	2.00
1½	1 29/32	8	7	6	1.01	1.11	1.23	1.882	.0580	1.766	11	1.000	2.00
1¾	2 5/32	8	7	6	1.17	1.29	1.38	2.116	.0580	2.000	11	1.125	2.25
2	2⅜	8	7	6	1.30	1.42	1.54	2.347	.0580	2.231	11	1.125	2.25

The above figures are based on British Standard Specification No. 21—1909.

ALUMINIUM PIPE SIZES

BRITISH STANDARD SIZES AND THREADS—(*Continued*)

Nominal Bore.	Approx. External Dia.	Thickness of Walls.			Approximate Weight/ft.			Dia. Top of Thread.	Depth of Thread.	Core Diameter.	No. of Threads per inch.	Length of Thread.	
		Gas.	Water.	Steam.	Gas.	Water.	Steam.					Pipe End	In Coupler Min.
in.	in.	s.w.g.	s.w.g.	s.w.g.	lb.	lb.	lb.	in.	in	in.	in.	in.	in.
2¼	2⅝	7	6	5	1.58	1.71	1.88	2.587	.0580	2.471	11	1.250	2.50
2½	3	7	6	5	1.84	1.98	2.17	2.960	.0580	2.844	11	1.250	2.50
2¾	3¼	7	6	5	1.99	2.16	2.37	3.210	.0580	3.094	11	1.375	2.75
3	3½	7	6	5	2.15	2.34	2.57	3.460	.0580	3.344	11	1.375	2.75
3¼	3¾	7	6	5	2.32	2.51	2.76	3.700	.0580	3.584	11	1.500	3.00
3½	4	7	6	5	2.48	2.69	2.96	3.950	.0580	3.834	11	1.500	3.00
3¾	4¼	7	6	5	2.64	2.87	3.15	4.200	.0580	4.084	11	1.500	3.00
4	4½	7	6	5	2.82	3.05	3.35	4.450	.0580	4.332	11	1.625	3.25
4½	5	7	6	5	3.13	3.40	3.75	4.950	.0580	4.834	11	1.625	3.25
5	5½	7	6	5	3.46	3.76	4.14	5.450	.0580	5.334	11	1.750	3.50
5½	6	7	6	5	3.77	4.13	4.58	5.950	.0580	5.834	11	1.875	3.75
6	6½	7	6	5	4.10	4.48	4.91	6.450	.0580	6.334	11	2.000	4.00

The above figures are based on British Standard Specification No. 21—1909.

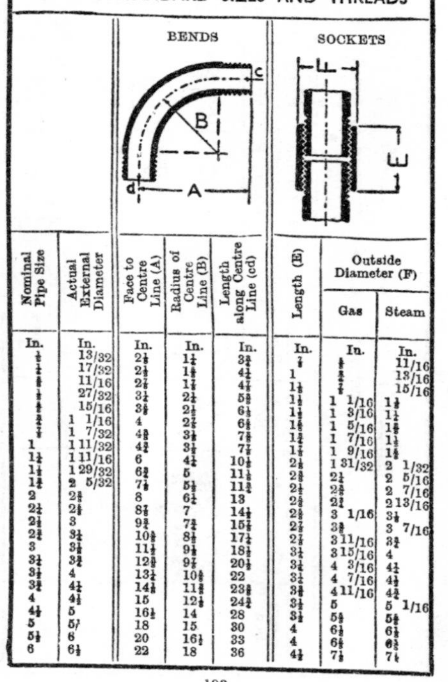

ALUMINIUM PIPE FITTINGS
BRITISH STANDARD SIZES AND THREADS

Nominal Pipe Size	Actual External Diameter	Bends: Face to Centre Line (A)	Bends: Radius of Centre Line (B)	Bends: Length along Centre Line (cd)	Sockets: Length (E)	Sockets: Outside Diameter (F) Gas	Sockets: Outside Diameter (F) Steam
In.	In.	In.	In.	In.	In.	In.	In.
1/8	13/32	2 1/8	1 1/4	3 3/4	7/8	5/8	11/16
1/4	17/32	2 1/2	1 3/8	4 1/4	1	3/4	13/16
3/8	11/16	2 7/8	1 7/8	4 7/8	1 1/8	7/8	15/16
1/2	27/32	3 1/4	2 1/4	5 5/8	1 1/4	1 1/16	1 1/8
5/8	15/16	3 5/8	2 1/2	6 1/2	1 1/4	1 3/16	1 1/4
3/4	1 1/16	4	2 7/8	6 5/8	1 3/8	1 5/16	1 3/8
7/8	1 7/32	4 3/8	3 1/8	7 5/8	1 1/2	1 7/16	1 1/2
1	1 11/32	4 3/4	3 1/2	7 7/8	1 7/8	1 9/16	1 5/8
1 1/4	1 11/16	6	4 1/4	10 1/8	2 1/8	1 31/32	2 1/32
1 1/2	1 29/32	6 3/4	5	11 1/8	2 3/8	2 1/4	2 5/16
1 3/4	2 5/32	7 1/8	5 1/2	11 3/4	2 1/2	2 3/8	2 7/16
2	2 3/8	8	6 1/4	13	2 5/8	2 3/4	2 13/16
2 1/4	2 5/8	8 7/8	7	14 1/2	2 5/8	3 1/16	3 1/8
2 1/2	3	9 3/4	7 3/4	15 7/8	2 7/8	3 3/8	3 7/16
2 3/4	3 1/4	10 5/8	8 1/2	17 1/4	2 7/8	3 11/16	3 3/4
3	3 1/2	11 1/2	9 1/8	18 1/2	3 1/4	3 15/16	4
3 1/4	3 3/4	12 3/8	9 7/8	20 1/2	3 1/4	4 3/16	4 1/4
3 1/2	4	13 1/4	10 5/8	22	3 1/2	4 7/16	4 1/2
3 3/4	4 1/4	14 1/8	11 3/8	23 5/8	3 5/8	4 11/16	4 3/4
4	4 1/2	15	12 1/8	24 3/4	3 1/2	5	5 1/16
4 1/2	5	16 1/2	14	28	3 1/2	5 5/8	5 5/8
5	5/	18	15	30	4	6 1/8	6 1/8
5 1/2	6	20	16 1/2	33	4	6 5/8	6 5/8
6	6 1/2	22	18	36	4 1/2	7 1/8	7 1/8

WROUGHT ALUMINIUM ALLOYS

British Standard Specifications. (Aircraft and General)

Common Designation	Nominal Composition *	Specification No.	Description and Heat Treatment †	Minimum Tensile Strength tons/sq. in.	0.1% Proof Stress tons/sq. in.	Elongation % in 2"
Duralumin	Cu 3.5–4.5% Mn. 0.4–0.7% Mg. 0.4–0.8%	5L3 (395)	Sheet. H.T.—Heated to 495 ± 10° C. and quenched. Then aged at room temp. for 5 days.	Over 12" wide 25 12" wide and under 25	Over .020" 14.5 15	Over .104" 15 15
		6L1 (477)	Bars for machining (up to 3"), Extruded Sections and Forgings. H.T.—As for 395 above.	Over ⅜" thick 25 Less than ⅜" thick 24	15 14	15 15

WROUGHT ALUMINIUM ALLOYS—(*Continued*)

British Standard Specifications. (Aircraft and General)

Common Designation	Nominal Composition *	Specification No.	Description and Heat Treatment †	Minimum Tensile Strength tons/sq. in.	0·2% Proof Stress tons/sq. in.	Elongation % in 2″
Duralumin	Cu. 3.5–4.5% Mn. 0.4–0.7% Mg. 0.4–0.8%	2L39	Bars for machining (3″–6″) H.T.—As for 395 above.	22	12	12
		5T4 (396)	Tubes. H.T.—Heated to 490 ± 10° C. and quenched. Aged at room temperature.	26	19	.064″ and under 8 .064″–.104″ 10 Over .104″ 12.5

WROUGHT ALUMINIUM ALLOYS—(*Continued*)

British Standard Specifications. (Aircraft and General)

Common Designation	Nominal Composition *	Specification No.	Description and Heat Treatment †	Minimum Tensile Strength tons/sq. in.	0.1% Proof Stress tons/sq. in.	Elongation % in 2″
Duralumin	Cu. 3·5–4.5% Mn. 0.4–0.7% Mg. 0.4–0.8%	532	Forgings H.T.—Heated to 480–490° C. and quenched. Then aged at room temperature for about 4 days.	2¼″ and under 25	14	15
				Over 2¼″–4″ 22	11.5	15
				Over 4″–6″ 20	10.5	15
				Over 6″ 18	9	15
		2L37	Rivets. Annealed condition. Anneal at 360–400° C. Test samples H.T. as for 395 above.	25	—	—

* In each of the above alloys, the remainder consists of aluminium and normal impurities.

† Where a material is covered by both an Aircraft and a general specification, the heat treatment given in the list applies to the Aircraft Specification.

WROUGHT ALUMINIUM ALLOYS—(*Continued*)

Air Ministry D.T.D. Specifications

Common Designation	Nominal Composition *	Specification No.	Description and Heat Treatment	Minimum Tensile Strength tons/sq. in.	0.1% Proof Stress tons/sq. in.	Elongation % in 2″
MG.7	Mg.6.5– 10.0% Mn. 0.6% Max. Si. 0.5% Max. Fe. 0.75% Max.	177A	Sheets (Hard).	25	17	Over .104″ 15
		182A	Sheets (Annealed).	Min. 20 Max. 23	10	Over .104″ 20
		186A	Tubes (Hard).	.104″ thick and under 26 Over .104″ 25	18 17	— —

WROUGHT ALUMINIUM ALLOYS—(*Continued*)

Air Ministry D.T.D. Specifications

Common Designation	Nominal Composition *	Specification No.	Description and Heat Treatment	Minimum Tensile Strength tons/sq. in.	0.1% Proof Stress tons/sq. in.	Elongation % in 2"
MG.7	Mg.6.5–10.0%	190	Tubes (Annealed).	Min. 20 Max. 23	10	—
	Mn. 0.6% Max. Si. 0.5% Max. Fe. 0.75% Max.	194	Bars.	Up to 1" 25 1"–3" 22 Over 3" 21	15 12 10	15 15 15
		198A	Wire and Rivets.	Min. 20 Max. 23	—	—
		297	Bars, Extruded Sections and Forgings (Softened).	20	18	15
		404	Hard drawn Wire and Rivets.	27	—	—

WROUGHT ALUMINIUM ALLOYS—(*Continued*)

Air Ministry D.T.D. Specifications

Common Designation	Nominal Composition *	Specification No.	Description and Heat Treatment	Minimum Tensile Strength tons/sq. in.	0.1% Proof Stress tons/sq. in.	Elongation % in 2"
Birmabright and RR.66	Mg. 3.0–6.0% Mn. 0.75% Max.	170A	Sheets (Hard).	20	15	Over .104" 3
		175A	Sheets (Half-Hard).	16	12	Over .104" 6
Birmabright 2, D.2, NA.4S and BA.20	Mg. 3.0% Max. Mn. 1.5% Max.	249	Sheets (Hard).	16	14	Over .104" 5
		266	Sheets (Half-Hard).	14	12	Over .104" 5

* In each of the above alloys, the remainder consists of aluminium and normal impurities.

WROUGHT ALUMINIUM ALLOYS—(*Continued*)

Air Ministry D.T.D. Specifications

Common Designation	Nominal Composition *	Specification No.	Description and Heat Treatment	Minimum Tensile Strength tons/sq. in.	0.1% Proof Stress tons/sq. in.	Elongation % in 2"
R.R.56	Cu. 1.5–3.0% Ni. 0.5–1.5% Mg. 0.4–1.0% Fe. 0.8–1.4%	179 and 184	Airscrew Forgings. H.T.—2 hours at 520–535° C. and quenched. 15–20 hours 165–175° C. and quenched.	27	20	10
		206	Sheets. H.T.—Heated at 530 ± 5° C. and quenched. Not less than 16 hours at 170 ± 5° C. and quenched.	27	21	Over .104" 10

WROUGHT ALUMINIUM ALLOYS—*(Continued)*

Air Ministry D.T.D. Specifications

Common Designation	Nominal Composition *	Specification No.	Description and Heat Treatment	Minimum Tensile Strength tons/sq. in.	0.1% Proof Stress tons/sq. in.	Elongation % in 2″
R.R.56	Cu. 1.5–3.0% Ni. 0.5–1.5% Mg. 0.4–1.0% Fe. 0.8 1.4%	220A	Tubes. H.T.—Heated to 530±5° C. and quenched. 10–20 hours at 170±5° C. and quenched.	27	(0.2% prf. stress) 23	Under .064″ 6 .064″-.104″ 8 Over .104″ 10
R.R.56 NS	Cu. 1.5–2.5% Ni. 0.5–1.5% Mg. 0.6–1.2% Fe. 0.8–1.5%	246A	Forgings (Softened). H.T.—2 hours at 515–525° C. and cooled in air. 5 hours at 200° C. and cooled in air.	16	7	16

WROUGHT ALUMINIUM ALLOYS—(*Continued*)

Air Ministry D.T.D. Specifications

Common Designation	Nominal Composition *	Specification No.	Description and Heat Treatment	Minimum Tensile Strength tons/sq. in.	0.1 % Proof Stress tons/sq. in.	Elongation % in 2″
Duralumin	Cu. 3.5–4.5% Mn. 0.4–0.7% Mg. 0.4–0.7%	147 and 150	Airscrew Forgings. H.T.—Heated to 480±10° C. quenched and aged for 4 days.	22.5	13.5	15
Duralumin F. and NA22S.	Cu. 3.5–4.7% Si. 0.7–1.5% Mg. 0.4–1.0% Mn. 0.4–1.5%	252	Bars for forging, etc. H.T.—Heated to 480-505° C. and quenched. Not less than 8 hours at 150–175° C. and quenched.	28	22	8
Duralumin G., Hiduminium 72 and NA.24S	Cu. 3.5–4.8% Mg. 0.8–1.8% Mn. 0.3–1.5%	270	Sheets. H.T.—Heated to 490±10° C. and quenched. Aged at room temp. for 5 days.	28	17.5	Over .104″ 15

WROUGHT ALUMINIUM ALLOYS—(*Continued*)

Air Ministry D.T.D. Specifications

Common Designation	Nominal Composition *	Specification No.	Description and Heat Treatment	Minimum Tensile Strength tons/sq. in.	0.1 % Proof Stress tons/sq. in.	Elongation % in 2"
Duralumin G., Hiduminium 72 and NA.24S	Cu. 3.5–4.8% Mg. 0.8–1.8% Mn. 0.3–1.5%	273	Tubes. H.T.—Heated to 490±10° C. and quenched. Aged at room temp.	¼" thick and under 29	22	7
				Over ¼" thick 29	22	10
		280	Bars up to 4" diam., or A/F and Extruded Sections up to 3" thick. H.T.—Heated to 490±10° C. and quenched. Aged at room temp. for 5 days.	28	18	10
		290	Material as covered by DTD.280, but larger sizes.	27	17	10

WROUGHT ALUMINIUM ALLOYS—(*Continued*)

Air Ministry D.T.D. Specifications

Common Designation	Nominal Composition *	Specification No.	Description and Heat Treatment	Minimum Tensile Strength tons/sq. in.	0.1% Proof Stress tons/sq. in.	Elongation % in 2"
BA60A	Mn. 1.5% Max. Cu. 0.15% Max. Fe. 0.75% Max.	213	Sheets.	11	—	—
Hiduminium S.12, NA.38S	Si. 11.5–13.0% Cu. 0.7–1.3% Ni. 0.7–1.3% Mg. 0.8–1.5%	324	Forgings for engine cylinders and pistons. H.T.—Heated to 520–535° C. and quenched. Not less than 5 hours at 130-160° C. After rough maching heated at 190–210° C. for not less than 2 hours.	20	—	3

WROUGHT ALUMINIUM ALLOYS—(*Continued*)

Air Ministry D.T.D. Specifications

Common Designation	Nominal Composition *	Specification No.	Description and Heat Treatment	Minimum Tensile Strength tons/sq. in.	0·1% Proof Stress tons/sq. in.	Elongation % in 2"
Hiduminium RR.75, NA.16ST	Cu. 1.5–3.0% Mg. 0.2–0.5%	327	Wire and Rivets. H.T.—Heated to 490±10° C. and quenched. Aged at room temp. for 5 days.	17	—	—
Duralumin H.	Si. 0.75–1.25% Mn. 0.5–1.0% Mg. 0.5–1.25%	346	Sheets and Strips (Soft).	11 max.	—	Over .104" 20
Duralumin E.	Cu. 3.0–4.5% Fe. 1.0% Max. Si. 1.0% ,,	356	Sheets and Strips— (a) Softened (b) Quenched	 — —	 — —	Over .104" 14 14

* In each of the above alloys, the remainder consists of aluminium and normal impurities.

WROUGHT ALUMINIUM ALLOYS—(*Continued*)

Air Ministry D.T.D. Specifications

Common Designation	Nominal Composition *	Specification No.	Description and Heat Treatment	Minimum Tensile Strength tons/sq. in.	0·1% Proof Stress tons/sq. in.	Elongation % in 2″
NA.26ST	Mn. 1.2% ,, Mg. 1.0% ,,		(c) Quenched and Aged. H.T.—Heated to 510±10° C. and quenched. Not less than 5 hours at 165±10° C.	27	21	
		364	Bars, Extruded Sections, and Forgings H.T.—Heated to 515±10° C. and quenched. Not less than 5 hours at 165±10° C.	30	26	8

WROUGHT ALUMINIUM ALLOYS—(*Continued*)

Air Ministry D.T.D. Specifications

Common Designation	Nominal Composition *	Specification No.	Description and Heat Treatment	Minimum Tensile Strength tons/sq. in.	0·1% Proof Stress tons/sq. in.	Elongation % in 2"
Hiduminium R.R.77	Cu. 1.5–3.0% Zn. 4.0–6.0% Mg. 2.0–4.0% Ni. 1.0% Max.	363	Bars and Extruded Sections. H.T.—Heated to 460±10° C. and quenched. 10–20 hours at 130±10° C.	33	27	8
Substitute for Duralumin	Mg. 0.5–1.25% Si. 0.75–1.25% Cu. 1.0% Max. Mn. 1.0% „ Fe. 0.75% „	423	Extruded Sections and Bars for machining. H.T.—Heated to 525±5° C. and quenched. Heated to 165±10° C. for the requisite period.	22	17	8

* In each of the above alloys, the remainder consists of aluminium and normal impurities.

DURALUMIN

CHEMICAL COMPOSITION

("B" Grade):

Copper: 3.5% to 4.5%
Manganese: 0.4% to 0.7%
Magnesium: 0.4% to 0.7%
Aluminium: About 94.5%

For special purposes other grades (A, M and H) are supplied. These vary somewhat in composition, according to the purpose for which they are intended.

PHYSICAL PROPERTIES ("B" Grade):

Specific Gravity: Aprox. 2.8

Specific Heat: 0.214 (Water=1).

Thermal Conductivity: 31 (Silver=100).

Electrical Conductivity: Normalised 33% to 35% (Copper=100). Annealed 39% to 41% (Copper=100).

Coefficient of Linear Expansion: 0.00001255 per deg. Fahrenheit. 0.0000226 per deg. Centigrade.

Young's Modulus of Elasticity (E): 4,500 tons per sq. inch.

Melting Range: 560 to 650 deg. Centigrade.

Annealing Range: 360 to 400 deg. Centigrade.

Heat Treatment or Normalising Temperature: 480 deg. Centigrade ±10 deg. C.

Forging Temperature: 400 to 470 deg. Centigrade.

MECHANICAL PROPERTIES:

Brinell Hardness: Annealed 60 (approx.).
Normalised 90 to 115.

Fatigue Range: ±9.5 tons per square inch.

Izod Impact Value: About 15ft. lb.

DURALUMIN—(*Continued*)

Those of the "B" grade in the normalised or fully heat-treated condition are best illustrated as follows:

SHEET AND STRIP

Thickness of Sheet or Strip	Ultimate Tensile Strength. Tons per sq. in.	0.1% Proof Stress. Tons per sq. in.	Elongation per cent. on 2in. Gauge Length.
Under 0.02in. (25 S.W.G.)	Not less than 25	Not less than 15	Not less than 8
0.02 to under 0.048in. (25 to 18 S.W.G.)	Not less than 25	Not less than 15	Not less than 12
0.048 in. (18 S.W.G) and thicker ..	Not less than 25	Not less than 15	Not less than 15

BARS, ETC.

Nominal Size of Bar (Diameter or Width across Flats).	0.1% Proof Stress. Tons per sq. in.	Ultimate Tensile Strength. Tons per sq. in.	Elongation per cent.	Reduction of Area per cent.
Inches.		Not less than	Not less than	Not less than
Up to $2\frac{5}{8}$	15	25	15	20
$2\frac{5}{8}$ to 4 ..	12	22	15	20
4 to 6 ..	10	20	15	20
Untreated Material	—	17	15	18

DURALUMIN—(*Continued*)

TUBES, ETC.

Thickness of Tube.	0.1% Proof Stress. Tons per sq. in.	Ultimate Tensile Strength. Tons per sq. in.	Elongation per cent. on 2 in. Gauge Length.
0.064 in. (16 S.W.G.) and under ..	Not less than 16.5	Not less than 26	Not less than 8
Over 0.064 in. (16 S.W.G.) and up to 0.104 in. (12 S.W.G.) ..	16.5	26	10
Over 0.104 in. (12 S.W.G.)..	16.5	26	12.5

NOTE.—These tests values apply to tubes in the condition in which they are supplied by the manufacturers, i.e., fully heat-treated and drawn to size. If the tubes are subsequently heat-treated by the purchaser they cannot be relied upon to have properties greater than the following :

Proof Stress : 18.0 tons per sq. in.
Ultimate Tensile Strength : 26.0 tons per sq. in.

VICKERS' DURALUMIN

The sizes shown indicate those for which tools are in existence.
Sections are not necessarily in stock.
Special sections are supplied to suit individual requirements.

ROUND RODS AND BARS

Sizes Inches	Decimal Equivalent	mm Equivalent	Approx. Weight	
			Lbs. per Ft. Run	Kilos per Lineal Metre
1/4	.250	6.35	.060	.089
5/16	.3125	7.94	.094	.134
11/12	.3437	8.70	.1125	.167
3/8	.875	9.52	.132	.20
7/16	.4375	11.11	.182	.27
15/32	.468	11.89	.208	.31
1/2	.500	12.70	.240	.36
17/32	.531	13.48	.266	.39
9/16	.5625	14.29	.302	.45
19/32	.593	15.06	.334	.50
5/8	.625	15.87	.372	.56
21/32	.656	16.65	.407	.61
11/16	.687	17.45	.452	.67
23/32	.718	18.25	.490	.72
3/4	.750	19.05	.538	.80
13/16	.812	20.62	.632	.94
7/8	.875	22.22	.729	1.08
29/32	.906	23.00	.782	1.16
15/16	.937	23.79	.835	1.23
1	1.000	25.40	.957	1.42
1 1/16	1.0625	26.98	1.08	1.61
1 3/32	1.093	27.77	1.13	1.68
1 1/8	1.125	28.57	1.21	1.80
1 3/16	1.187	30.16	1.35	2.01
1 1/4	1.250	31.75	1.49	2.22
1 5/16	1.312	33.34	1.64	2.45
1 3/8	1.375	34.42	1.81	2.70
1 7/16	1.437	36.51	1.96	2.92

DURALUMIN
ROUND RODS AND BARS—(*Continued*)

Sizes Inches	Decimal Equivalent	mm Equivalent	Approx. Weight	
			Lbs. per Ft. Run	Kilos per Lineal Metre
1½	1.500	38.10	2.15	3.20
1 9/16	1.562	39.69	2.32	3.45
1⅝	1.625	41.27	2.53	3.77
1 11/16	1.687	42.85	2.71	4.03
1¾	1.750	44.45	2.93	4.36
1 13/16	1.812	46.02	3.12	4.65
1⅞	1.875	47.62	3.36	5.00
1 29/32	1.906	48.40	3.46	5.14
1 15/16	1.937	49.19	3.57	5.30
2	2.000	50.80	3.83	5.70
2 1/16	2.0625	52.50	4.10	6.10
2⅛	2.125	54.00	4.32	6.45
2 3/16	2.187	55.56	4.55	6.78
2¼	2.250	57.15	4.85	7.22
2 5/16	2.312	58.74	5.05	7.51
2⅜	2.375	60.32	5.40	8.05
2 7/16	2.437	61.90	5.65	8.40
2½	2.500	63.50	3.99	8.91
2 9/16	2.5625	65.09	6.30	9.40
2⅝	2.625	66.67	6.55	9.75
2 11/16	2.687	68.25	6.92	10.30
2¾	2.750	69.85	7.20	10.75
2⅞	2.875	73.02	7.95	11.85
3	3.000	76.20	8.65	12.89
3⅛	3.125	79.37	9.38	13.98
3 3/16	3.187	81.20	9.60	14.30
3¼	3.250	82.55	10.01	14.90
3 5/16	3.312	84.14	10.5	15.65
3⅜	3.375	85.72	10.83	16.13
3½	3.500	88.90	11.71	17.45
3⅝	3.625	92.07	12.73	18.95
3¾	3.750	95.25	13.40	20.00
3⅞	3.875	98.42	14.30	21.37
4	4.00	101.60	15.30	22.81
4⅛	4.125	104.60	16.20	24.15

DURALUMIN
ROUND RODS AND BARS—(*Continued*)

Sizes Inches	Decimal Equivalent	mm Equivalent	Approx. Weight	
			Lbs. per Ft. Run	Kilos per Lineal Metre
4¼	4.250	107.95	17.20	25.68
4⅜	4.375	111.12	18.2	27.13
4½	4.50	114.30	19.30	28.76
4⅝	4.625	117.47	20.4	30.41
4¾	4.750	120.65	21.5	32.05
4⅞	4.875	123.82	21.6	32.20
5	5.00	127.00	24.0	35.35
5⅛	5.125	130.20	25.0	37.50
5¼	5.250	133.35	26.14	38.65
5⅜	5.375	136.52	27.6	40.05
5½	5.50	139.7	28.87	43.00
5⅝	5.625	142.7	30.2	44.90
5¾	5.750	146.05	31.25	46.50
6	6.00	152.40	34.15	50.60
Metric Sizes				
	.315	8	.093	133
	.394	10	.148	.22
	.472	12	.213	.32
	.512	13	.250	.37
	.590	15	.332	.49
	.787	20	.590	.88
	.890	22.5	.755	1.13
	1.181	30	1.325	1.99
	1.290	33	1.59	2.36
	1.378	35	1.81	2.70
	1.575	40	2.35	3.52
	1.772	45	3.05	4.55
	1.968	50	3.68	5.50
	2.165	55	4.51	6.72
	2.362	60	5.31	7.92
	2.441	62	5.65	8.40
	2.756	70	7.22	10.76
	2.953	75	8.58	12.35
	3.150	80	9.50	14.00

DURALUMIN HEXAGON RODS

(Whitworth Standard)

Sizes of Bolts Inches	Across Flats	mm Equivalent	Approx. Weight	
			Lbs. per Ft. Run	Kilos per Lineal Metre
1/8	.338	8.58	.124	.18
3/16	.448	11.38	.211	.31
1/4	.525	13.33	.291	.43
5/16	.601	15.26	.375	.56
3/8	.709	18.01	.521	.78
7/16	.820	20.83	.708	1.05
1/2	.919	23.34	.875	1.30
9/16	1.011	25.68	1.062	1.58
5/8	1.101	27.96	1.250	1.86
11/16	1.201	30.50	1.521	2.26
3/4	1.301	33.04	1.791	2.67
13/16	1.390	35.30	2.062	3.07
7/8	1.479	37.57	2.333	3.32
15/16	1.574	39.98	2.625	3.91
1	1.670	42.42	2.916	4.34
1 1/8	1.960	47.42	3.586	5.34
1 1/4	2.048	52.02	4.541	6.76
1 1/2	2.413	61.29	5.860	8.72
Odd Sizes				
	.190/.193	4.82	.038	.056
	.250	6.35	.066	.098
	.280	7.11	.082	.122
	.312	7.92	.100	.149
	.324	8.22	.111	.165
	.375	9.52	.147	.219
	.410/.413	10.42	.177	.263
	.656	16.66	.447	.546
	.750	19.05	.590	.879
	.875	22.22	.812	1.249
	1.468	37.29	2.300	3.427
	2.219	56.36	5.200	7.818

DURALUMIN SQUARE RODS

Sizes Inches	mm Equivalent	Approx. Weight	
		Lbs. per Ft. Run	Kilos per Lineal Metre
3/8	9.52	.171	.25
7/16	11.11	.233	.35
.450	11.43	.246	.36
1/2	12.70	.305	.45
9/16	14.29	.386	.57
5/8	15.87	.475	.70
3/4	19.05	.685	1.02
7/8	22.22	.933	1.39
1	25.40	1.220	1.82
1 1/8	28.57	1.540	2.30
1 1/4	31.75	1.901	2.83
1 3/8	34.92	2.23	3.28
1 1/2	38.10	2.75	4.10
1.5748	40.00	2.98	4.44
1 5/8	41.27	3.27	4.85
1 3/4	44.45	3.73	5.56
1 7/8	47.62	4.26	6.35
*2	50.80	4.86	7.23
2 1/8	53.98	5.49	8.18
2 1/4	57.15	6.15	9.14
2 3/8	60.32	6.82	10.16
2 1/2	63.50	7.57	11.29
*2 3/4	69.85	9.00	13.35
2.7559	70.00	9.24	13.82
3	76.20	10.75	16.00
3.3464	85.00	13.56	20.20
*3 1/2	88.90	14.82	22.20
*4	101.60	19.00	28.20

*—Round Corners.

FLUXES FOR SOLDERING

Metals.	*Fluxes.*	*Fluxes generally used.*
Iron	Chloride of zinc	Chloride of zinc (killed spirit)
Steel	Sal-ammoniac	
Copper	Chloride of zinc	
Brass	Resin / Sal-ammoniac	Resin
Zinc (new) / Zinc (old)	Chloride of zinc	
Lead (with fine solder)	Hydrochloric acid	
Lead (with coarse solder)	Tallow and resin	
Tin	Tallow	
Pewter	Resin and sweet oil	

COMPOSITION OF SOFT SOLDERS

Solder.	*Composition.*	*Melting-point*
Fine	1½ parts tin, 1 part lead	334°F.
Tinman's	1 part tin, 1 part lead	370°F.
Plumber's	1 part tin, 2 parts lead	440°F.
Pewterer's	1 part tin, 1 part lead and 2 parts bismuth	203°F.
Wood's Metal	1 part tin, 2 parts lead 4 parts bismuth, 1 part cadmium	165°F.

A mixture of 1½ parts tin and 1 part lead fuses at a lower temperature than any other mixed proportion of these metals.

COMPOSITION OF HARD SOLDERS

Solder.	*Composition.*
Hard brazing	3 parts copper, 1 part zinc
Hard brazing	1 part copper, 1 part zinc
Softer brazing	4 parts copper, 3 parts zinc, and 1 part tin

TEMPERATURES OF TEMPERING COLOURS

Tint of Oxide on Surface of Steel	Centigrade	Fahrenheit	Suitable for
Dark Blue ..	316°	600°	Hand Saws
Blue	293°	560°	Fine Saw Blades, Augers
Bright Blue ..	288°	550°	Watch Springs, Swords
Purple ..	277°	530°	Table Knives, Large Shears, Wood Turning Tools
Brown, beginning to show Purple ..	266°	510°	Axes, Planes and Wood - working Tools
Brown ..	254°	490°	Scissors, Shears, Cold Chisels
Golden Yellow	243°	470°	Penknives, Hammers Taps, Reamers
Straw	230°	446°	Razor Blades
Pale Yellow ..	221°	430°	Small Edge Tools

HEAT COLOUR TEMPERATURES

Colour	Centigrade	Fahrenheit
Just Visible Red	500°—600°	932°—1112°
Dull Cherry Red	700°—750°	1300°—1385°
Cherry Red ..	750°—825°	1385°—1517°
Bright Cherry Red	825°—875°	1517°—1600°
Brightest Red ..	900°—950°	1652°—1750°
Orange.. ..	950°—1000°	1750°—1835°
Light Orange ..	1000°—1050°	1835°—1925°
Lemon	1100°—1200°	2012°—2200°
White	1200°—1300°	2200°—2372°

NOTE.—The above Colours and Temperatures are, of course, only roughly approximate.

WALTHAM WATCH SCREW TAPS

No. of tap	No. of threads to an inch	Diameter of thread on screws		Diameter of drill	
		Inches	Millimetres	Inches	Millimetres
1	110	0.05906	1.50	0.05197	1.32
3	110	0.04724	1.20	0.04016	1.02
5	120	0.04331	1.10	0.03741	0.95
7	140	0.03937	1.00	0.03347	0.85
9	160	0.03661	0.93	0.02796	0.71
11	170	0.05276	1.34	0.04803	1.22
13	180	0.03937	1.00	0.03347	0.85
15	180	0.03268	0.83	0.02796	0.71
17	200	0.02560	0.65	0.02126	0.54
19	220	0.02166	0.55	0.01772	0.45
21	240	0.01772	0.45	0.01339	0.34
23	254	0.01379	0.35	0.01064	0.27

PENDANT TAPS

Size	Diameter of tap		Threads per inch	Diameter of drill	
	Inches	Millimetres		Inches	Millimetres
18	.236	5.90	50	.211	5.28
16	.200	5.00	60	.180	4.50
12–6	.176	4.40	66	.158	3.95
0	.156	3.90	66	.138	3.45
5/0	.128	3.20	80	.114	2.85
10/0	.103	2.58	90	.086	2.15

WATCH CROWN TAPS

Size	Diameter of tap		Threads per inch	Diameter of Drill	
	Inches	Milli-metres		Inches	Milli-metres
18	.091	2.28	60	.071	1.78
16	.077	1.93	72	.063	1.58
12-6-0	.061	1.53	80	.048	1.20
5/0-10/0	.048	1.20	110	.038	0.95

WATCHMAKERS' MEASUREMENTS

Showing the comparative values of the standards of measurements commonly used by watchmakers.

One douzieme = 0.0074 inch.

One millimetre = 0.03937 inch.

One ligne = 0.0888 inch.

One ligne = 2.256 millimetres.

The Elgin Watch Company has two gauges, one known as the "upright," the other as the "fine."

One degree "upright" gauge = $\frac{1}{500}$ or 0.002 inch.

One degree "fine" gauge = $\frac{1}{2500}$ or 0.0004 inch.

ELGIN WATCH SCREW TAPS

Diameter of tap		Threads per Inch	Diameter of drill	
Inches	Millimetres		Inches	Millimetres
.0132	.33	360	.0112	.28
.0148	.37	320	.0120	.30
.0168	.42	260	.0132	.33
.0208	.52	220	.0168	.42
.0228	.57	260	.0188	.47
.0248	.62	220	.0200	.50
.0268	.67	180	.0220	.55
.0288	.72	220	.0248	.62
.0308	.77	180	.0248	.62
.0308	.77	220	.0268	.67
.0368	.92	140	.0280	.70
.0368	.92	220	.0268	.67
.0408	1.02	120L	.0300	.75
.0408	1.02	200	.0348	.87
.0428	1.07	120	.0328	.82
.0448	1.12	110	.0340	.85
.0468	1.17	110	.0348	.87
.0488	1.22	140	.0400	1.00
.0488	1.22	200	.0436	1.09
.0508	1.27	110L	.0388	.97
.0548	1.37	180	.0488	1.22
.0608	1.52	110	.0488	1.22
.0608	1.52	110L	.0488	1.22
.0708	1.77	180L	.0648	1.62
.0768	1.92	110L	.0708	1.77
.0772	1.93	80L	.0612	1.53
.0892	2.23	80L	.0712	1.78

TABLE OF JEWEL-SETTING DIAMETERS

Showing their sizes in thousandths of an inch and equivalent diameters in millimetres.

Inches	Milli-metres	Inches	Milli-metres	Inches	Milli-metres
.054	1.37	.088	2.23	.126	3.20
.060	1.52	.092	2.33	.140	3.55
.070	1.78	.096	2.43	.144	3.65
.076	1.93	.106	2.69	.156	3.96
.080	2.03	.116	2.94	.160	4.06
.084	2.13	.122	3.09	.170	4.31
.086	2.18				

THICKNESS OF PALLET STONES

In thousandths of an inch and equivalent thickness in millimetres.

	Inches	Millimetres
18 and 16 Old Style (thick)	0.0168	=0.4267
18 and 16 New Style (thin)	0.0148	=0.3759
6 Old Style (thick)	0.0146	=0.3708
6 New Style (thin)	0.0128	=0.3251

ENGLISH MUSIC WIRE GAUGE

Music Wire Gauge No.	*Decimal Diameter.*	*Sectional Area.*	*No. of Feet in 1 Lb.*	*Weight. Lb. per 100 Feet.*
9/0	0·005	0·0000196	14,705	0·0069
8/0	0·0055	0·0000237	12,048	0·0083
7/0	0·006	0·0000283	10,204	0·0098
6/0	0·0065	0·0000332	8,697	0·0115
5/0	0·007	0·0000385	7,462	0·0134
4/0	0·0075	0·0000442	6,493	0·0154
3/0	0·008	0·0000503	5,714	0·0175
2/0	0·0085	0·0000567	5,263	0·019
1/0	0·009	0·0000636	4,545	0·022
1	0·010	0·0000785	3,700	0·027
2	0·011	0·0000950	3,033	0·033
3	0·012	0·0001131	2,560	0·039
4	0·013	0·0001327	2,170	0·046
5	0·014	0·0001539	1,886	0·053
6	0·016	0·0002011	1,428	0·070
7	0·018	0·0002545	1,136	0·088
8	0·020	0·0003142	917	0·109
9	0·022	0·0003801	757	0·132
10	0·024	0·0004524	636	0·157
11	0·026	0·0005309	540	0·185
12	0·028	0·0006158	467	0·214
13	0·030	0·0007069	406	0·246
14	0·032	0·0008042	357	0·28
15	0·034	0·0009079	322	0·31
16	0·036	0·0010179	285	0·35
17	0·038	0·0011341	256	0·39
18	0·040	0·0012566	232	0·43

ENGLISH MUSIC WIRE GAUGE

Music Wire Gauge No.	*Decimal Diameter.*	*Sectional Area.*	*No. of Feet in 1 Lb.*	*Weight. Lb. per 100 Feet.*
19	0·042	0·0013854	208	0·48
20	0·044	0·0015205	188	0·53
21	0·046	0·0016619	172	0·58
22	0·048	0·0018095	158	0·63
23	0·051	0·0020428	140	0·71
24	0·055	0·0023758	121	0·82
25	0·059	0·0027340	105	0·95
26	0·063	0·0031173	92	1·08
27	0·067	0·0035257	81	1·23
28	0·071	0·0039592	72	1·38
29	0·074	0·0043009	66	1·5
30	0·078	0·0047784	60	1·66
31	0·082	0·0052810	54	1·84
32	0·086	0·0054106	49	2·02
33	0·090	0·0063617	45	2·22
34	0·094	0·0069398	41	2·42
35	0·098	0·0075430	38	2·63
36	0·102	0·0081710	35	2·85
37	0·106	0·0088250	32	3·07
38	0·112	0·0098520	29	3·43
39	0·118	0·0109359	26	3·81
40	0·125	0·0122718	23	4·28
41	0·132	0·0136848	21	4·77
42	0·139	0·0151747	18·5	5·29
43	0·146	0·0167415	17	5·84
44	0·153	0·0183854	16·6	6·41
45	0·160	0·0201062	14·2	7·01

APPROXIMATE PERCENTAGE COMPOSITION OF SOME ORDINARY METALS AND ALLOYS.

	Copper.	Zinc.	Tin.	Lead.	Manganese.	Iron.	Aluminium.	Antimony.	Phosphorus.
Copper Sheets, Strip, Rods, Ordinary quality	99.3	—	—	—	—	—	—	—	—
Copper Sheets, Strip, Rods, H.C. quality	99.9	—	—	—	—	—	—	—	—
Muntz Metal	60	40	—	—	—	—	—	—	—
Yellow Metal	60	40	—	—	—	—	—	—	—
Brass Rods, Common screwing quality	58	41	—	1	—	—	—	—	—
Brass Rods, Riveting quality	62	38	—	—	—	—	—	—	—
Brass Wire	62	38	—	—	—	—	—	—	—
Brass Sheets and Strip—									
Ordinary quality	62	38	—	—	—	—	—	—	—
Deep stamping quality ..	65	35	—	—	—	—	—	—	—
Spinning quality (a)	67	33	—	—	—	—	—	—	—
" " (b)	70	30	—	—	—	—	—	—	—
Brazing quality (a)	67	33	—	—	—	—	—	—	—
" " (b)	70	30	—	—	—	—	—	—	—
Dipping quality (a)	70	30	—	—	—	—	—	—	—
" " (b)	73	27	—	—	—	—	—	—	—
Cartridge quality (a)	70	30	—	—	—	—	—	—	—
" " (b)	72	28	—	—	—	—	—	—	—

APPROXIMATE PERCENTAGE COMPOSITION OF SOME ORDINARY METALS AND ALLOYS—*Continued.*

	Copper.	Zinc.	Tin.	Lead.	Manganese.	Iron.	Aluminium.	Antimony.	Phosphorus.
Princes Metal (*a*)	75	25	—	—	—	—	—	—	—
,, (*b*)	80	20	—	—	—	—	—	—	—
Ordinary Gilding Metal (6 and 1 quality)	86	14	—	—	—	—	—	—	—
Best Gilding Metal (9 and 1 quality)	90	10	—	—	—	—	—	—	—
Tombac	85	15	—	—	—	—	—	—	—
Pinchbeck (light)	87	13	—	—	—	—	—	—	—
,, (dark)	90	10	—	—	—	—	—	—	—
Ordinary Brass Brazing Solder	50	50	—	—	—	—	—	—	—
Brass Solder Strip	59	41	—	—	—	—	—	—	—
Naval Brass	62	37	1	—	—	—	—	—	—
Admiralty Brass	70	29	1	—	—	—	—	—	—
Tobin Bronze	60	38	2	—	—	—	—	—	—
Manganese Bronze Sheets and Rods	59	40	—	—	1	—	—	—	—
Manganese Bronze Castings (high tensile)	57	36	—	—	3	2	2	—	—
Manganese Copper Rods (*a*)..	94	—	—	—	6	—	—	—	—
,, ,, ,, (*b*)..	94	—	—	—	5	1	—	—	—

APPROXIMATE PERCENTAGE COMPOSITION OF SOME ORDINARY METALS AND ALLOYS—*Continued.*

	Copper.	Zinc.	Tin.	Lead.	Manganese.	Iron.	Aluminium.	Antimony.	Phosphorus.
Phosphor Bronze Ingots, Chill-bars and Castings (best quality)	88	—	$11\frac{1}{2}$	—	—	—	—	—	$\frac{1}{2}$
Phosphor Bronze Strip, Sheets	94	—	6	—	—	—	—	—	.2
Phosphor Bronze Tubes, Wire	95	—	5	—	—	—	—	—	.2
Coinage Bronze	95	1	4	—	—	—	—	—	—
Gun Metal Ingots, Chill-bars and Castings (Admiralty quality)	88	2	10	—	—	—	—	—	—
Aluminium Bronze Sheets and Strip	90	—	—	—	—	—	10	—	—
Aluminium Bronze Sheets and Strip	95	—	—	—	—	—	5	—	—
Aluminium Bronze Wire ..	95	—	—	—	—	—	5	—	—
Pewter	—	—	50	50	—	—	—	—	—
Britannia Metal	1	—	94	—	—	—	—	5	—
Meter Metal	—	—	95	—	—	—	—	5	—
Plumbers' Solder	—	—	33	67	—	—	—	—	—
Tinmen's Solder	—	—	50	50	—	—	—	—	—

PHYSICAL AND MECHANICAL DATA OF VARIOUS MATERIALS

	Atomic Weight, O=16	Specific Gravity	Weight, lb. per cu. in.	Melting-point, ° C.	Thermal Conductivity, cgs. units at 20° C.	Specific Heat, cal. per g. per ° C. at 20° C.	Co-efficient of Linear Expansion per ° C. × 10–6 20°–100° C.	Young's Modulus of Elasticity, lbs. per sq. in. × 106	Specific Resistance, microhms per cu. cm. at 20° C.	Electro-Chemical Equivalent g. per amp.-hr. at valency ( )
Commercially pure Aluminium	26.97	2.706	0.0978	658	0.503	0.225	23.8	10.0	2.828	0.3354 (3)
Brass	—	8.47	0.306	935	0.24	0.091	18.8	15.0	7.2	—
Bronze	—	8.78	0.317	1.000	0.12	0.086	18.2	15.0	16.0	—
Copper	63.57	8.93	0.322	1,083	0.92	0.092	16.5	18.0	1.72	1.186 (2)
Iron, cast	—	7.2	0.261	1,230	0.11	0.118	11.0	16.0	14.0	—
Iron, wrought	—	7.7	0.278	1,510	0.14	0.115	12.0	28.0	11.0	—
Lead	207.22	11.34	0.410	327	0.083	0.031	29.4	2.5	20.8	3.865 (2)
Magnesium	24.32	1.74	0.063	651	0.38	0.246	25.8	6.3	4.45	0.454 (2)
Nickel	58.69	8.9	0.322	1,455	0.14	0.105	12.8	30.0	7.2	1.095 (2)
Steel, cast	—	7.85	0.283	1,350	0.12	0.118	13.0	29.0	19.0	—
Steel, structural	—	7.85	0.283	1,350	0.12	0.118	12.6	29.0	19.0	—
Steel, stainless	—	7.89	0.285	1,400	0.055	0.118	17.0	28.0	74.0	—
Tin	118.7	7.3	0.264	232	0.15	0.054	21.6	5.9	11.5	2.214 (2)
Zinc	65.38	7.19	0.260	419	0.26	0.092	26.3	13.4	6.0	1.220 (2)

PROPERTIES OF ELEMENTS

Metal	Symbol	Colour	International Atomic Weight (1934)	Specific Gravity	Specific Heat		Electrical Conductivity Silver at 0° C. = 100	Heat Conductivity Silver = 100	Melting Point		Boiling Point	
					Temp.	Mean			°F.	°C.	°F.	°C.
Aluminium ..	Al	Tin-white ..	26.97	2.70	20—400	0.24	57	35	1218	659	3272	1800
Antimony ..	Sb	Silver-white	121.76	6.62	0—100	0.0495	4	4.02	1166	630	2975 ±15	1635 ±8
Arsenic	As	Steel-grey ..	74.91	5.73	21—268	0.0830	5		1562	850	Sub-limes.	Sub-limes.
Barium	Ba	Yellowish-white	137.36	3.66	—185 to 20	0.0680			1299	704		
Beryllium ..	Be or Gl	Steel-coloured	9.02	1.83	0—100	0.4246			2338	1281		
Bismuth ..		White ..	209.00	9.82	9—102	0.0298	1.3	1.8	514	268	2840	1560
Cadmium ..	Cd	White with blue tinge	112.41	8.64	0—100	0.0548	20	20	610	321	1411	765
Caesium ..	Cs	Silver-white	132.91	1.87	0	0.0522			83	28.25	1238	670
Calcium.. ..	Ca	Yellowish-white	40.08	1.55 (at 29°)	0—100	0.1490	18	25.4	1564	851	2264	1240
Cerium ..	Ce	Steel-grey ..	140.13	6.92	0—100	0.0450			1175	635		
Chromium ..	Cr	Greyish-white	52.01	7.14	22—51	0.1000			3326	1830	3992	2200
Cobalt	Co	Steel-grey ..	58.94	8.79	15—100	0.1030	15	17.2	2673	1467	4379	2415
Columbium (Niobium) ..	Cb Nb	Steel-grey ..	93.3	12.75	0	0.065			3542	1950		

PROPERTIES OF ELEMENTS (*Continued*)

Metal	Symbol	Colour	International Atomic Weight (1934)	Specific Gravity	Specific Heat		Electrical Conductivity Silver at 0° C. = 100	Heat Conductivity Silver = 100	Melting Point		Boiling Point	
					Temp.	Mean			°F.	°C.	°F.	°C.
Copper	Cu	Reddish-yellow	63.57	8.93	15—238	0.0951	94	92	1981	1082.6	4190	2310
Dysprosium ..	Dy		162.46									
Erbium ..	Er		167.64	4.77(?)								
Europium ..	Eu		152.0									
Gadolinium ..	Gd		157.3									
Gallium	Ga	Silver-white	69.72	5.95	0	0.079			86	30	3632	2000
Germanium ..	Ge	Greyish-white	72.60	5.47	0	0.0737			1652	900		
Gold	Au	Yellow ..	197.2	19.32	0	0.0316	67	70	1945.5	1063	3992	2200
Hafnium ..	Hf		178.6	13.08								
Indium	In	Silver-white	114.76	7.12	0	0.0570			311	155		
Iridium	Ir	Grey .	193.1	22.41	0—100	0.0323			4451	2454		
Iron	Fe	Greyish-white	55.84	7.87	20—100	0.1190	17	16	2780.6 ±5	1527 ±3	4442	2450
Lanthanum ..	La	White ..	138.9	6.12	0	0.0449			1490	810		
Lead	Pb	Blue-grey ..	207.22	11.37	18—100	0.0310	7.2	8.5	621	327	3132	1740
Lithium	Li	Silver-white	6.94	0.534	0—100	0.9600	16		367	186	2437	1336
Magnesium ..	Mg	Silver-white	24.32	1.74	18—99	0.2460	34	34.3	1200	649	2048	1120
Manganese ..	Mn	White-grey	54.93	7.39	0	0.1217			2268	1242	3452	1900
Mercury	Hg	White ..	200.61	13.56 (at 15° C)	20—50	0.0331	1.5	5.3	—38	—38.5	673	356.7

PROPERTIES OF ELEMENTS (*Continued*)

Metal	Symbol	Colour	International Atomic Weight (1934)	Specific Gravity	Specific Heat		Electrical Conductivity Silver at 0° C. = 100	Heat Conductivity Silver = 100	Melting Point		Boiling Point	
					Temp.	Mean			°F.	°C.	°F.	°C.
Molybdenum ..	Mo	Dull silver ..	96.0	10.0	15—91	0.0723			4532	2500		
Neodymium ..	Nd		144.27	6.96					1544	840		
Nickel	Ni	White ..	58.69	8.90	0—100	0.1147	20.5	14	2651	1455	4244	2340
Osmium ..	Os	Blue-white	191.5	22.48	18—98	0.0311			4532	2500		
Palladium ..	Pd	White ..	106.7	12.16	18—100	0.059			2831	1555	3992	2200
Platinum ..	Pt	White ..	195.23	21.4	0—100	0.0323	13.5	37.9	3224	1773.5	7772	4300
Potassium ..	K	Silver-white	39.096	0.862	0	0.1728	17	45	144	62.5	1404	762.2
Praseodymium..	Pr	White ..	140.92	6.48	0	0.0314			1724	940		
Radium.. ..	Ra	Brilliant-white	225.97						1292	700		
Rhenium ..	Re		186.31	21.4					6224 ±108	3440 ±60		
Rhodium ..	Rh	Bluish-white	102.91	12.41	0	0.0580			3571 ±5	1966 ±3		
Rubidium ..	Rb	White ..	85.44	1.532	0	0.0802			100	38	1285	696
Ruthenium ..	Ru	White ..	101.7	12.3	0	0.0611						
Samarium ..	Sm		150.43	7.8					2462	1350		
Scandium ..	Sc		45.10									
Selenium ..	Se	Steel-grey ..	78.96	4.5	15—217	0.084	Varies		423	217	1274	690
Silver	Ag	White ..	107.88	10.5	17—517	0.059	100	100	1760	960	3550	1955
Sodium	Na	Silver-white	22.997	0.971	0	0.2811	28	36.5	207	97.5	1622	882.9

PROPERTIES OF ELEMENTS (*Continued*)

Metal	Symbol	Colour	International Atomic Weight (1934)	Specific Gravity	Specific Heat		Electrical Conductivity Silver at 0° C. = 100	Heat Conductivity Silver = 100	Melting Point		Boiling Point	
					Temp.	Mean			°F.	°C.	°F.	°C.
Strontium ..	Sr	Yellowish-white	87.63	2.74	0	0.0735			1420	771		
Tantalum ..	Ta	Iron-grey ..	181.4	16.6	—185 to 20	0.0326			5270	2910		
Tellurium ..	Te	Shining-white	127.61	6.25	15—100	0.0483	0.077		842	450	2534	1390
Terbium ..	Tr		159.2									
Thallium ..	Tl	White ..	204.39	11.9	20—100	0.0326	8		574	301	2654	1457
Thorium ..	Th	Greyish-white	232.12	11.3	0	0.0276			3090	1700		
Thulium ..	Tm		169.4									
Tin ..	Sn	Silver-white	118.70	7.29	0—100	0.0559	11.3	15.2	450	232	4118	2270
Titanium ..	Ti	Dark-grey ..	47.90	4.8	0—100	0.1125			3632	2000		
Tungsten ..	W	Steel-grey ..	184.0	19.2	15—93	0.0340			6107	3375		
Uranium ..	U	Silver-white	238.14	18.7	0	0.028			3360	1800		
Vanadium ..	V	Light-grey .	50.95	5.5	0—100	0.1153			3128	1720		
Ytterbium ..	Yb		173.04									
Yttrium ..	Y	Dark-grey ..	88.92	3.8(?)								
Zinc ..	Zn	Bluish-white	65.38	7.1	20—100	0.0931	25.5	26.3	786.9	419.4	1662	905.7
Zirconium ..	Zr	Grey ..	91.22	6.53	0	0.0662			3501	1927		

(These tables are by courtesy of "Metal Industry.")

IMPERIAL STANDARD WIRE GAUGE

TABLE OF SIZES, WEIGHTS, AND LENGTHS OF STEEL WIRE

As adopted by the Iron and Steel Wire Manufacturers' Association—January 1904

Diameter in.	Size on Standard Wire Gauge	DIAMETER		Sectional Area in sq. in.	APPROXIMATE WEIGHT OF		
		Decimal of an in.	mm.		100ft.	Mile	Kilometre
	7/0	.500	12.7	.19635	66.7 lb.	3522 lb.	2188 lb.
15/32	6/0	.464	11.8	.16910	57.44	3033	1885
7/16	5/0	.432	11.0	.14657	49.79	2629	1634
13/32	4/0	.400	10.2	.12568	42.69	2254	1400
3/8	3/0	.372	9.4	.10869	36.93	1950	1211
11/32	2/0	.348	8.8	.09510	32.31	1706	1060
	1/0	.324	8.2	.08244	28.01	1479	919
	1	.300	7.6	.07069	24.01	1268	788
	2	.276	7.0	.05982	20.32	1073	667
	3	.252	6.4	.04987	16.85	895	556
	4	.232	5.9	.04227	14.36	758	471
	5	.212	5.4	.03530	12.00	633	393
3/16	6	.192	4.9	.02896	9.81	518	323
	7	.176	4.5	.02432	8.26	436	271
	8	.160	4.1	.02011	6.82	360	224
	9	.144	3.7	.01628	5.53	292	182
1/8	10	.128	3.3	.01287	4.37	231	143

IMPERIAL STANDARD WIRE GAUGE (Continued)

Diameter in.	Size on Standard Wire Gauge	DIAMETER		Sectional Area in sq. in.	APPROXIMATE WEIGHT OF		
		Decimal of an in.	mm.		100ft.	Mile	Kilometre
	11	.116	3.0	.01057	3.60 lb.	190 lb.	118 lb.
	12	.104	2.6	.00850	2.88	152	95
3/32	13	.092	2.3	.00665	2.25	119	74
	14	.080	2.0	.00503	1.70	90	56
	15	.072	1.8	.00407	1.38	73	45
1/16	16	.064	1.6	.00322	1.10	58	36
	17	.056	1.4	.00246	.83	44	27.5
3/64	18	.048	1.2	.00181	.61	32.5	20.2
	19	.040	1.0	.00126	.42	22.54	14.0
	20	.036	0.9	.00102	.34	18.25	11.34
1/32	21	.032	0.8	.00080	.273	14.42	8.96
	22	.028	0.7	.00062	.209	11.04	6.86
	23	.024	0.6	.00045	.154	8.11	5.04
3/128	24	.022	0.55	.00038	.129	6.82	4.24
	25	.020	0.5	.00031	.107	5.63	3.5
	26	.018	0.45	.00025	.086	4.56	2.84
	27	.0164	0.4	.00021	.072	3.79	2.35
1/64	28	.0148	0.37	.00017	.058	3.09	1.92
	29	.0136	0.35	.00014	.050	2.61	1.62
	30	.0124	0.32	.00012	.041	2.17	1.35

IMPERIAL STANDARD WIRE GAUGE (Continued)

Diameter in.	Size on Standard Wire Gauge	DIAMETER		Sectional Area in sq. in.	APPROXIMATE WEIGHT OF		
		Decimal of an in.	mm.		100ft.	Mile	Kilometre
	31	.0116	0.28	.00010	.036 lb.	1.89 lb.	1.16 lb.
	32	.0108	0.27	.000091	.031	1.64	1.02
	33	.0100	0.254	.000078	.026	1.40	.875
	34	.0092	0.230	.000066	.022	1.19	.744
	35	.0084	0.203	.000055	.019	.901	.563
1/128	36	.0076	0.177	.000045	.015	.813	.508
	37	.0068	0.172	.000036	.012	.651	.407
	38	.0060	0.152	.000028	.0096	.507	.317
	39	.0052	0.127	.000021	.0072	.380	.238
	40	.0048	0.122	.000018	.0061	.324	.202
	41	.0044	0.112	.000015	.0051	.272	.170
	42	.004	0.101	.000012	.0042	.225	.140
	43	.0036	0.091	.000010	.0034	.182	.114
	44	.0032	0.081	.000008	.0027	.144	.090
	45	.0028	0.071	.000006	.0021	.110	.070
	46	.0024	0.061	.000004	.0015	.081	.050
	47	.002	0.050	.000003	.00106	.056	.035
	48	.0016	0.040	.000002	.00082	.036	.0225
	49	.0012	0.030	.000001	.000266	.020	.0125
	50	.001	0.025	.0000007	.000259	.014	.0097

STEEL SHEET AND WIRE GAUGES

Gauge No.	B.G.			S.W.G.			
	Thickness.		Weight per Sq. Foot.	Thickness.		Weight.	
						Wire per 100 yds.	Sheets per sq. ft.
	Ins.	Mm.	Lb.	Ins.	Mm.	Lb.	Lb.
1	·353	8·97	14·41	·300	7·62	72·0	12·24
2	·315	7·99	12·84	·276	7·01	61·0	11·26
3	·280	7·12	11·44	·252	6·40	50·8	10·28
4	·250	6·35	10·20	·232	5·89	43·1	9·47
5	·222	5·65	9·08	·212	5·38	36·0	8·65
6	·198	5·03	8·08	·192	4·88	29·4	7·83
7	·176	4·48	7·20	·176	4·47	24·8	7·18
8	·157	3·99	6·41	·160	4·06	20·4	6·53
9	·140	3·55	5·70	·144	3·66	16·6	5·87
10	·125	3·17	5·10	·128	3·25	13·1	5·22
11	·111	2·83	4·54	·116	2·95	10·8	4·73
12	·099	2·52	4·04	·104	2·64	8·63	4·24
13	·088	2·24	3·60	·092	2·34	6·76	3·75
14	·078	1·99	3·20	·080	2·03	5·11	3·26
15	·070	1·77	2·85	·072	1·83	4·15	2·94
16	·062	1·59	2·55	·064	1·63	3·29	2·61
17	·056	1·41	2·27	·056	1·42	2·50	2·28
18	·049	1·26	2·02	·048	1·22	1·83	1·96
19	·044	1·12	1·79	·040	1·02	1·27	1·63
20	·039	·996	1·60	·033	·914	1·03	1·47
21	·035	·886	1·42	·032	·813	·819	1·31
22	·031	·794	1·27	·028	·711	·628	1·14
23	·028	·707	1·13	·024	·610	·461	·979
24	·025	·629	1·01	·022	·559	·387	·898
25	·022	·560	·899	·020	·508	·320	·816

STEEL SHEET AND WIRE GAUGES
(Continued)

Gauge No.	B.G.			S.W.G.			
	Thickness.		Weight per Sq. Foot.	Thickness.		Weight.	
						Wire per 100 yds.	Sheet per sq. ft.
	Ins.	Mm.	Lb.	Ins.	Mm.	Lb.	Lb.
26	·020	·498	·800	·018	·457	·259	·734
27	·017	·443	·712	·016	·417	·215	·669
28	·016	·397	·637	·015	·376	·175	·604
29	·0139	·353	·567	·0136	·345	·148	·555
30	·0123	·312	·502	·0124	·315	·123	·506
31	·0110	·279	·449	·0116	·295	..	·473
32	·0098	·249	·400	·0108	·274	..	·441
33	·0087	·221	·335	·0100	·254	..	·408
34	·0077	·196	·314	·0092	·234	..	·375
35	·0069	·175	·282	·0084	·213	..	·343
36	·0061	·155	·249	·0076	·193	..	·310
37	·0054	·137	·220	·0068	·173	..	·277
38	·0048	·122	·196	·0060	·152	..	·245
39	·0043	·109	·175	·0052	·132	..	·212
40	·0039	·098	·157	·0048	·122	..	·196
41	·0034	·087	·140	·0044	·112	..	·180
42	·0031	·078	·125	·0040	·102	..	·163
43	·0027	·069	·111	·0036	·091	..	·147
44	·0024	·061	·099	·0032	·081	..	·131
45	·0021	·055	·088	·0028	·071	..	·144
46	·0019	·049	·078	·0024	·061	..	·098
47	·0017	·043	·069	·0020	·051	..	·082
48	·0015	·039	·062	·0016	·041	..	·065
49	·0013	·034	·055	·0012	·030	..	·049
50	·0012	·030	·049	·0010	·025	..	·041

Instrument-wire Gauges

No. (S.W.G.)	Dia. In.	No. (S.W.G.)	Dia. In.	No. (S.W.G.)	Dia. In.
4/0	.400	15	.072	33	.0100
3/0	.372	16	.064	34	.0092
2/0	.348	17	.056	35	.0084
0	.324	18	.048	36	.0076
1	.300	19	.040	37	.0068
2	.276	20	.036	38	.0060
3	.252	21	.032	39	.0052
4	.232	22	.028	40	.0048
5	.212	23	.024	41	.0044
6	.192	24	.022	42	.0040
7	.176	25	.020	43	.0036
8	.160	26	.018	44	.0032
9	.144	27	.0164	45	.0028
10	.128	28	.0148	46	.0024
11	.116	29	.0136	47	.0020
12	.104	30	.0124	48	.0016
13	.092	31	.0116	49	.0012
14	.080	32	.0108	50	.0010

SHEET ZINC TRADE GAUGE

No.	*Inch.*	*No.*	*Inch.*
1	0·00395	**14**	0·0323
2	0·00554	**15**	0·0375
3	0·0067	**16**	0·0426
4	0·0082	**17**	0·0478
5	0·0097	**18**	0·0526
6	0·0114	**19**	0·0577
7	0·0132	**20**	0·0632
8	0·0149	**21**	0·0699
9	0·0177	**22**	0·0768
10	0·0196	**23**	0·0843
11	0·0228	**24**	0·0915
12	0·0260	**25**	0·0980
13	0·0292	**26**	0·1052

BELT DRIVES

Finding the Length of a Belt.—When the radius of the pulleys and the distance between two shafts are known the following formula applies in which:

R = radius of the large pulley.
r = radius of the small pulley.
c = the centre distance.
L = the total length.

Open Belt—

$$L = \pi(R + r) + 2\sqrt{c^2 + (R - r)^2}.$$

For Equal Pulleys—

$$L = \pi(R + r) + 2c.$$

For Crossed Belt—

$$L = \pi(R + r) + 2\sqrt{c^2 + (R + r)^2}.$$

Belt Speed.—The usual velocity of belts varies from 1,000ft. to 1,600ft. per minute for the main driving belts in a workshop. The belts driving the machine tools vary in speed from 1,000ft. to 4,000ft. per minute.

Belt Tension.—Belt tension (working side) should not exceed 420lb. per sq. in. when they have cemented and sewn joints, and 280lb. per sq. in. when they are laced. Working tension is usually expressed per inch of width. The following table is a useful guide:

Single Belts	50lb.
Light Double Belts	70lb.
Heavy Double Belts	90lb.
$\frac{3}{8}$in. Link Belts	45lb.
$\frac{1}{2}$in. Link Belts	50lb.
$\frac{5}{8}$in. Link Belts	60lb.
$\frac{3}{4}$in. Link Belts	68lb.
$\frac{7}{8}$in. Link Belts	80lb.
1in. Link Belts	95lb.

BELT DRIVES (Continued)

Various Forms of Belt Drive

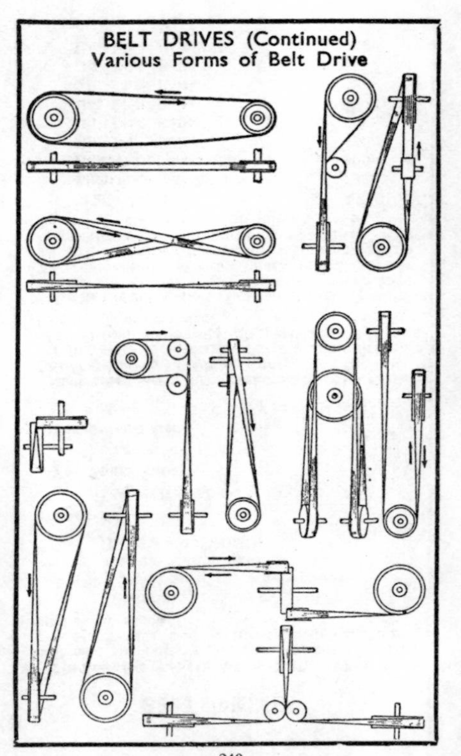

PULLEY CALCULATIONS.

In the workshop, calculations regarding speeds of machine countershafts, etc., are frequent, and the following formulæ apply:

R.P.M. of Driven Pulley or Gear

$$= \frac{\text{Diam. of Driver} \times \text{R.P.M. of Driver.}}{\text{Diam. of Driven.}}$$

Diam. of Driven Pulley or Gear

$$= \frac{\text{Diam. of Driver} \times \text{R.P.M. of Driver.}}{\text{R.P.M. of Driven.}}$$

R.P.M. of Driver Pulley or Gear

$$= \frac{\text{Diam. of Driven} \times \text{R.P.M. of Driven.}}{\text{Diam. of Driver.}}$$

Diam. of Driver

$$= \frac{\text{Diam. of Driven} \times \text{R.P.M. of Driven.}}{\text{R.P.M. of Driver}}$$

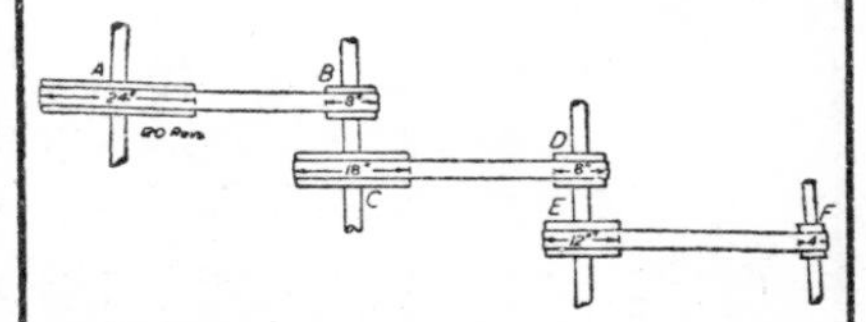

The illustration shows an example of triple belt drive, in which the main shaft revolves at 120 r.p.m. In this example

$$F = \frac{120 \times 24 \times 18 \times 12}{8 \times 8 \times 4} = 2{,}430 \text{ r.p.m.}$$

PULLEY PROPORTIONS

A = width of face = B + $\frac{1}{4}$ in. to $\frac{1}{2}$ in.
B = width of belt.
C = thickness of belt.
D = diameter of pulley.
E = thickness of rim $= 0.005 \times D + \frac{1}{4}$ in.
F = crown of face = $\frac{1}{4}$ in. per 12 in. face.
G = $\frac{1}{2}$ of J.
H = $\frac{1}{2}$ of I.
I = width of arm = $(0.04 \times D) + \frac{7}{8}$.
J = thickness of arm = $\frac{1}{2}$ of I.
K = taper of rim = $\frac{1}{2}$ of E.
L = metal around bore = 7/16 of bore
O = $\frac{1}{2}$ of I.

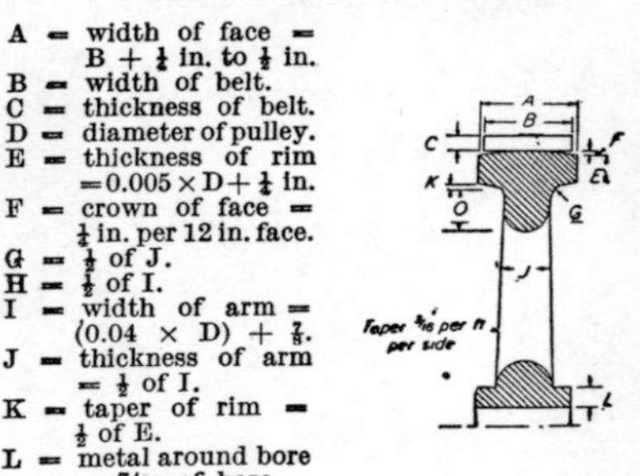

Diagrams illustrating pulley proportions.

Pulley Arms

(see diagrams.)

I = width of arm.
J = thickness of arm = $\frac{1}{2}$ of I
M = radius = $\frac{3}{4}$ of I.
N = radius = $\frac{1}{8}$ of I.

Number of Arms in Pulleys

6 to 24 in., 4 arms.
24 to 36 in., 5 arms.
36 to 96 in., 6 arms.
8 to 16 ft., 8 arms.
16 to 24 ft., 10 arms.

SCREW-CUTTING

The principle of screw-cutting, that is, as understood by the term when applied to centre-lathes, consists of gearing the lead-screw to the headstock mandrel in such a manner that by revolving the mandrel one turn the lead-screw will rotate sufficiently to carry the saddle forward a distance exactly equal to the pitch of the thread to be cut.

For screw-cutting purposes the saddle is generally connected to the lead-screw by means of a lever-operated split-nut. To cut a thread of, say, 24 threads to the inch, the saddle, and consequently the tool, must travel a distance of 1/24 of an inch for each revolution of the lathe. Thus, the factors governing the gearing ratio are—number of threads per inch required, and the number of threads per inch of the leading screw.

Lead-screws are commonly cut either 8, 6, 4 or 2 threads per inch, but on the class of lathe that the reader is likely to use, will not be finer than ¼in. pitch. To cut a thread of 24 threads per inch on a lathe with a screw of this pitch it will follow that while the mandrel is making one turn the lead-screw, in order to advance 1/24 inch, must revolve only ⅙ of a turn—in other words, the mandrel must revolve six times as fast as the lead-screw. Obviously, to do this the gear train connecting the mandrel to the screw must give a speed reduction of six to one.

Therefore, the first step in ascertaining which change-wheels to use is to find the ratio of the gearing to employ by dividing the number of threads per inch to be cut by the number of threads per inch of the lead-screw.

Change-Wheels.—As supplied with the lathe, a set of change-wheels may consist of 22 wheels, all identically bored and keywayed, ranging from 20 to 120 teeth, each wheel having 5 teeth more

Screw-cutting

(continued)

than the next smaller, one of the smaller gears, usually a 40, being in duplicate. With lathes having a lead-screw of 6 or 8 threads per inch, the wheels may be run from 24 to 100 teeth in increasing stages of 4 teeth.

Having found the ratio, all that needs to be done is to select a pair of wheels having numbers of teeth in the same ratio. As an example, what wheels are required to cut 20 threads per inch on a lathe having a lead-screw of 4 threads per inch ? 20 divided by 4 equals 5. The wheel on the lead-screw, therefore, requires to have five times as many teeth as that on the spindle, or 100 and 20 teeth respectively. As these two gears run on centres that are fixed, some means is necessary to transmit the drive from one gear to the other. Provision is made for this in the slotted quadrant plate which is pivoted off the centre of the lead-screw. Into the slot is fitted an adjustable stud, working on which is a bush having a key to suit the change-wheels. Any convenient-sized wheel is selected as an intermediate gear, and the stud is raised in the slot sufficiently to allow the wheel to pass on to the bush clear of the lead-screw wheel. A smaller gear is put on in front as packing and locked by the nut provided, the intermediate wheel is then dropped into mesh with the lead-screw wheel and the stud locked in the slot, after which the quadrant plate is swung over, until the gear engages the one on the spindle, and locked. In Fig. 1 a simple gear train is illustrated. The top left-hand wheel represents the spindle and the lowest one the screw, the smaller wheels on the screw and in front of the intermediate or idler gear being used as packing.

Compound Trains.—Obviously, with the standard wheels the limits of a simple train of gears are represented by a ratio of 6 to 1, 20 being the smallest gear available and 120 the largest;

Screw-cutting

(continued)

these, by the way, will cover all standard Whitworth pitches from 5/16th in. diameter and B.S.F. from 5/16in. diameter up to the largest size likely to be handled, without needing a compound train.

For pitches finer than 24 threads per inch, use will have to be made of compounding the gear train. As, for instance, a screw having a pitch of 30 threads per inch is required, $30 \div 4 = 7\frac{1}{2}$ to 1

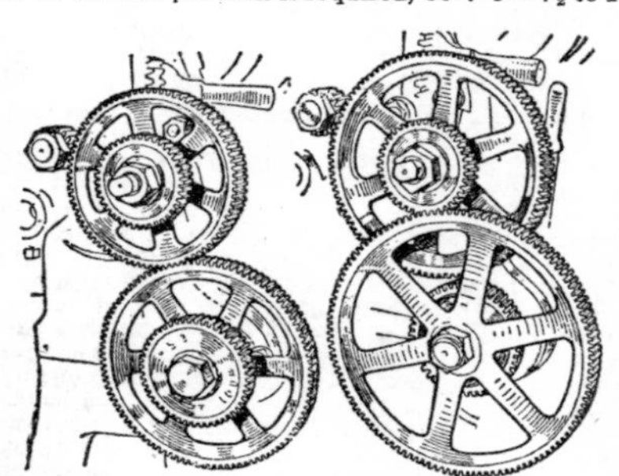

Figs. 1 & 2.—(Left) Showing a simple gear train. (Right) A compound train. The gear wheel behind the lead-screw wheel is used as packing.

ratio. To cut this with a simple train would require wheels of 20 and 150 teeth; as the larger size is not available, recourse is made to a compound train of gears. This consists of splitting the gearing up into two units, as it were; one gear on the spindle drives a gear on the stud; a second

Screw-cutting

(continued)

gear having a different number of teeth is also mounted on the stud in front of the first and by virtue of the key in the bush is driven at the same speed. This front or second gear meshes with the wheel on the screw.

Thus, by using a 20 wheel on the spindle to drive a 100 wheel on the stud, giving a reduction of 5 to 1, and the second wheel on the stud having 30 teeth driving a wheel with 120 teeth on the screw, a total reduction of 20 to 1 would be obtained. In the case under review the gears could be split into two trains to give a first reduction of $3\frac{3}{4}$ to 1 and a second of 2 to 1; or, a first of $2\frac{1}{2}$ to 1 and a second of 3 to 1. To cut 30 threads, then, the following gears could be used: spindle 20, driving stud 75. Stud 50, driving screw 100, or in the same order 30, 75, 40, 120. This may be expressed as follows:

$$\frac{\text{Lead-screw, threads per in.}}{\text{Threads required per in.}} = \frac{4}{30} = \frac{4 \times 1}{15 \times 2} = \frac{4 \times 5}{15 \times 5} \times$$

$$\frac{1 \times 50}{2 \times 50} = \frac{20}{75} \times \frac{50}{100} \text{ or } \frac{4}{30} = \frac{2 \times 2}{6 \times 5} = \frac{2 \times 20}{6 \times 20} \times \frac{2 \times 15}{5 \times 15} =$$

$$\frac{40 \times 30}{120 \times 75}$$

It will be seen that the numerator and denominator in each factor are multiplied by the same number to give suitable wheels, and, further, either of the wheels indicated by the numerator may be used as a driver to driven wheel with the same result. Thus, in the last example, 40 could drive the 75, giving a reduction of 1-35/40 to 1 and the 30 driving the 120 giving a second reduction of 4 to 1, so that $1\frac{7}{8}$ multiplied by 4 is equal to $7\frac{1}{2}$ to 1 required. The examples instanced do not exhaust the possible combination that could

Screw-cutting
(continued)

be used, and in selecting the gears the only point to watch is that those selected will be large enough to permit meshing when the quadrant is swung into position.

A compound train is shown in Fig. 2; the gear behind the lead-screw wheel is used as packing. It is hoped that the explanation has made it clear that the working out of change-wheels is only a question of simple mental arithmetic. For the benefit of those who have or may get a lathe having a change-wheel plate, and are unfamiliar with the terms thereon, spindle means the headstock spindle or shaft connected thereto by tumbler gearing, stud is the stud on the quadrant plate, and screw is the lead-screw. Alternative markings meaning the same things in the order named are spindle-driven, driver-screw or driver-driven. In both of these cases where only the first and last columns are marked it means that only a simple train is needed, and as before stated any wheel can be utilised as an intermediate.

Cutting a Thread.—Having turned the work ready for threading and mounted the wheels, set the screw-cutting tool for centre height and the flanks of the tool square with the work by means of a centre or screw-cutting gauge of the standard type, bring the saddle back so that the tool is well clear of the front of the work. The nut is engaged with the lead-screw and the tailstock locked hard up against the saddle to form a stop. A cut is put on, noting the position or reading on the cross slide index. When the tool has travelled a distance along the work equal to the length of thread required, disengage the nut and at the same time recede the tool smartly. Return the saddle up to the stop and put further cuts on as before until the thread fits the female part. It will be noticed that the tool is actually cutting and not merely scraping. During the cutting particularly with deep threads, the tool is advanced

Screw-cutting

(continued)

slightly several times during the operations, but allowing the tool to cut all over on the last one or two cuts to obtain a threa l to correct form.

Where the thread washes out, as on a stud, care must be taken not to allow the tool to travel beyond the point of withdrawal on the previous cut, or the nose of the tool will break. Aim at withdrawing each succeeding cut slightly in advance of the previous one to obtain a gradual wash-out.

This procedure is adopted for all threads per inch that are multiples of the thread on the lead-screw. When threads such as 9, 10 or 11 are to be cut the nut is not disengaged, but the cut is withdrawn and the lathe reversed by pulling the belt backwards, or, when the saddle is first brought back against the stop and the nut engaged, a mark is made on the face of the headstock cone or on the gear wheel and a corresponding mark made on the front bearing housing or gear guard. Similar lines are made on the lead-screw and lead-screw bracket. The nut is disengaged at the end of the cut as before and the saddle returned to the stop, which in this case is essential, and the lathe run until both sets of lines coincide at the same time when the nut is dropped in. It should be mentione l that to cut left-hand threads, where the lathe is not fitted with a tumbler gear, two intermediate wheels will be required in a simple train and one intermediate wheel in a compound train. This is necessary to reverse the direction of the lead-screw.

As distinct from the screw-cutting gauge referred to is the screw-pitch with serrated blades covering the various screw pitches. The blades represent sections of threads of different pitches and are correctly formed according to the standard represented, and on this account, apart from finding pitches, are useful for checking thread-form when cutting.

Screw-cutting
(Continued)

Cutting Screws of English Pitch with Lathe having Metric Lead Screw.—If a wheel of 127 teeth is too large to use, apply the ratio $\frac{2160}{85}$. The ratio is $\frac{127}{5} = \frac{2152}{85}$, so the error per inch is less than 0.0005. The change wheels supplied with the lathe usually advance in fives from 20 to 100, and the pitch of the lead screw is 6, usually mm. These two items enhance the value of the quantity 2160, as both 6 and 10 are factors of it. The 85 wheel is always placed on the lead screw, so the ratio which remains is $\frac{2160}{6 \times N}$, N being the number of threads per inch. As an example, to cut 12 threads per inch $\frac{2160}{6 \times 12} = 30$ and wheels $\frac{45 \times 40}{60}$ to give this value.

Measurements and Identification of Screw Threads.—The length of a bolt or screw is the measurement taken from under the head to the end of the thread, excepting in the case of countersunk screws, in which the length is the overall measurement.

The threads in common use are British Standard Whitworth, British Standard Fine, British Standard Pipe, and British Association, these being, of course, the standards of this country. Those of America are United States Standard or Sellers, Society of Automobile Engineers, and American Society of Mechanical Engineers. The International System Metric Thread is the standard of most Continental countries. In the order named, the threads mentioned, in an abbreviated form, are designated as follows: B.S.W. or Whit., B.S.F., B.S.P., B.A., U.S.S., S.A.E., A.S.M.E., and S.I.

Change Wheels for Screw Cutting

Threads per Inch to be Cut.	Lead Screw, ¼-in. pitch.		Lead Screw, ½-in. pitch.	
	Drivers.	Driven.	Drivers.	Driven.
50	20 30	75 100	20 20	100 100
	20 40	80 125	20 30	120 125
48	20 25	60 100	20 25	100 120
	25 30	75 120	20 20	80 120
45	20 30	75 90	20 20	75 120
	20 40	90 100	20 25	90 125
40	20 55	100 110	20 30	100 120
	20 40	80 100	20 25	100 100
35	30 40	100 105	20 30	100 105
	20 40	70 100	25 30	105 125
80	20 60	90 100	20 40	100 120
	20 50	75 100	20 35	100 105
28	20 30	40 105	20 25	70 100
	20 30	60 70	20 45	105 120
26	20 30	60 65	20 25	65 100
	25 40	65 100	20 30	65 120
25	30 40	75 100	20 30	75 100
	20 60	75 100	20 60	120 125
24	20	120	25 30	75 120
	20 40	60 80	20 25	60 100
23	20	115	20 50	100 115
	30 40	60 115	20 30	60 115
22	20	110	20 30	60 110
	30 50	76 115	20 40	80 110
21	20 40	60 70	20 40	70 120
	30 40	70 90	20 30	70 90
20	20	100	20 40	80 100
	20 40	50 80	20 35	70 100
19	20	95	25 40	95 100
	30 40	60 95	20 60	95 120
18	20	90	25 40	75 120
	30 40	60 90	35 40	105 120
17	20	85	20 60	85 120
	30 40	60 85	20 45	85 90
16	20	80	25 30	50 120
	35 40	70 80	30 45	90 120
15	20	75	20 80	100 120
	20 40	30 100	20 70	100 105

Change Wheels for Screw Cutting
(Continued)

Threads per Inch to be Cut.	*Lead Screw, ¼-in. pitch.*		*Lead Screw, ½-in. pitch.*	
	Drivers.	*Driven.*	*Drivers.*	*Driven.*
14	20	70	20 75	100 105
	30 40	60 70	20 50	70 100
13	20	65	20 50	65 100
	40 45	65 90	20 60	65 120
12	20	60	20	120
	30 50	60 75	25 60	90 100
11	40	110	20	110
	30 40	55 60	30 60	90 110
10	40	100	20	100
	30 40	50 60	35 60	100 105
9	40	90	20	90
	30 40	45 60	30 70	90 105
8	40	80	20	80
	20 75	50 60	35 60	70 120
7½	40	75	20	75
	20 80	50 60	30 80	75 120
7	40	70	20	70
	30 80	60 70	30 80	70 120
6½	40	65	20	65
	30 60	45 65	30 80	65 120
6	30	45	30	90
	20 60	40 45	35 80	70 120
5½	40	55	20	55
	40 60	30 110	40	110
5	40	50	30	75
	60	75	40	100
4½	40	45	40	90
	40 100	75 60	20	45
4	40	40	30	60
	30 105	90 35	40	80
3½	40	35	40	70
	40 60	30 70	30 90	45 105
3¼	80	65	40	65
	70 40	35 65	50 80	65 100
3	80	60	40	60
	40	30	30	45
2⅞	40 100	115 25	20 100	115 25
	40 120	115 30	40 100	115 50
2¾	80	55	40	55
	60 100	55 75	80	110

Change Wheels for Screw Cutting
(Continued)

Threads per Inch to be Cut	Lead Screw, ¼-in. pitch. Drivers.		Driven.		Lead Screw, ½-in. pitch. Drivers.		Driven.	
$2\frac{5}{8}$	40	100	105	25	80		105	
	40	120	105	30	40	100	105	50
$2\frac{1}{2}$	80		50		40		50	
	40	90	75	30	40	120	100	60
$2\frac{3}{8}$	40	100	95	25	80		95	
	40	120	95	30	40	100	95	50
$2\frac{1}{4}$	80		45		40		45	
	40	100	75	30	40	100	90	50
2	80		40		60		60	
	40	75	50	30	30	75	90	25
$1\frac{7}{8}$	40	80	50	30	80		75	
	40	80	75	20	40	80	100	30
$1\frac{3}{4}$	80		35		80		70	
	80	100	70	50	60	90	105	45
$1\frac{5}{8}$	60	80	65	30	60	100	75	65
	50	80	65	25	40	90	65	45
$1\frac{1}{2}$	80		30		80		60	
	60	100	75	30	60	110	90	55
$1\frac{3}{8}$	80	120	110	30	80		55	
	80	70	55	35	80	70	110	35
$1\frac{1}{4}$	80		25		80		50	
	80	120	75	40	40	120	100	30
$1\frac{1}{8}$	60	80	45	30	80		45	
	80	100	50	45	80	100	90	50
1	100		25		60		30	
	80	100	50	40	80		40	
One Thread in $1\frac{1}{4}$	100	120	60	40	100		40	
	75	50	30	25	80	75	60	40
$1\frac{1}{2}$	80	90	40	30	90		30	
	70	75	35	25	60	70	40	35
$1\frac{3}{4}$	70	75	30	25	75	105	90	25
	80	105	40	30	70	105	60	35
2	80	100	40	25	80	60	40	30
	75	80	30	25	70	110	55	35
$2\frac{1}{4}$	75	90	30	25	90	105	60	35
	90	100	40	25	70	90	40	35
$2\frac{1}{2}$	100	75	30	25	75	100	50	30
	100	120	40	30	75	110	55	30
$2\frac{3}{4}$	100	110	40	25	100	110	50	40
	110	75	30	25	90	110	45	40

Change Wheels for mm. Pitches

Pitch of Screw to be Cut.	$\frac{1}{4}$*-in. Pitch Lead Screw.*		$\frac{1}{2}$*-in. Pitch Lead Screw.*	
Millimetres.	*Drivers.*	*Driven.*	*Drivers.*	*Driven.*
1 (.039 in.)	63 × 20	80 × 100	21 × 30	80 × 100
	35 × 45	100 × 100	21 × 45	100 × 120
2 (.079 in.)	63 × 30	60 × 100	63 × 20	80 × 100
	63 × 40	80 × 100	63 × 30	100 × 120
3 (.118 in.)	63 × 30	40 × 100	63 × 30	80 × 100
	63 × 45	60 × 100	63 × 45	100 × 120
4 (.157 in.)	63 × 30	50 × 60	63 × 30	60 × 100
	63 × 20	40 × 50	63 × 20	50 × 80
5 (.197 in.)	63 × 30	40 × 60	63 × 30	60 × 80
	45 × 70	50 × 80	45 × 70	80 × 100
6 (.236 in.)	63 × 45	50 × 60	63 × 30	50 × 80
	63 × 60	50 × 80	63 × 45	60 × 100
7 (.275 in.)	63 × 35	40 × 50	63 × 35	50 × 80
	63 × 70	50 × 80	63 × 70	80 × 100
8 (.315 in.)	63	50	63	100
	63 × 45	50 × 70	63 × 45	60 × 75
9 (.354 in.)	63 × 90	50 × 80	63 × 45	50 × 80
	63 × 45	40 × 50	63 × 90	80 × 100
10 (.393 in.)	63	40	63	80
	70 × 90	50 × 80	70 × 90	80 × 100
11 (.433 in.)	63 × 55	25 × 80	63 × 55	50 × 80
	63 × 110	50 × 80	63 × 110	80 × 100
12 (.474 in.)	63 × 30	20 × 50	63 × 30	40 × 50
	63 × 60	40 × 50	63 × 60	50 × 80
13 (.512 in.)	63 × 65	40 × 50	63 × 65	40 × 100
	63 × 65	25 × 80	63 × 65	50 × 80
14 (.551 in.)	63 × 70	40 × 50	63 × 70	50 × 80
	63 × 105	40 × 75	63 × 70	40 × 100
15 (.591 in.)	63 × 75	40 × 50	63 × 75	50 × 80
	63 × 90	40 × 60	63 × 90	60 × 80
16 (.630 in.)	63 × 80	40 × 50	63 × 80	50 × 80
	63 × 60	30 × 50	63 × 60	50 × 60
17 (.669 in.)	63 × 85	20 × 100	63 × 85	40 × 100
	63 × 85	40 × 50	63 × 85	50 × 80
18 (.708 in.)	63 × 90	40 × 50	63 × 90	50 × 80
	63 × 45	20 × 50	63 × 45	40 × 50
19 (.748 in.)	63 × 95	40 × 50	63 × 95	50 × 80
	63 × 95	20 × 100	63 × 95	40 × 100

Change Wheels for mm. Pitches
(Continued)

Pitch of Screw to be Cut.	¼-in. Pitch Lead Screw.		½-in. Pitch Lead Screw.	
Millimetres.	Drivers.	Driven.	Drivers.	Driven.
20 (.787in.)	63 × 75	25 × 60	63 × 75	50 × 60
	63 × 60	30 × 40	63 × 60	40 × 60
21 (.826in.)	63 × 63	20 × 60	63 × 63	40 × 60
	63 × 63	30 × 40	63 × 63	30 × 80
22 (.866in.)	63 × 55	20 × 50	63 × 55	40 × 50
	63 × 100	40 × 50	63 × 100	50 × 80
23 (.905in.)	63 × 46	20 × 40	63 × 46	40 × 40
	63 × 115	40 × 50	63 × 115	50 × 80
24 (.945in.)	63 × 90	30 × 50	63 × 90	50 × 60
	63 × 60	20 × 50	63 × 60	40 × 50
25 (.984in.)	63 × 50	20 × 40	63 × 50	20 × 80
	70 × 90	40 × 40	70 × 90	40 × 80
26 (1.023in.)	63 × 65	20 × 50	63 × 65	25 × 80
	63 × 65	25 × 40	63 × 65	40 × 50
27 (1.063in.)	63 × 54	20 × 40	63 × 54	20 × 80
	63 × 81	30 × 40	63 × 81	40 × 60
28 (1.102in.)	63 × 70	20 × 50	63 × 70	40 × 50
	63 × 70	25 × 40	63 × 70	25 × 80
29 (1.140in.)	63 × 58	20 × 40	63 × 58	20 × 80
	63 × 145	20 × 100	63 × 145	40 × 100
30 (1.180in.)	63 × 60	20 × 40	63 × 60	20 × 80
	63 × 90	30 × 40	63 × 90	40 × 60
31 (1.220in.)	63 × 62	20 × 40	62 × 62	20 × 80
	63 × 62	25 × 32	63 × 62	40 × 40
32 (1.260in.)	63 × 60	25 × 30	63 × 60	30 × 50
	63 × 80	25 × 40	63 × 80	40 × 50
33 (1.300in.)	33 × 63	20 × 20	66 × 63	20 × 80
	66 × 63	20 × 40	63 × 99	40 × 60
34 (1.338in.)	63 × 85	20 × 50	63 × 85	40 × 50
	63 × 85	25 × 40	63 × 85	25 × 80
35 (1.378in.)	63 × 70	20 × 40	63 × 70	40 × 40
	63 × 105	30 × 40	63 × 70	20 × 80
36 (1.417in.)	63 × 90	20 × 50	63 × 90	40 × 50
	63 × 90	25 × 40	63 × 90	25 × 80
37 (1.456in.)	63 × 37	20 × 20	63 × 37	20 × 40
	63 × 74	20 × 40	63 × 74	20 × 80
38 (1.496in.)	63 × 95	25 × 40	63 × 95	40 × 50
	63 × 95	20 × 50	63 × 95	20 × 100

Change Wheels for mm. Pitches
(Continued)

Pitch of Screw to be Cut.	*¼-in. Pitch Lead Screw.*		*½-in Pitch Lead Screw.*	
Millimetres.	*Drivers.*	*Driven.*	*Drivers.*	*Driven.*
39 (1.535in.)	63 × 78	20 × 40	63 × 78	40 × 40
	63 × 78	25 × 32	63 × 78	20 × 80
40 (1.575in.)	63 × 80	20 × 40	63 × 80	40 × 40
	63 × 40	20 × 20	70 × 90	40 × 50
42 (1.653in.)	63 × 84	20 × 40	63 × 105	40 × 50
	63 × 105	20 × 40	63 × 81	20 × 80
44 (1.732in.)	63 × 55	20 × 25	63 × 55	20 × 50
	63 × 110	25 × 40	63 × 110	40 × 50
45 (1.770in.)	63 × 45	20 × 20	63 × 45	20 × 40
	63 × 90	20 × 40	63 × 90	40 × 40
46 (1.811in.)	63 × 115	20 × 50	63 × 115	40 × 50
	63 × 115	25 × 40	63 × 115	20 × 100
48 (1.890in.)	63 × 60	20 × 25	63 × 60	25 × 40
	63 × 90	25 × 30	63 × 90	30 × 50
50 (1.968in.)	63 × 50	20 × 20	63 × 50	20 × 40
	63 × 75	20 × 30	63 × 75	30 × 40
55 (2.165in.)	63 × 55	20 × 20	63 × 55	20 × 40
	63 × 110	20 × 40	63 × 110	40 × 40
60 (2.362in.)	63 × 60	20 × 20	63 × 60	20 × 40
	63 × 90	20 × 30	63 × 75	20 × 50
65 (2.560in.)	63 × 65	20 × 20	63 × 65	20 × 40
	63 × 78	20 × 24	63 × 78	20 × 48
70 (2.756in.)	63 × 70	20 × 20	63 × 70	20 × 40
	63 × 105	20 × 30	63 × 105	20 × 60
75 (2.953in.)	63 × 75	20 × 20	63 × 75	20 × 40
	63 × 90	20 × 24	63 × 90	20 × 48
80 (3.149in.)	63 × 80	20 × 20	63 × 80	20 × 40
	63 × 100	20 × 25	63 × 100	25 × 40
85 (3.346in.)	63 × 85	20 × 20	63 × 85	20 × 40
	63 × 102	20 × 24	63 × 102	24 × 40
90 (3.543in.)	63 × 90	20 × 20	63 × 90	20 × 40
	63 × 108	20 × 24	63 × 108	40 × 24
95 (3.740in.)	63 × 95	20 × 20	63 × 95	20 × 40
	63 × 76	20 × 16	63 × 76	20 × 32
100 (3.930in.)	63 × 100	20 × 20	63 × 100	20 × 40
	70 × 90	20 × 20	70 × 90	20 × 40

TAPPING SIZES

Whitworth Standard Bolt Threads

Diameter, Nominal Size ins.	1/16	3/32	1/8	5/32	3/16	7/32	1/4	5/16	3/8	7/16	1/2	9/16	5/8	11/16	3/4
Threads per inch ..	60	48	40	32	24	24	20	18	16	14	12	12	11	11	10
Diameter of Tap Drill .. ,, Nos.	56	48	40	31	28	11/64	5 m/m.	1/4	5/16	23/64	13/32	15/32	17/32	37/64	41/64

Diameter, Nominal Size ins.	13/16	7/8	15/16	1	1⅛	1¼	1⅜	1½	1⅝
Threads per inch ..	10	9	9	8	7	7	6	6	5
Diameter of Tap Drill .. ,,	45/64	3/4	13/16	55/64	24.5 m/m.	1 3/32	1 3/16	1 5/16	35.5 m/m.

Diameter, Nominal Size ins.	1¾	1⅞	2	2¼	2½	2¾	3	3¼
Threads per inch ..	5	4½	4½	4	4	3½	3½	3¼
Diameter of Tap Drill ,,	1 33/64	1⅝	1¾	1 31/32	2 7/32	2 7/16	68 m/m.	74 m/m.

Whitworth Standard Pipe Threads

Nominal Inside Diam. ins.	1/8	1/4	3/8	1/2	5/8	3/4	7/8	1	1¼	1½	1¾
External Diameter of Pipe,,	.383	.518	.656	.825	.902	1.041	1.189	1.309	1.650	1.882	2.116
Threads per inch ..	28	19	19	14	14	14	14	11	11	11	11
Diameter of Tap Drill ,,	11/32	15/32	19/32	3/4 or 19 m/m.	53/64 or 21 m/m.	31/32	1 7/64	1 7/32	1 35/64	1 25/32	2 1/32

Nominal Inside Diameter .. ins.	2	2¼	2½	2¾	3	3¼	3½	3¾	4
External Diameter of Pipe .. ,,	2.347	2.587	2.960	3.210	3.460	3.700	3.950	4.200	4.450
Threads per inch	11	11	11	11	11	11	11	11	11
Diameter of Tap Drill .. .. ,,	2¼	2½	2⅞	3 9/64	3 25/64	3 19/32	3 27/32	4 3/32	4 11/32

TAPPING SIZES—(*Continued*)

British Association Standard Threads

Designating No.		0	1	2	3	4	5	6	7	8	9	10	11	12	13	14	15	16
Diameter	m/m.	6.0	5.3	4.7	4.1	3.6	3.2	2.8	2.5	2.2	1.9	1.7	1.5	1.3	1.2	1.0	.90	.79
Pitch	,,	1.0	.90	.81	.73	.66	.59	.53	.48	.43	.39	.35	.31	.28	.25	.23	.21	.19
Diameter of Tap Drill : Steel	Nos.	8-9	15-16	23	28	31	36-37	41-42	45	48	51	53	54-55	61	63	69	71	74
Diameter of Tap Drill : Cast Iron, Brass or Ebonite	Nos.	10	16-17	24-25	29	32	37-38	42-43	46	49-50	52-53	54-55	56	62	64	70	71	74

British Standard Fine Threads

Diameter, Nominal Size	7/32	1/4	5/16	3/8	7/16	1/2	9/16	5/8	11/16	3/4	13/16	7/8	1	1 1/8	1 1/4	1 3/8
Threads per inch	28	26	22	20	18	16	16	14	14	12	12	11	10	9	9	8
Diameter of Tap Drill	No. 15	No. 5	G	O	3/8	27/64	12.5 m/m.	35/64	39/64	21/32	23/32	25/32	57/64	1	1 1/8	1 15/64

Diameter, Nominal Size	1 1/2	1 5/8	1 3/4	2	2 1/4	2 1/2	2 3/4	3 ins
Threads per inch	8	8	7	7	6	6	6	5
Diameter of Tap Drill	34.5 m/m.	1 31/64	1 19/32	1 27/32	2 1/16	2 5/16	2 9/16	70 m/m.

TAP DRILLS. **A simple rule for ascertaining the size of tap drill required and which in many cases is good enough is:**

Subtract the pitch of one thread from the diameter of the tap.

Example:

A 3/8 inch tap 16 threads would be 3/8 minus 1/16 = 5/16 in. drill.

A 3/4 inch tap 10 threads would be 3/4 minus 1/10 = 75/100 − 10/100 or 0.75 − 0.10 = 65/100 or 0.65, or a little over 5/8 of an inch, so a 5/8 inch drill will do nicely. With a 1 inch tap we have 1 − 1/8 = 7/8 in. drill, which is a little large but leaves enough thread for most cases.

A tap should be between 0.002 and 0.003 inch large for clearance between top and bottom of threads.

Standard Wood Screws

No. of Screw Gauge	Diameter. A in Dec.	Dia. A in Fractions.	B In.	Depth of C'sink	Slot.	
					Width	Depth.
0	.05784	1/16	7/64	1/32	1/64	1/64
1	.07100	5/64	9/64	3/64	1/64	1/32
2	.08416	5/64	11/64	3/64	1/64	1/32
3	.09732	3/32	3/16	3/64	1/64	1/32
4	.11048	7/64	7/32	1/16	1/32	1/32
5	.12364	1/8	15/64	1/16	1/32	1/32
6	.13680	9/64	17/64	5/64	1/32	3/64
7	.14996	5/32	19/64	5/64	1/32	3/64
8	.16312	5/32	21/64	3/32	3/64	3/64
9	.17628	11/64	23/64	3/32	3/64	3/64
10	.18944	3/16	3/8	7/64	3/64	1/16
11	.20260	13/64	13/32	7/64	3/64	1/16
12	.21576	7/32	7/16	1/8	3/64	1/16
13	.22892	15/64	29/64	1/8	1/16	1/16
14	.24208	1/4	31/64	9/64	1/16	1/16
15	.25524	1/4	33/64	9/64	1/16	1/16
16	.26840	17/64	17/32	5/32	1/16	5/64
17	.28156	9/32	9/16	5/32	1/16	5/64
18	.29472	19/64	19/32	11/64	5/64	5/64
19	.30788	5/16	39/64	11/64	5/64	5/64
20	.32104	21/64	41/64	11/64	5/64	5/64
21	.33420	21/64	43/64	3/16	5/64	3/32
22	.34736	11/32	11/16	3/16	3/32	3/32
23	.36052	23/64	28/32	13/64	3/32	3/32
24	.37368	3/8	3/4	13/64	3/32	3/32

WOOD SCREW PROPORTIONS

TWIST DRILLS FOR WOOD SCREWS

No. (or) size) of Screw.	*Diameter of Neck or Shank.*	*For Wood or Metal.*			*With Side Lips and Centre for Wood only.*	
			No., etc.	*Diameter.*	*Size.*	*Diameter.*
1	.066	Stubb's Wire Gauge Drills.	51	.067	—	—
2	.080		46	.081	—	—
3	.094		41	.096	—	—
4	.108		35	.110	—	—
5	.122		30	.128	1/8	.125
6	.136		28	.140	—	—
7	.150		23	.154	5/32	.156
8	.164		18	.169	—	—
9	.178		14	.182	3/16	.187
10	.192		9	.196	—	—
11	.206		4	.209	7/32	.218
12	.220		1	.228	—	—
13	.234	Letter Gauge Drills.	B	.238	—	—
14	.248		E	.250	1/4	.250
15	.262		H	.266	—	—
16	.276		K	.281	9/32	.281
17	.290		M	.295	—	—
18	.304		O	.316	5/16	.312
19	.318		P	.323	—	—
20	.332		R	.339	11/32	.343
21	.346		{ S	.348	—	—
			{ T	.358		
22	.360		U	.368	3/8	.375
23	.374		V	.377	3/8	.375
24	.388		X	.397	—	—
25	.402		Z	.413	13/32	.406
26	.416	27/64		.421	—	—
27	.430	7/16		.437	7/16	.437
28	.444	29/64		.453	—	—
29	.458	15/32		.468	15/32	.468
30	.472	31/64		.484	—	—
31	.486	1/2		.500	1/2	.500
32	.500	33/64		.515	1/2	.500

All dimensions in parts of an inch.

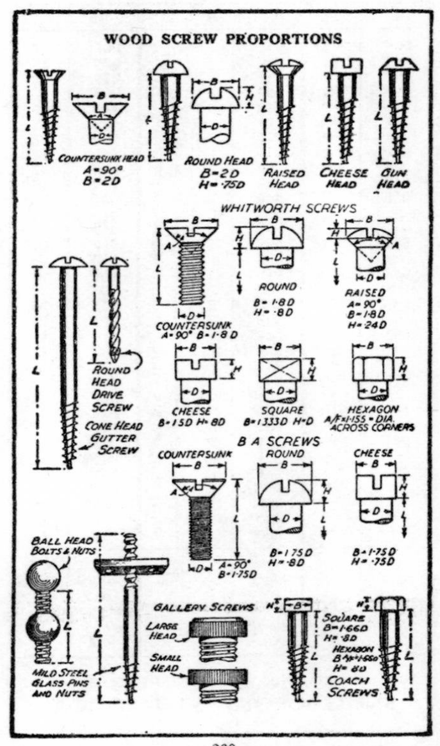
WOOD SCREW PROPORTIONS
COUNTERSUNK HEAD
A=90°
B=2D
ROUND HEAD
B=2D
H=·75D
RAISED HEAD
CHEESE HEAD
GUN HEAD
WHITWORTH SCREWS
COUNTERSUNK
A=90° B=1·8D
ROUND
B=1·8D
H=·8D
RAISED
A=90°
B=1·8D
H=·24D
CHEESE
B=1·5D H=·8D
SQUARE
B=1·333D H=D
HEXAGON
A/F×1·155 = DIA. ACROSS CORNERS
ROUND HEAD DRIVE SCREW
CONE HEAD GUTTER SCREW
B A SCREWS
COUNTERSUNK
A=90°
B=1·75D
ROUND
B=1·75D
H=·8D
CHEESE
B=1·75D
H=·75D
BALL HEAD BOLTS & NUTS
MILD STEEL GLASS PINS AND NUTS
GALLERY SCREWS
LARGE HEAD
SMALL HEAD
SQUARE
B=1·66D
H=·8D
HEXAGON
H=·8D
COACH SCREWS

Rules and Formulæ for Worm Gears

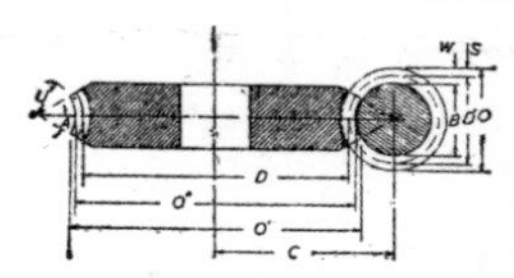

To compute the necessary dimensions for a worm gear drive the following formulas should be used in connection with the above figure.

P = circular pitch of wheel and linear pitch of worm.

L = lead of worm.

N′ = number of threads in worm.

S = addendum.

D′ = pitch diameter of worm.

D = pitch diameter of wormwheel.

O = outside diameter of worm.

O″ = throat diameter of wormwheel.

O′ = diameter of wormwheel over sharp corners.

B = bottom diameter of worm.

N = number of teeth in wormwheel.

W = whole depth of worm tooth.

T = width of thread tool at end.

B′ = helix angle of worm.

90°—B′ = gashing angle of wormwheel.

U = radius of curvature of wormwheel throat.

C = centre distance.

A = face angle of wormwheel.

RULES AND FORMULÆ FOR WORM GEARS

To Find	Rule	Formula
Linear Pitch	Divide the lead by the number of threads. It is understood that by the number of threads is meant, not number of threads per inch, but the number of threads in the whole worm—one, if it is single-threaded; four if it is quadruple-threaded, etc. ...	$P = \frac{L}{N'}$
Addendum of worm tooth	Multiply the linear pitch by 0.3183	$S = 0.3183\ P$
Pitch diameter of worm	Subtract twice the addendum from the outside diameter	$D' = 0 - 2\ S$
Pitch diameter of wormwheel	Multiply the number of teeth in the wheel by the linear pitch of the worm, and divide the product by 3.1416	$D = \frac{NP}{3.1416}$
Centre distance between worm and wormwheel	Add together the pitch diameter of the worm and the pitch diameter of the wormwheel, and divide the sum by 2	$C = \frac{D+D'}{2}$
Whole depth of worm tooth	Multiply the linear pitch by 0.6866	$W = 0.6866 \times P$
Bottom diameter of worm	Subtract twice the whole depth of tooth from the outside diameter	$B = O - 2W$

RULES AND FORMULÆ FOR WORM GEARS (*Contd.*)

To Find	Rule	Formula
Helix angle of worm	Multiply the pitch diameter of the worm by 3.1416, and divide the product by the lead; the quotient is the cotangent of the helix angle of the worm	$\text{Cotan B} = \frac{3.1416\ D'}{L}$
Width of thread tool at end	Multiply the linear pitch by 0.31	$T = 0.31\ P$
Throat diameter of wormwheel	Add twice the addendum of the worm tooth to the pitch diameter of the wormwheel	$O'' = D + 2S$
Radius of wormwheel throat	Subtract twice the addendum of the worm tooth from half the outside diameter of the worm	$U = \frac{O}{2} - 2S$
Outside diameter of worm	Add together the pitch diameter and twice the addendum	$O = D' + 2S$
Pitch diameter of worm	Subtract the pitch diameter of the wormwheel from twice the centre distance	$D' = 2C - D$
Diameter of wormwheel to sharp corners	Multiply the radius of curvature of the wormwheel throat by the cosine of half the face angle, subtract this quantity from the radius of curvature, multiply the remainder by 2, and add the product to the throat diameter of the wormwheel	$O' = 2(U - U \times \cos A) + O''$
Gashing angle of wormwheel	Divide the lead of the worm by the circumference of the pitch circle. The result will be the tangent of the gashing angle	$\text{Tan}\ (90° - B) = \frac{L}{\pi D'}$

INDEX TABLE FOR MILLING MACHINES

40 Turns of Worm to 1 Revolution of Wormwheel

Division	*Circle*	*Turns*	*Holes*
2	any	20	..
3	39	13	13
	33	13	11
	27	13	9
	21	13	7
	18	13	6
	15	13	5
4	any	10	..
5	any	8	..
6	39	6	26
	33	6	22
	27	6	18
	21	6	14
	18	6	12
	15	6	10
7	49	5	35
	21	5	15
8	any	5	..
9	27	4	12
	18	4	8
10	any	4	..
11	33	3	21
12	39	3	13
	33	3	11
	27	3	9
	21	3	7
	18	3	6
12	15	3	5
13	39	3	3
14	49	2	42
	21	2	18
15	39	2	26
	33	2	22
	27	2	18
	21	2	14
	18	2	12
	15	2	10
16	20	2	10
	18	2	9
	16	2	8
17	17	2	6

Division	*Circle*	*Turns*	*Holes*
18	27	2	6
	18	2	4
19	19	2	2
20	any	2	..
21	21	1	19
22	33	1	27
23	23	1	17
24	39	1	26
	33	1	22
	27	1	18
	21	1	14
	18	1	12
	15	1	10
25	20	1	12
26	39	1	21
27	27	1	13
28	49	1	21
	21	1	9
29	29	1	11
30	39	1	13
	33	1	11
	27	1	9
	21	1	7
	18	1	6
	15	1	5
31	31	1	9
32	20	1	5
	16	1	4
33	33	1	7
34	17	1	3
35	49	1	7
	21	1	3
36	27	1	3
	18	1	2
37	37	1	3
38	19	1	1
39	39	1	1
40	any	1	..

This Table relates to Simple Indexing

RIVET SPACING.

The diagrams show limiting positions for the centres of adjacent rivets of various diameters.

The circular arcs in Fig. **A** below show the closest allowable positions of the centre of a rivet in relation to an adjacent rivet (shown in the top left-hand corner of the diagram) when the distance centre to centre is to be not less than 3 times the diameter—this being the minimum usually observed in structural work.

If the rivets are staggered as in the inset Fig. B, and the foregoing rule is to be observed, the diagram will show at a glance the smallest allowable value for s for a given longitudinal half pitch p, and conversely.

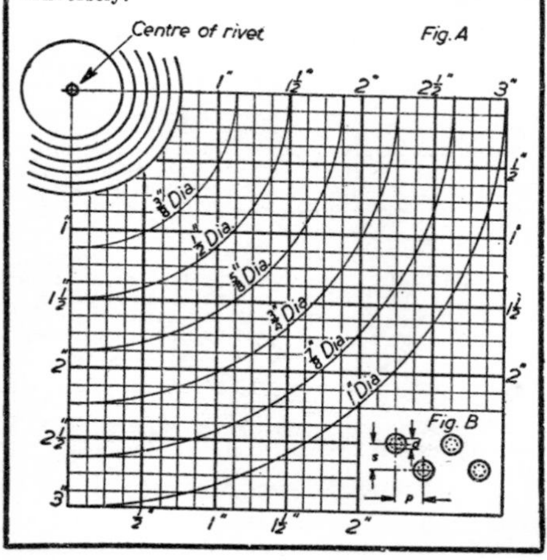

RIVET SPACING (Continued)

The diagrams show limiting positions for the centres of adjacent rivets of various diameters.

When rivets are used in the same plane through two flanges of an angle or tee as in Fig. C, $\frac{1}{4}''$ clearance, as shown therein, must be allowed for machine driving.

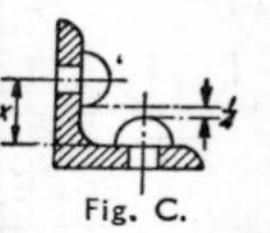

Fig. C.

The diameter of the rivets being known, the lowest permissible value for x can be read at a glance from Fig. D below. This diagram shows the closest permissible position for the centre of the second rivet being along the circular arcs and straight lines continuing them.

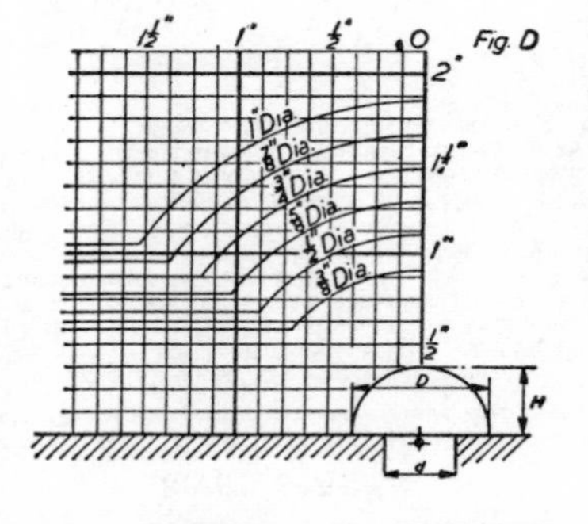

DIMENSIONS FOR SQUARE SHAFT AND FITTING

HOLE — PERMANENT FIT AND SLIP FIT — SHAFT

Nom. Dia.	PERMANENT FIT					SLIP FIT				
	W	d′	d	D′	D	W	d′	d	D′	D
¼	.193	.1895 .1885	.1875 .1865	.250 .245	.260 .252	.257	.248 .247	.250 .249	.3437 .3387	.3537 .3457
⅜	.290	.2832 .2822	.2812 .2802	.375 .370	.385 .377	.386	.373 .372	.375 .374	.5156 .5106	.5256 .5176
½	.386	.3770 .3760	.3750 .3740	.500 .495	.516 .502	33/64	.498 .497	.500 .599	.6875 .6825	.6975 .6895
⅝	33/64	.5020 .5010	.5000 .4990	.625 .620	.635 .627	41/64	.623 .622	.625 .624	.8437 .8387	.8537 .8457
¾	37/64	.5645 .5635	.5625 .5615	.750 .745	.760 .752	49/64	.748 .747	.750 .749	1.031 1.026	1.051 1.036
⅞	45/64	.6895 .6885	.6875 .6865	.875 .870	.885 .877	29/32	.873 .872	.875 .874	1.187 1.182	1.207 1.192
1	27/32	.8155 .8145	.8125 .8115	1.000 .995	1.020 1.005	1 1/32	.998 .997	1.000 .999	1.375 1.370	1.395 1.380
1⅛	29/32	.8780 .8770	.8750 .8740	1.125 1.120	1.145 1.130	1 5/32	1.123 1.122	1.125 1.124	1.562 1.557	1.582 1.567
1¼	1 1/32	1.003 1.002	1.000 .999	1.250 1.245	1.270 1.255	1 9/32	1.248 1.247	1.250 1.249	1.687 1.682	1.707 1.692

DIMENSIONS FOR SQUARE SHAFT AND FITTING (*Contd.*)

Nom. Dia.	PERMANENT FIT					SLIP FIT				
	W	d′	d	D′	D	W	d′	d	D′	D
1⅜	1 5/32	1.128 1.127	1.125 1.124	1.375 1.370	1.395 1.380	1 27/64	1.373 1.372	1.375 1.374	1.875 1.870	1.895 1.880
1½	1 5/32	1.128 1.127	1.125 1.124	1.500 1.495	1.520 1.505	1 35/64	1.498 1.497	1.500 1.499	2.062 2.057	2.082 2.067
1¾	1 27/64	1.378 1.377	1.375 1.374	1.750 1.745	1.770 1.755	1 13/16	1.748 1.747	1.750 1.749	2.375 2.370	2.395 2.380
2	1 35/64	1.504 1.503	1.5000 1.4985	2.000 1.995	2.020 2.005	2 1/16	1.9975 1.9965	2.0000 1.9985	2.750 2.745	2.770 2.755
2¼	1 13/16	1.754 1.753	1.7500 1.7485	2.250 2.245	2.270 2.255	2 5/16	2.2475 2.2465	2.2500 2.2485	3.062 3.057	3.082 3.067
2½	2 1/16	2.004 2.003	2.0000 1.9985	2.500 2.495	2.520 2.505	2 37/64	2.4975 2.4965	2.5000 2.4985	3.437 3.432	3.457 3.442
2¾	2 5/16	2.254 2.253	2.2500 2.2485	2.750 2.745	2.770 2.755	2 55/64	2.7475 2.7465	2.7500 2.7485	3.750 3.745	3.770 3.755
3	2 37/64	2.504 2.503	2.500 2.498	3.000 2.995	3.020 3.005	3 3/32	2.997 2.996	3.000 2.998	4.125 4.120	4.145 4.130
3½	2 55/64	2.754 2.753	2.750 2.748	3.500 3.495	3.520 3.505	3 39/64	3.497 3.496	3.500 3.498	4.750 4.745	4.770 4.755
4	3 23/64	3.254 3.253	3.250 3.248	4.000 3.995	4.020 4.005	4⅛	3.997 3.996	4.000 3.998	5.500 5.495	5.520 5.505

DIMENSIONS FOR 4-KEY SPLINED HOLES & SHAFTS

(PERMANENT FIT)

HOLE (All dimensions in inches) SHAFT

Normal Diam.	D	d	W	D′	d′	W′
¾	.750	.637	.181	.742	.6395	.1815
	.749	.636	.179	.738	.6375	.1805
⅞	.875	.744	.211	.867	.7465	.2115
	.874	.743	.209	.863	.7445	.2105
1	1.000	.850	.241	.992	.8525	.2415
	.999	.849	.239	.988	.8505	.2405
1⅛	1.125	.956	.271	1.117	.9585	.2715
	1.124	.955	.269	1.113	.9565	.2705
1¼	1.250	1.062	.301	1.242	1.0645	.3015
	1.249	1.061	.229	1.238	1.0625	.3005
1⅜	1.375	1.169	.331	1.367	1.1715	.3315
	1.374	1.168	.329	1.363	1.1695	.3305
1½	1.500	1.275	.361	1.490	1.2785	.362
	1.499	1.274	.359	1.486	1.2755	.360
1⅝	1.625	1.381	.391	1.615	1.3845	.392
	1.624	1.380	.389	1.611	1.3815	.390
1¾	1.750	1.487	.422	1.740	1.4905	.423
	1.749	1.486	.420	1.736	1.4875	.421
2	2.000	1.700	.482	1.988	1.7035	.483
	1.998	1.698	.479	1.984	1.7005	.481
2¼	2.250	1.912	.542	2.238	1.9155	.543
	2.248	1.910	.539	2.234	1.9125	.541
2½	2.500	2.125	.602	2.488	2.1285	.603
	2.498	2.123	.599	2.484	2.1255	.601
3	3.000	2.550	.723	2.988	2.5535	.724
	2.998	2.548	.720	2.984	2.5505	.722

4-KEY SPLINED HOLES & SHAFTS
(PERMANENT FIT)

POSSIBLE FIT BETWEEN HOLE AND SHAFT

+ = CLEARANCE — = INTERFERENCE

(See diagrams in preceding table)

NORMAL DIAM.	D	d	W
3/4	+ .012 + .007	− .0005 − .0035	+ .0005 − .0025
7/8	+ .012 + .007	− .0005 − .0035	+ .0005 − .0025
1	+ .012 + .007	− .0005 − .0035	+ .0005 − .0025
1 1/8	+ .012 + .007	− .0005 − .0035	+ .0005 − .0025
1 1/4	+ .012 + .007	− .0005 − .0035	+ .0005 − .0025
1 3/8	+ .012 + .007	− .0005 − .0035	+ .0005 − .0025
1 1/2	+ .014 + .009	− .0005 − .0045	+ .001 − .003
1 5/8	+ .014 + .009	− .0005 − .0045	+ .001 − .003
1 3/4	+ .014 + .009	− .0005 − .0045	+ .001 − .003
2	+ .016 + .010	− .0005 − .0055	+ .001 − .004
2 1/4	+ .016 + .010	− .0005 − .0055	+ .001 − .004
2 1/2	+ .016 + .010	− .0005 − .0055	+ .001 − .004
3	+ .016 + .010	− .0005 − .0055	+ .001 − .004

DIMENSIONS FOR 4-KEY SPLINED HOLES AND SHAFTS

(SLIDING FIT WHEN NOT UNDER LOAD)

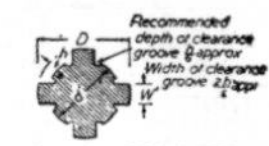

HOLE (All dimensions in inches) SHAFT

NORMAL DIAM.	D	d	W	D′	d′	W′
3/4	.750 .749	.562 .561	.181 .179	.742 .738	.560 .558	.178 .177
7/8	.875 .874	.656 .655	.211 .209	.867 .863	.654 .652	.208 .207
1	1.000 0.999	.750 .749	.241 .239	.992 .988	.748 .746	.238 .237
1 1/8	1.125 1.124	.844 .843	.271 .269	1.117 1.113	.842 .840	.268 .267
1 1/4	1.250 1.249	.937 .936	.301 .299	1.242 1.238	.935 .933	.298 .297
1 3/8	1.375 1.374	1.031 1.030	.331 .329	1.367 1.363	1.029 1.027	.328 .327
1 1/2	1.500 1.499	1.125 1.124	.361 .359	1.490 1.486	1.123 1.120	.358 .356
1 5/8	1.625 1.624	1.219 1.218	.391 .389	1.615 1.611	1.217 1.214	.388 .386
1 3/4	1.750 1.749	1.312 1.311	.422 .420	1.740 1.736	1.310 1.307	.419 .417
2	2.000 1.998	1.500 1.498	.482 .479	1.988 1.984	1.497 1.494	.478 .476
2 1/4	2.250 2.248	1.687 1.685	.542 .539	2.238 2.234	1.684 1.681	.538 .536
2 1/2	2.500 2.498	1.875 1.873	.602 .599	2.488 2.484	1.872 1.869	.598 .596
3	3.000 2.998	2.250 2.248	.723 .720	2.988 2.984	2.247 2.244	.719 .717

4-KEY SPLINED HOLES AND SHAFTS

(SLIDING FIT WHEN NOT UNDER LOAD)

POSSIBLE FIT BETWEEN HOLE AND SHAFT

+ = CLEARANCE — = INTERFERENCE

NORMAL DIAM.	D	d	W
$\frac{3}{4}$	+ .012 / + .007	+ .004 / + .001	+ .004 / + .001
$\frac{7}{8}$	+ .012 / + .007	+ .004 / + .001	+ .004 / + .001
1	+ .012 / + .007	+ .004 / + .001	+ .004 / + .001
$1\frac{1}{8}$	+ .012 / + .007	+ .004 / + .001	+ .004 / + .001
$1\frac{1}{4}$	+ .012 / + .007	+ .004 / + .001	+ .004 / + .001
$1\frac{3}{8}$	+ .012 / + .007	+ .004 / + .001	+ .004 / + .001
$1\frac{1}{2}$	+ 0.14 / + .009	+ .005 / + .001	+ .005 / + .001
$1\frac{5}{8}$	+ .014 / + .009	+ .005 / + .001	+ .005 / + .001
$1\frac{3}{4}$	+ .014 / + .009	+ .005 / + .001	+ .005 / + .001
2	+ .016 / + .010	+ .006 / + .001	+ .006 / + .001
$2\frac{1}{4}$	+ .016 / + .010	+ .006 / + .001	+ .006 / + .001
$2\frac{1}{2}$	+.016 / + .010	+ .006 / + .001	+ .006 / + .001
3	+ .016 / + .010	+ .006 / + .001	+ .006 / + .001

DIMENSIONS FOR 6-KEY SPLINED HOLES AND SHAFTS
(Permanent Fit)

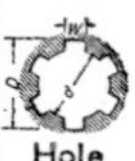

Hole

All dimensions in inches

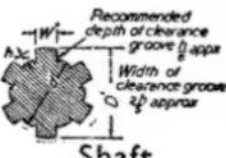

Shaft

Normal Diam.	D	d	W	D′	d′	W′
3/4	.750 .749	.675 .674	.188 .186	.742 .738	.6775 .6755	.1885 .1875
7/8	.875 .874	.788 .787	.219 .217	.867 .863	.7905 .7885	.2195 .2185
1	1.000 .999	.900 .899	.250 .248	.992 .988	.9025 .9005	.2505 .2495
1 1/8	1.125 1.124	1.013 1.012	.281 .279	1.117 1.113	1.0155 1.0135	.2815 .2805
1 1/4	1.250 1.249	1.125 1.124	.313 .311	1.242 1.238	1.1275 1.1255	.3135 .3125
1 3/8	1.375 1.374	1.238 1.237	.344 .342	1.367 1.363	1.2405 1.2385	.3445 .3435
1 1/2	1.500 1.499	1.350 1.349	.375 .373	1.490 1.486	1.3535 1.3505	.376 .374
1 5/8	1.625 1.624	1.463 1.462	.406 .404	1.615 1.611	1.4665 1.4635	.407 .405
1 3/4	1.750 1.749	1.575 1.574	.438 .436	1.740 1.736	1.5785 1.5755	.439 .437
2	2.000 1.998	1.800 1.798	.500 .497	1.988 1.984	1.8035 1.8005	.501 .499
2 1/4	2.250 2.248	2.025 2.023	.563 .560	2.238 2.234	2.0285 2.0255	.564 .562
2 1/2	2.500 2.498	2.250 2.248	.625 .622	2.488 2.484	2.2535 2.2505	.626 .624
3	3.000 2.998	2.700 2.698	.750 .747	2.988 2.984	2.7035 2.7005	.751 .749

6-KEY SPLINED HOLES AND SHAFTS

(Permanent Fit)

Possible Fit Between Hole and Shaft

+ = Clearance — = Interference

Normal Diam.	D	d	W
$\frac{3}{4}$	+.012 +.007	—.0005 —.0035	+.0005 —.0025
$\frac{7}{8}$	+.012 +.007	—.0005 —.0035	+.0005 —.0025
1	+.012 +.007	—.0005 —.0035	+.0005 —.0025
$1\frac{1}{8}$	+.012 +.007	—.0005 —.0035	+.0005 —.0025
$1\frac{1}{4}$	+.012 +.007	—.0005 —.0035	+.0005 —.0025
$1\frac{3}{8}$	+.012 +.007	—.0005 —.0035	+.0005 —.0025
$1\frac{1}{2}$	+.014 +.009	—.0005 —.0045	+.001 —.003
$1\frac{5}{8}$	+.014 +.009	—.0005 —.0045	+.001 —.003
$1\frac{3}{4}$	+.014 +.009	—.0005 —.0045	+.001 —.003
2	+.016 +.010	—.0005 —.0055	+.001 —.004
$2\frac{1}{4}$	+.016 +.010	—.0005 —.0055	+.001 —.004
$2\frac{1}{2}$	+.016 +.010	—.0005 —.0055	+.001 —.004
3	+.016 +.010	—.0005 —.0055	+.001 —.004

DIMENSIONS FOR 6-KEY SPLINED HOLES AND SHAFTS
(Sliding Fit When Not Under Load)

Hole

All dimensions in inches

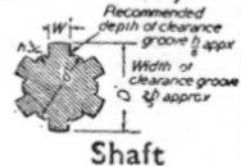

Shaft

Normal Diam.	D	d	W	D′	d′	W′
¾	.750 .749	.638 .637	.188 .186	.742 .738	.636 .635	.185 .184
⅞	.875 .874	.744 .743	.219 .217	.867 .863	.742 .741	.216 .215
1	1.000 .999	.850 .849	.250 .248	.992 .988	.848 .847	.247 .246
1⅛	1.125 1.124	.956 .955	.281 .279	1.117 1.113	.954 .953	.278 .277
1¼	1.250 1.249	1.063 1.062	.313 .311	1.242 1.238	1.061 1.060	.310 .309
1⅜	1.375 1.374	1.169 1.168	.344 .342	1.367 1.363	1.167 1.166	.341 .340
1½	1.500 1.499	1.275 1.274	.375 .373	1.490 1.486	1.273 1.271	.372 .370
1⅝	1.625 1.624	1.381 1.380	.406 .404	1.615 1.611	1.379 1.377	.403 .401
1¾	1.750 1.749	1.488 1.487	.438 .436	1.740 1.736	1.486 1.484	.435 .433
2	2.000 1.998	1.700 1.698	.500 .497	1.988 1.984	1.697 1.694	.496 .494
2¼	2.250 2.248	1.913 1.911	.563 .560	2.238 2.234	1.910 1.907	.559 .557
2½	2.500 2.498	2.125 2.123	.625 .622	2.488 2.484	2.122 2.119	.621 .619
3	3.000 2.998	2.550 2.548	.750 .747	2.988 2.984	2.547 2.544	.746 .744

6-KEY SPLINED HOLES AND SHAFTS
(Sliding Fit When Not Under Load)

Possible Fit Between Hole and Shaft

+ = Clearance — = Interference

Normal Diam.	D	d	W
3/4	+.012 / +.007	+.001 / +.003	+.001 / +.004
7/8	+.012 / +.007	+.001 / +.003	+.001 / +.004
1	+.012 / +.007	+.001 / +.003	+.001 / +.004
1 1/8	+.012 / +.007	+.001 / +.003	+.001 / +.004
1 1/4	+.012 / +.007	+.001 / +.003	+.001 / +.004
1 3/8	+.012 / +.007	+.001 / +.003	+.001 / +.004
1 1/2	+.014 / +.009	+.001 / +.004	+.001 / +.005
1 5/8	+.014 / +.009	+.001 / +.004	+.001 / +.005
1 3/4	+.014 / +.009	+.001 / +.004	+.001 / +.005
2	+.016 / +.010	+.001 / +.006	+.001 / +.006
2 1/4	+.016 / +.010	+.001 / +.006	+.001 / +.006
2 1/2	+.016 / +.010	+.001 / +.006	+.001 / +.006
3	+.016 / +.010	+.001 / +.006	+.001 / +.006

DIMENSIONS FOR 6-KEY SPLINED HOLES & SHAFTS
(Sliding Fit When Under Load)

Hole

All dimensions in inches

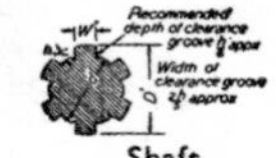

Shaft

Normal Diam.	D	d	W	D′	d′	W′
¾	.750 .749	.600 .599	.188 .186	.742 .738	.598 .597	.185 .184
⅞	.875 .874	.700 .699	.219 .217	.867 .863	.698 .697	.216 .215
1	1.000 .999	.800 .799	.250 .248	.992 .988	.798 .797	.247 .246
1⅛	1.125 1.124	.900 .899	.281 .279	1.117 1.113	.898 .897	.278 .277
1¼	1.250 1.249	1.000 .999	.313 .311	1.242 1.238	.998 .997	.310 .309
1⅜	1.375 1.374	1.100 1.099	.344 .342	1.367 1.363	1.098 1.097	.341 .340
1½	1.500 1.499	1.200 1.199	.375 .373	1.490 1.486	1.198 1.196	.372 .370
1⅝	1.625 1.624	1.300 1.299	.406 .404	1.615 1.611	1.298 1.296	.403 .401
1¾	1.750 1.749	1.400 1.399	.438 .436	1.740 1.736	1.398 1.396	.435 .433
2	2.000 1.998	1.600 1.598	.500 .497	1.988 1.984	1.597 1.594	.496 .494
2¼	2.250 2.248	1.800 1.798	.563 .560	2.238 2.234	1.797 1.794	.559 .557
2½	2.500 2.498	2.000 1.998	.625 .622	2.488 2.484	1.997 1.994	.621 .619
3	3.000 2.998	2.400 2.398	.750 .747	2.988 2.984	2.397 2.394	.746 .744

6-KEY SPLINED HOLES & SHAFTS

(Sliding Fit When Under Load)

Possible Fit Between Hole and Shaft

+ = Clearance — = Interference

Normal Diameter	D	d	W
$\frac{3}{4}$	+.012 / +.007	+.001 / +.003	+.001 / +.004
$\frac{7}{8}$	+.012 / +.007	+.001 / +.003	+.001 / +.004
1	+.012 / +.007	+.001 / +.003	+.001 / +.004
$1\frac{1}{8}$	+.012 / +.007	+.001 / +.003	+.001 / +.004
$1\frac{1}{4}$	+.012 / +.007	+.001 / +.003	+.001 / +.004
$1\frac{3}{8}$	+.012 / +.007	+.001 / +.003	+.001 / +.004
$1\frac{1}{2}$	+.014 / +.009	+.001 / +.004	+.001 / +.005
$1\frac{5}{8}$	+.014 / +.009	+.001 / +.004	+.001 / +.005
$1\frac{3}{4}$	+.014 / +.009	+.001 / +.004	+.001 / +.005
2	+.016 / +.010	+.001 / +.006	+.001 / +.006
$2\frac{1}{4}$	+.016 / +.010	+.001 / +.006	+.001 / +.006
$2\frac{1}{2}$	+.016 / +.010	+.001 / +.006	+.001 / +.006
3	+.016 / +.010	+.001 / +.006	+.001 / +.006

DIMENSIONS FOR 10-KEY SPLINED HOLES AND SHAFTS
(PERMANENT FIT)

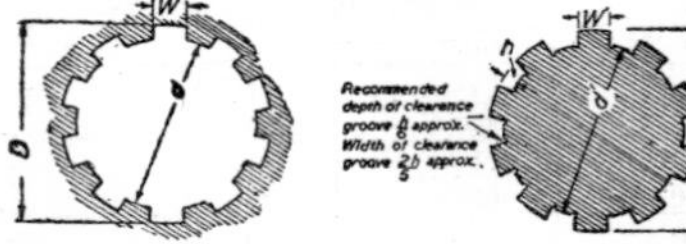

Hole (All dimensions in inches) Shaft

Nor. Dia.	D	d	W	D'	d'	W'
$\frac{3}{4}$	.750 .749	.683 .682	.117 .115	.742 .738	.6850 .6835	.1175 .1165
$\frac{7}{8}$	.875 .874	.796 .795	.137 .135	.867 .863	.7980 .7965	.1375 .1365
1	1.000 .999	.910 .909	.156 .154	.992 .988	.9120 .9105	.1565 .1555
$1\frac{1}{8}$	1.125 1.124	1.024 1.023	.176 .174	1.117 1.113	1.0260 1.0245	.1765 .1755
$1\frac{1}{4}$	1.250 1.249	1.138 1.137	.195 .193	1.242 1.238	1.1400 1.1385	.1955 .1945
$1\frac{3}{8}$	1.375 1.374	1.251 1.250	.215 .213	1.367 1.363	1.2530 1.2515	.2155 .2145
$1\frac{1}{2}$	1.500 1.499	1.365 1.364	.234 .232	1.490 1.486	1.3675 1.3655	.235 .234
$1\frac{5}{8}$	1.625 1.624	1.479 1.478	.254 .252	1.615 1.611	1.4815 1.4795	.255 .254
$1\frac{3}{4}$	1.750 1.749	1.593 1.592	.273 .271	1.740 1.736	1.5955 1.5935	.274 .273
2	2.000 1.998	1.820 1.818	.312 .309	1.988 1.984	1.8225 1.8205	.313 .312
$2\frac{1}{4}$	2.250 2.248	2.048 2.046	.351 .348	2.238 2.234	2.0505 2.0485	.352 .351
$2\frac{1}{2}$	2.500 2.498	2.275 2.273	.390 .387	2.488 2.484	2.2775 2.2755	.391 .390
3	3.000 2.998	2.730 2.728	.468 .465	2.988 2.984	2.7325 2.7305	.469 .468

10-KEY SPLINED HOLES AND SHAFTS

(PERMANENT FIT)

POSSIBLE FIT BETWEEN HOLE AND SHAFT

+ = CLEARANCE — = INTERFERENCE

NORMAL DIA.	D	d	W
3/4	+.012 / +.007	−.0005 / −.0030	+.0005 / −.0025
7/8	+.012 / +.007	−.0005 / −.0030	+.0005 / −.0025
1	+.012 / +.007	−.0005 / −.0030	+.0005 / −.0025
1 1/8	+.012 / +.007	−.0005 / −.0030	+.0005 / −.0025
1 1/4	+.012 / +.007	−.0005 / −.0030	+.0005 / −.0025
1 3/8	+.012 / +.007	−.0005 / −.0030	+.0005 / −.0025
1 1/2	+.014 / +.009	−.0005 / −.0035	.000 / −.003
1 5/8	+.014 / +.009	−.0005 / −.0035	.000 / −.003
1 3/4	+.014 / +.009	−.0005 / −.0035	.000 / −.003
2	+.016 / +.010	−.0005 / −.0045	.000 / −.004
2 1/4	+.016 / +.010	−.0005 / −.0045	.000 / −.004
2 1/2	+.016 / +.010	−.0005 / −.0045	.000 / −.004
3	+.016 / +.010	−.0005 / −.0045	.000 / −.004

DIMENSIONS FOR 10-KEY SPLINED HOLES AND SHAFTS
(SLIDING FIT WHEN NOT UNDER LOAD)

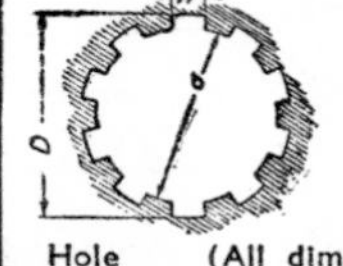

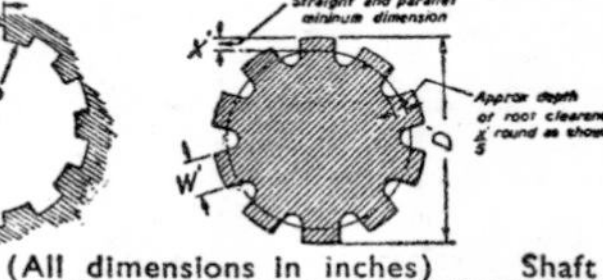

Hole (All dimensions in inches) Shaft

Nor. Dia.	D	d	W	D′	X′	W′
3/4	.750 / .749	.645 / .644	.117 / .115	.746 / .745	.051	.1145 / .1135
7/8	.875 / .874	.753 / .752	.137 / .135	.871 / .870	.059	.1345 / .1335
1	1.000 / .999	.860 / .859	.156 / .154	.996 / .995	.068	.1535 / .1525
1 1/8	1.125 / 1.124	.968 / .967	.176 / .174	1.121 / 1.120	.077	.1735 / .1725
1 1/4	1.250 / 1.249	1.075 / 1.074	.195 / .193	1.246 / 1.245	.086	.1925 / .1915
1 3/8	1.375 / 1.374	1.183 / 1.182	.215 / .213	1.371 / 1.370	.094	.2125 / .2115
1 1/2	1.500 / 1.499	1.290 / 1.289	.234 / .232	1.495 / 1.494	.103	.231 / .230
1 5/8	1.625 / 1.624	1.398 / 1.397	.254 / .252	1.620 / 1.619	.112	.251 / .250
1 3/4	1.750 / 1.749	1.505 / 1.504	.273 / .271	1.745 / 1.744	.120	.270 / .269
2	2.000 / 1.998	1.720 / 1.718	.312 / .309	1.994 / 1.993	.138	.3080 / .3060
2 1/4	2.250 / 2.248	1.935 / 1.933	.351 / .348	2.244 / 2.243	.155	.3470 / .3450
2 1/2	2.500 / 2.498	2.150 / 2.148	.390 / .387	2.494 / 2.493	.173	.3860 / .3840
3	3.000 / 2.998	2.580 / 2.578	.468 / .465	2.994 / 2.993	.208	.4640 / .4620

10-KEY SPLINED HOLES AND SHAFTS

(SLIDING FIT WHEN NOT UNDER LOAD)

POSSIBLE FIT BETWEEN HOLE AND SHAFT

+ = CLEARANCE — = INTERFERENCE

NORMAL DIAM.	D		W
$\frac{3}{4}$	+.005 +.003		+.0005 +.0035
$\frac{7}{8}$	+.005 +.003		+.0005 +.0035
1	+.005 +.003		+.0005 +.0035
$1\frac{1}{8}$	+.005 +.003		+.0005 +.0035
$1\frac{1}{4}$	+.005 +.003		+.0005 +.0035
$1\frac{3}{8}$	+.005 +.003		+.0005 +.0035
$1\frac{1}{2}$	+.006 +.004		+.001 +.004
$1\frac{5}{8}$	+.006 +.004		+.001 +.004
$1\frac{3}{4}$	+.006 +.004		+.001 +.004
2	+.007 +.004		+.001 +.006
$2\frac{1}{4}$	+.007 +.004		+.001 +.006
$2\frac{1}{2}$	+.007 +.004		+.001 +.006
3	+.007 +.004		+.001 +.006

DIMENSIONS FOR 10-KEY SPLINED HOLES AND SHAFTS

(SLIDING FIT WHEN UNDER LOAD)

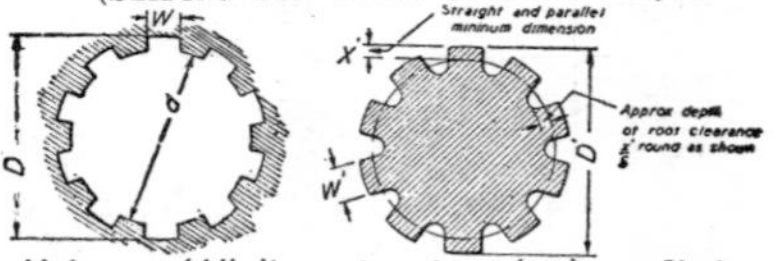

Straight and parallel minimum dimension

Approx depth of root clearance $\frac{X'}{4}$ round as shown

Hole (All dimensions in inches) Shaft

Nor. Dia	D	d	W	D′	X′	W′
$\frac{3}{4}$	.750	.608	.117	.746	.069	.1145
	.749	.607	.115	.745		.1135
$\frac{7}{8}$	.875	.709	.137	.871	.081	.1345
	.874	.708	.135	.870		.1335
1	1.000	.810	.156	.996	.093	.1535
	.999	.809	.154	.995		.1525
$1\frac{1}{8}$	1.125	.911	.176	1.121	.105	.1735
	1.124	.910	.174	1.120		.1725
$1\frac{1}{4}$	1.250	1.013	.195	1.246	.117	.1925
	1.249	1.012	.193	1.245		.1915
$1\frac{3}{8}$	1.375	1.114	.215	1.371	.129	.2125
	1.374	1.113	.213	1.370		.2115
$1\frac{1}{2}$	1.500	1.215	.234	1.495	.140	.231
	1.499	1.214	.232	1.494		.230
$1\frac{5}{8}$	1.625	1.316	.254	1.620	.152	.251
	1.624	1.315	.252	1.619		.250
$1\frac{3}{4}$	1.750	1.418	.273	1.745	.164	.270
	1.749	1.417	.271	1.744		.269
2	2.000	1.620	.312	1.994	.188	.3080
	1.998	1.618	.309	1.993		.3060
$2\frac{1}{4}$	2.250	1.823	.351	2.244	.211	.3470
	2.248	1.821	.348	2.243		.3450
$2\frac{1}{2}$	2.500	2.025	.390	2.494	.235	.3860
	2.498	2.023	.387	2.493		.3840
3	3.000	2.430	.468	2.994	.282	.4640
	2.998	2.428	.465	2.993		.4620

10-KEY SPLINED HOLES AND SHAFTS

(SLIDING FIT WHEN UNDER LOAD)

POSSIBLE FIT BETWEEN HOLE AND SHAFT

+ = CLEARANCE — = INTERFERENCE

NORMAL DIAM.	D		W
$\frac{3}{4}$	+.005 +.003		+.0005 +.0035
$\frac{7}{8}$	+.005 +.003		+.0005 +.0035
1	+.005 +.003		+.0005 +.0035
$1\frac{1}{8}$	+.005 +.003		+.0005 +.0035
$1\frac{1}{4}$	+.005 +.003		+.0005 +.0035
$1\frac{3}{8}$	+.005 +.003		+.0005 +.0035
$1\frac{1}{2}$	+.006 +.004		+.001 +.004
$1\frac{5}{8}$	+.006 +.004		+.001 +.004
$1\frac{3}{4}$	+.006 +.004		+.001 +.004
2	+.007 +.004		+.001 +.006
$2\frac{1}{4}$	+.007 +.004		+.001 +.006
$2\frac{1}{2}$	+.007 +.004		+.001 +.006
3	+.007 +.004		+.001 +.006

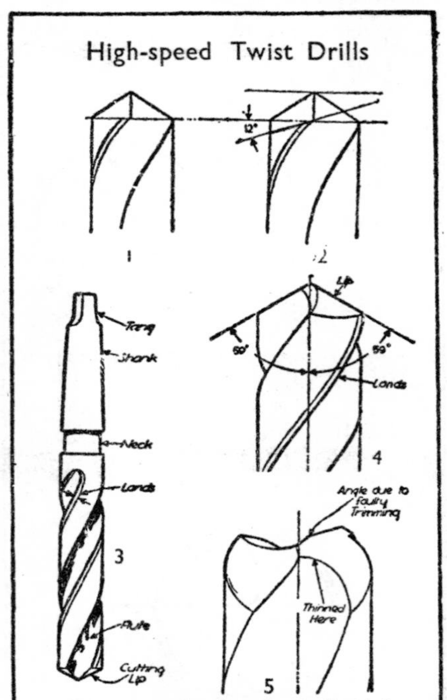

1.—Drill without lip clearance. 2.—Correct lip clearance. 3.—Taper-shank twist drill. 4.—Correct angles of cutting edges. 5.—Correct trimming after regrinding is an important operation.

HIGH-SPEED TWIST DRILLS (Contd.)

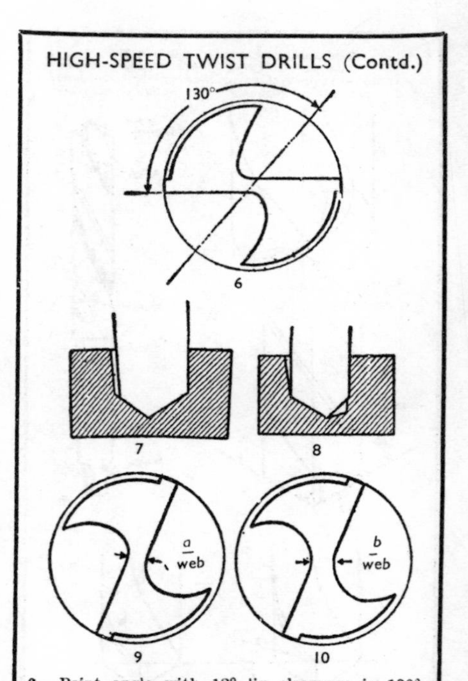

6.—Point angle with 12° lip clearance is 130°. 7.—Fault due to unequal length of lip. 8.—Fault due to unequal length of lip. 9 and 10.—The web should be thicker at the shank than at the web. (Compare these two diagrams.)

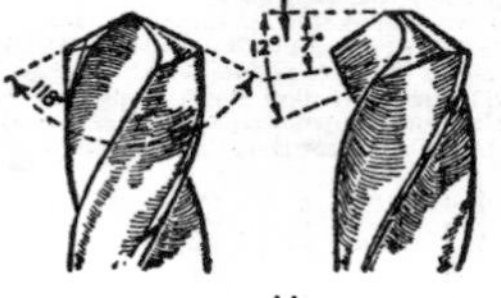

11

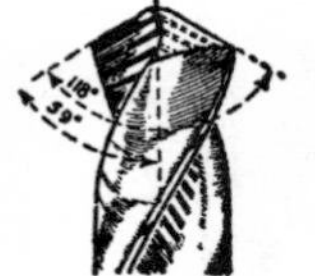

12

11.—Modified clearance angle for drilling stainless steels. 12.—Chip control grooves for stainless steel.

The above twelve drill shapes are typical of those manufactured by various firms, but variations of them inevitably occur over the years.

Cutting Angles of Tools for Various Materials

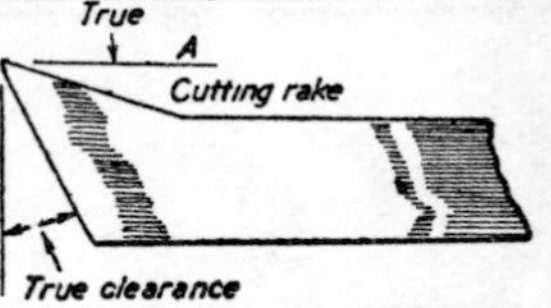

Material.	Degrees Angle B.	Degrees Angle A.
Electron	5–7	5–10
Duralumin	5–7	30–45
Aluminium	6–8	30–45
Copper	6–8	25–35
Brass	8–10	0–5
Silicon Aluminium	5–7	30–45
Mild Steel	5–7	15–35
Machinery Steel	5–7	10–20
Manganese Bronze	6–8	0–5
Monel Metal	8–10	15–25
Gunmetal	6–8	0–5
S.1 Steel	5–7	15–35
S.2 "	5–7	5–15
S.11 "	5–7	5–15
S.14 "	5–7	15–35
S.15 "	5–7	10–25
S.21 "	5–7	15–35
S.28 "	5–7	5–15
S.61 "	5–7	10–20
S.62 "	5–7	10–20
S.65 "	5–7	5–15
S.67 "	5–7	10–20
S.68 "	5–7	10 20
S.69 "	5–7	10–25
S.70 "	5–7	5–15
S.71 "	5–7	15–25
S.76 "	5–7	10–20
S.77 "	5–7	15–25
S.79 "	5–7	5–15
S.80 "	5–7	5–15
S.81 "	5–7	0–10

NOTE.—A and B are diagrammatic only, and are not intended to differentiate between side, front, or a combination of the two, both as regards rake angle and clearance angle. The most efficient cutting angles can only be determined when all the conditions are known, and it will be appreciated that the angles given are only approximate, as the best cutting rake is influenced by (a) feed ; (b) whether cut is continuous or not ; (c) type of operation, e.g., turning or planing ; (d) condition of machine and character of work. If chatter is experienced, a reduction of rake is often necessary ; (e) power of machine tool. It sometimes happens that keener tools than normal are used to obviate stalling of machines : (f) finish of work.

Standard Planing, Slotting and Boring Tools

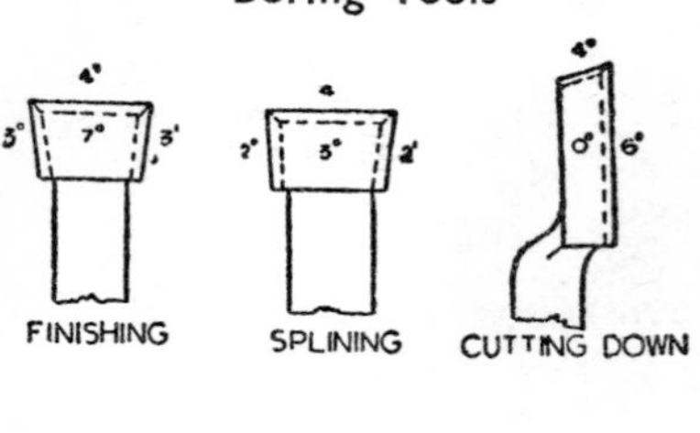

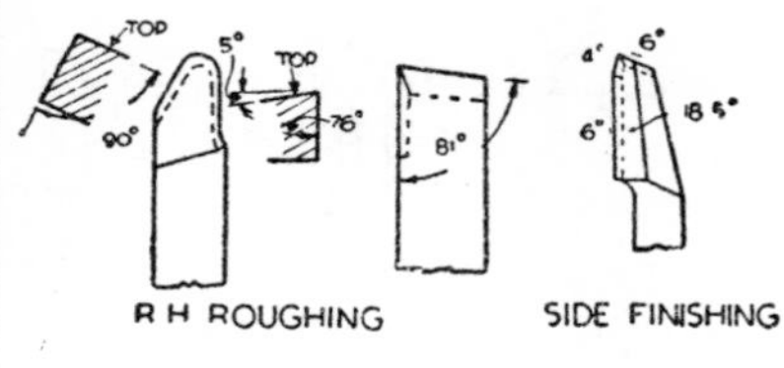

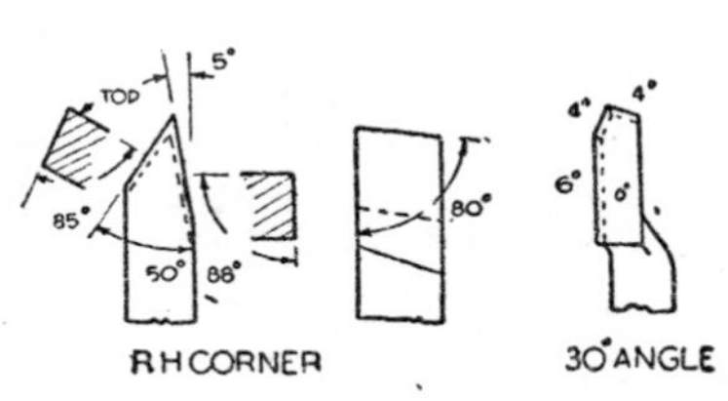

Standard Planing, Slotting and Boring Tools—(Continued)

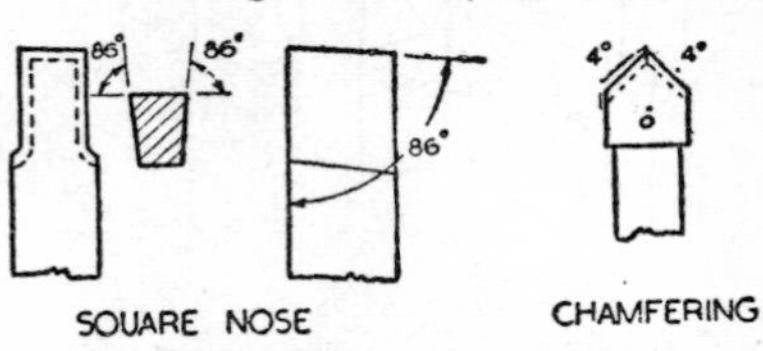

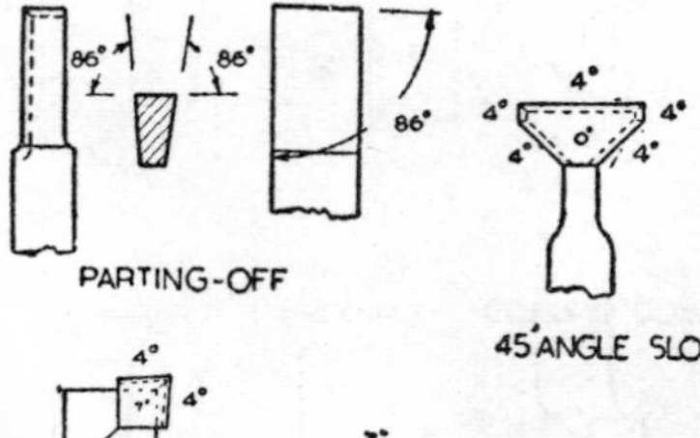

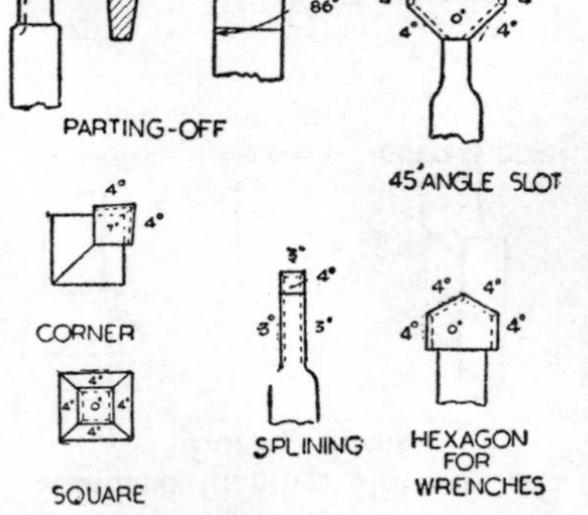

Standard Planing, Slotting and Boring Tools—(Continued)

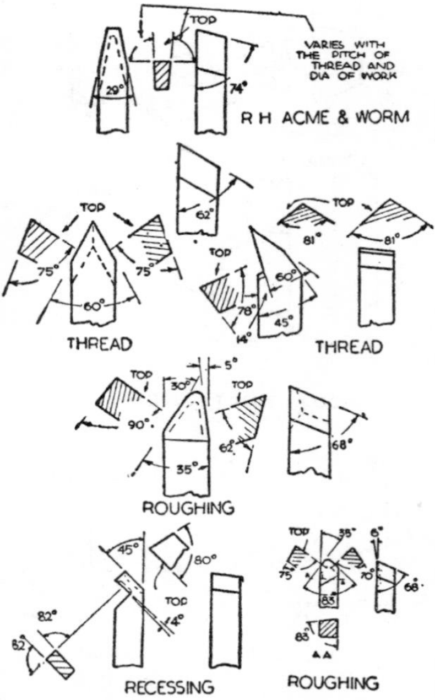
TOP
VARIES WITH THE PITCH OF THREAD AND DIA OF WORK
29°
74°
R H ACME & WORM
62°
TOP
TOP
81°
81°
75°
75°
60°
THREAD
TOP
60°
78°
45°
14°
THREAD
5°
TOP
30°
TOP
90°
62°
68°
35°
ROUGHING
45°
80°
TOP
4°
82°
82°
RECESSING
TOP
35°
6°
75
70°
68°
83
83
A A
ROUGHING

Standard Planing, Slotting and Boring Tools—(Continued)

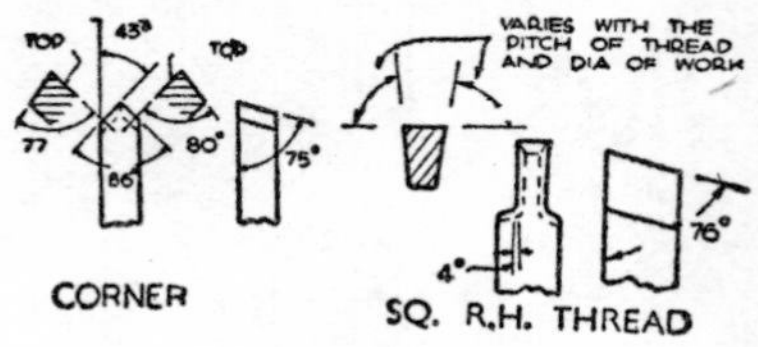

Standard Lathe Tools

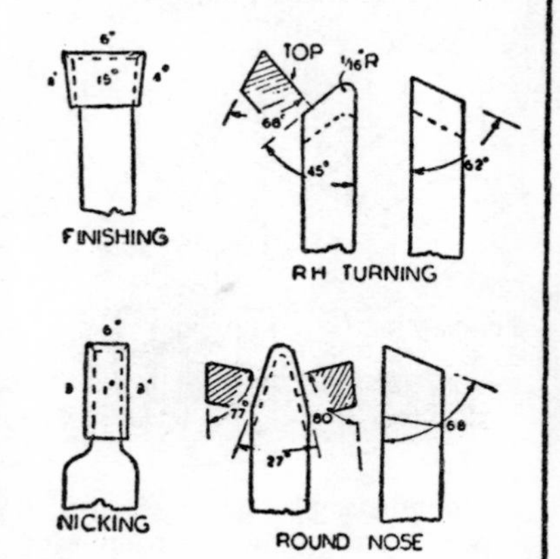

Standard Lathe Tools

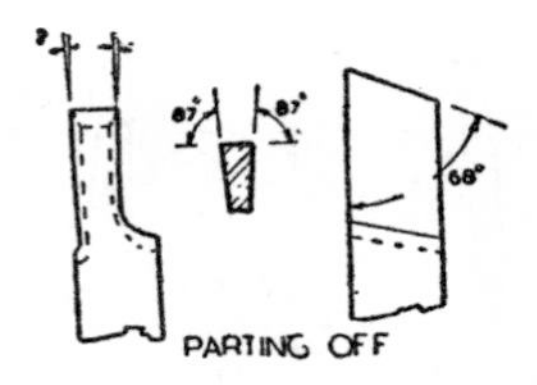

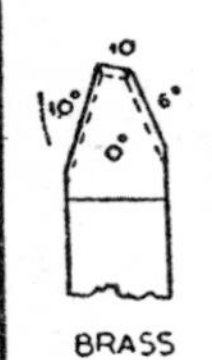

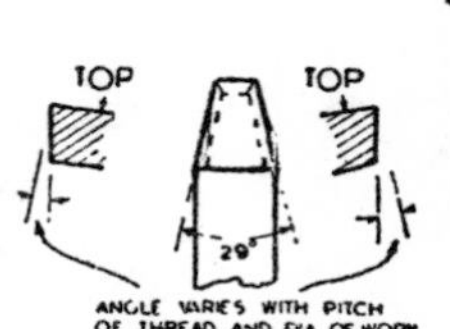

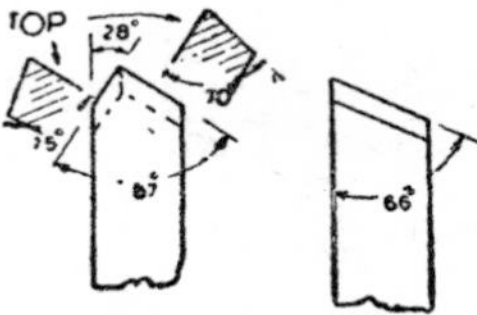

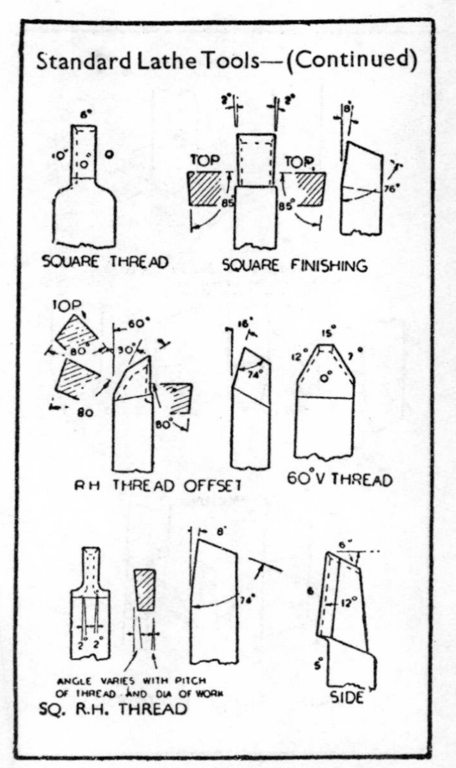
Standard Lathe Tools—(Continued)
SQUARE THREAD
SQUARE FINISHING
TOP
TOP
R H THREAD OFFSET
60°V THREAD
ANGLE VARIES WITH PITCH
OF THREAD AND DIA OF WORK
SQ. R.H. THREAD
SIDE

Standard Lathe Tools—(Continued)

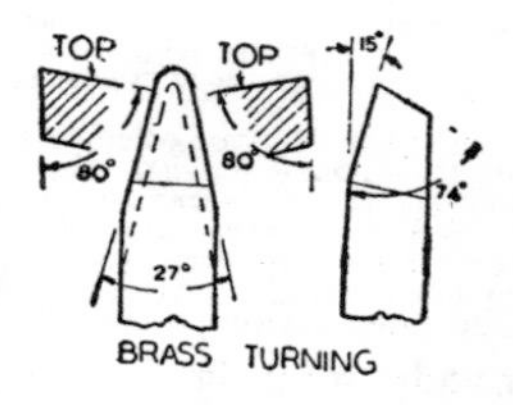

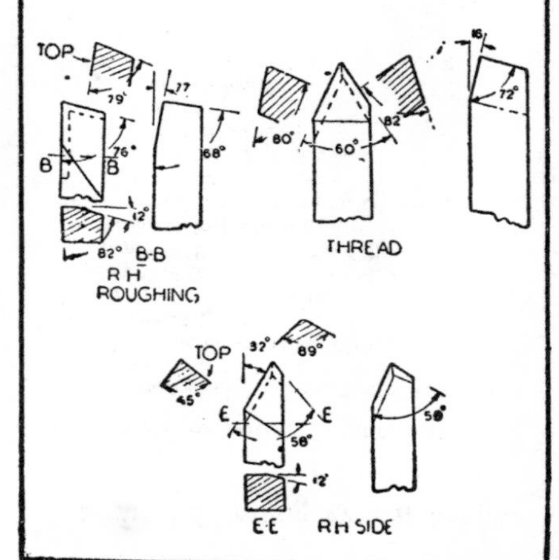

Standard Grinding Wheel Shapes

Key to Letter Dimensions

(*See illustrations, pages* 297, 298 *and* 299)

A—Flat Spot of Bevelled Wall.
D—Diameter (overall).
E—Centre or Back Thickness.
F—Depth of Recess (see Type 5).
G—Depth of Recess (see Type 7).
H—Arbor Hole Diameter.
J—Diameter of Flat or Small Diameter.
K—Diameter of Flat Inside.
M—Large Diameter of Bevel.
P—Diameter of Recess.
R—Radius.
T—Thickness (overall).
U—Width of Face.
V—Angle of Bevel.
W—Thickness of Wall.

Polishing Spindle Speeds

The speed at which brushing wheels are revolved is an important matter. It should be remembered that wire wheels must always run more slowly than bristle or fibre wheels; also that the larger the wheel and the coarser the wire, the more slowly it must revolve. The following speeds are recommended for average working conditions:

Small bristle and fibre wheels	2500 r.p.m.
Large bristle and fibre wheels .	2000 r.p.m.
Fine-wire scratch wheels . .	1600–1700 r.p.m.
Medium-wire scratch wheels .	1200–1500 r.p.m.
Coarse-wire scratch wheels .	700–1000 r.p.m.
Extra heavy-wire scratch wheels	500–600 r.p.m.

Standard Grinding Wheel Shapes
(Continued)

(For letter references see page 296)

Standard Grinding Wheel Shapes
(Continued)

(For letter references see page 296)

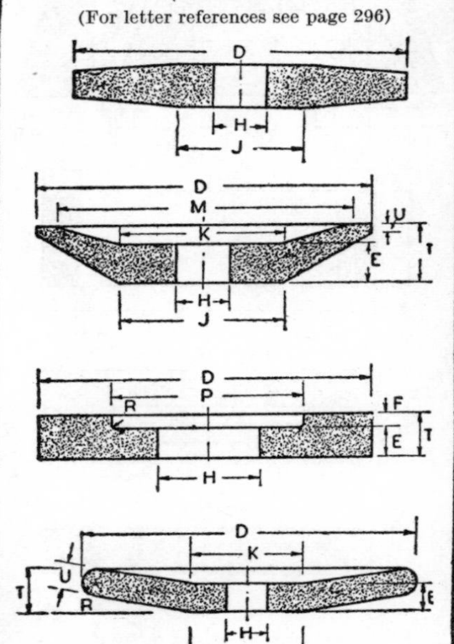

Standard Grinding Wheel Shapes

(For letter references see page 296)

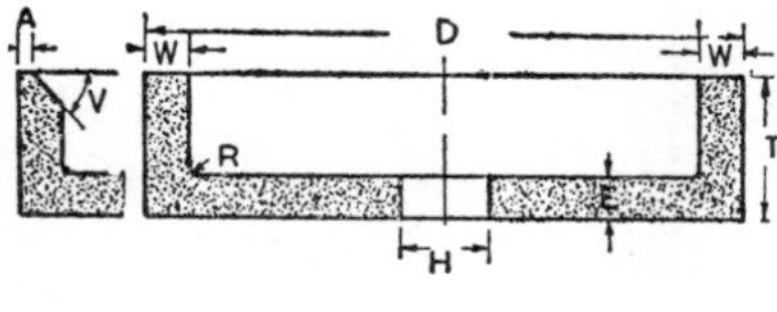

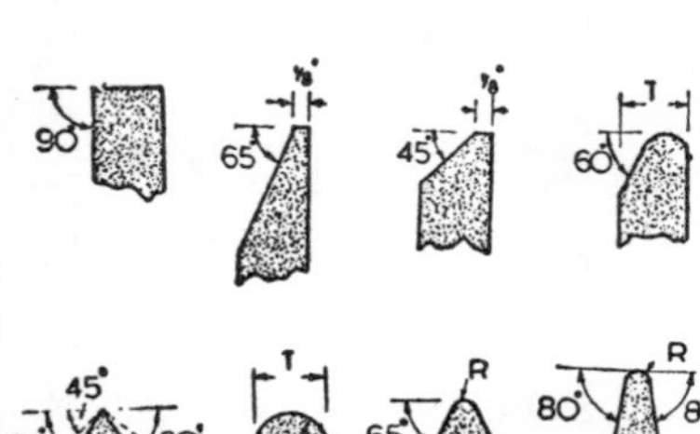

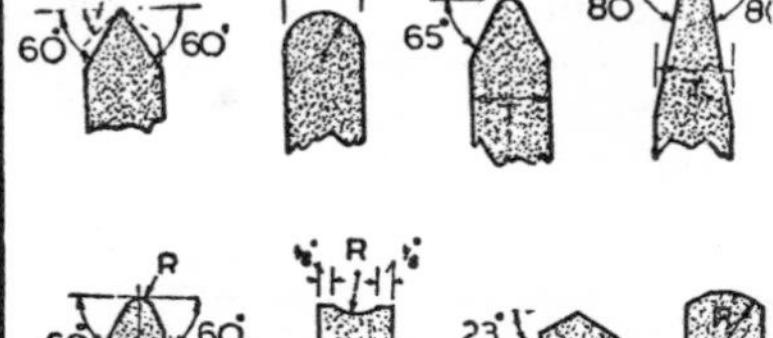

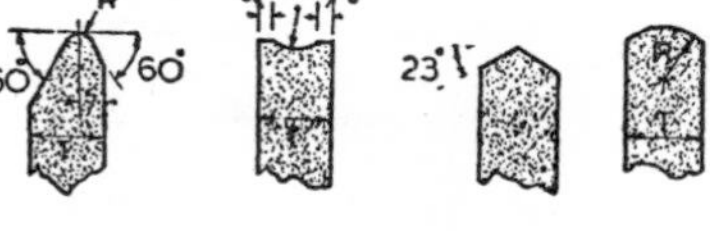

GRINDING WHEEL SPEEDS

High Speed Table

Diameter of Wheels	7,000 S.F.P.M.	8,000 S.F.P.M.	9,000 S.F.P.M.	10,000 S.F.P.M.
mm	r.p.m.	r.p.m.	r.p.m.	r.p.m.
25.4	26,738	30,558	34,377	38,197
50.8	13,369	15,279	17,189	19,098
76.2	8,913	10,186	11,459	12,732
101.6	6,684	7,639	8,594	9,549
127	5,347	6,111	6,875	7,639
152	4,456	5,093	5,729	6,366
178	3,820	4,365	4,911	5,457
203	3,342	3,820	4,297	4,775
254	2,674	3,056	3,439	3,820
305	2,228	2,546	2,865	3,183
356	1,910	2,183	2,455	2,728
406	1,671	1,910	2,148	2,387
457	1,485	1,698	1,910	2,122
508	1,337	1,528	1,719	1,910
559	1,215	1,389	1,563	1,736
609	1,114	1,273	1,432	1,591
660	1,028	1,175	1,322	1,469
711	955	1,091	1,228	1,364
762	891	1,018	1,146	1,273
813	835	955	1,074	1,194
863	786	899	1,101	1,123
914	743	849	955	1,061

GRINDING WHEEL SPEEDS

Low Speed Table

Diameter of Wheels	4,000 S.F.P.M.	5,000 S.F.P.M.	6,000 S.F.P.M.	6,500 S.F.P.M.
mm	r.p.m.	r.p.m.	r.p.m.	r.p.m.
25.4	15,279	19,098	22,918	24,828
50.8	7,639	9,549	11,459	12,414
76.2	5,093	6,366	7,639	8,276
101.6	3,820	4,775	5,729	6,207
127	3,056	3,820	4,584	4,966
152	2,546	3,183	3,820	4,138
178	2,183	2,728	3,274	3,547
203	1,910	2,387	2,865	3,103
254	1,528	1,910	2,292	2,483
305	1,273	1,591	1,910	2,069
356	1,091	1,364	1,637	1,773
406	955	1,194	1,432	1,552
457	849	1,061	1,273	1,379
508	764	955	1,146	1,241
559	694	868	1,042	1,128
609	637	796	955	1,034
660	588	734	881	955
711	546	682	818	887
762	509	637	764	828
813	477	597	716	776
863	449	562	674	730
914	424	530	637	690

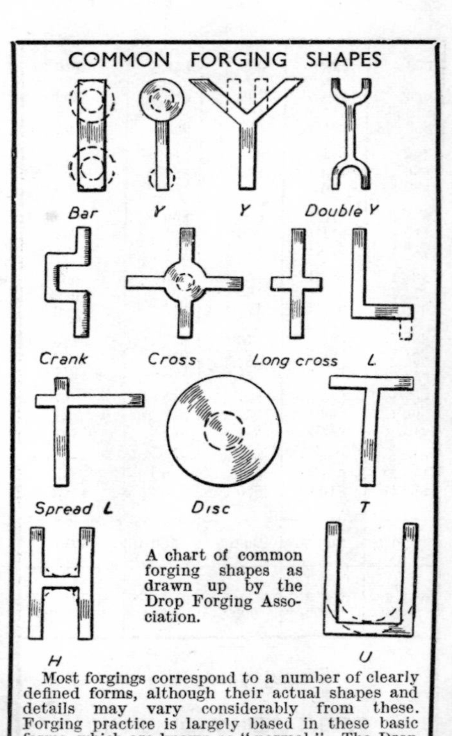

A chart of common forging shapes as drawn up by the Drop Forging Association.

Most forgings correspond to a number of clearly defined forms, although their actual shapes and details may vary considerably from these. Forging practice is largely based in these basic forms, which are known as " normal." The Drop Forging Association drew up a chart of common

COMMON FORGING SHAPES
(continued).

forging shapes, which is shown on page 302, and these they defined in principle in the following terms:

Bar. Those pieces of bar or cylinder shape, approximately of uniform weight throughout their length, are bar-shaped pieces. The cross section of the piece may be of any shape. The piece may be drawn or fullered in the centre portion, but it must not be necessary materially to spread the stock to forge the ends.

Y. This class includes banjo-shaped pieces, connecting rods, and three-pointed stars. It is meant to include those forgings for which the stock is large enough (or too large) for the small end, and the large end requires spreading.

Double Y. Two Y-shaped pieces joined together make a double Y. One or both ends may be closed, with or without holes.

Crank. The sketch shows a single-throw crank. Crank shapes are always designated with the number of throws as "one throw crank shape," "two throw crank shape," "three throw crank shape."

Cross. These pieces are in the shape of a cross or four-pointed star, with or without a hole in the centre.

Long Cross. A cross shape with two opposite arms much longer than the other two becomes a long cross shape.

L. These pieces are either right-angle pieces (or nearly so), or shaped like a crank arm.

Spread L. When an L-shape has projections at the angle, or when a cross-shape has two adjacent arms much longer than the other two, the piece becomes a spread L shape.

Disc. This includes discs with or without holes, such as gear blanks, hubs and rings.

T. This includes pieces of the general shape of the letter T.

H. This includes pieces of the general shape of the letter H.

U. All pieces of the general shape of the letter U, whether with a flat or round bottom, fall into this class.

PRESS SELECTION

Correct press type.—Here experience only counts.

Whilst a job might be done on any one of several types, in most cases there is a best type on which production will be highest.

Generalisations are always dangerous, but below is given a broad summary of common press types and suitable jobs for them.

Single acting open-fronted presses.—These are usually made inclinable, so that pressings, etc., may fall away by gravity. They are used as general purpose presses for the smaller work. Suitable for light blanking, not usually over 50 tons. For all manner of raising, forming, bending, etc., dies.

Single acting double-sided presses.—Used for all heavy blanking, raising, etc., above the open-fronted range.

Made in all sizes up to 80 inches or more between standards, and tonnages up to 500 tons and occasionally beyond this.

Smaller sizes usually have solid frames, and the larger sizes are built up, i.e., sides, bed and bridge of separate castings held together by tie rods through all the parts.

When fitted with air cushion have superseded the larger toggle double-acting presses, at least on comparatively shallow draws.

Double acting presses.—For all drawing work. Smaller sizes have blank holder cam operated. Heavier types blank holder operated by toggle levers; combination cut and draw work may be done in lighter types.

STANDARD TYPES OF BOTTOM SEAMS

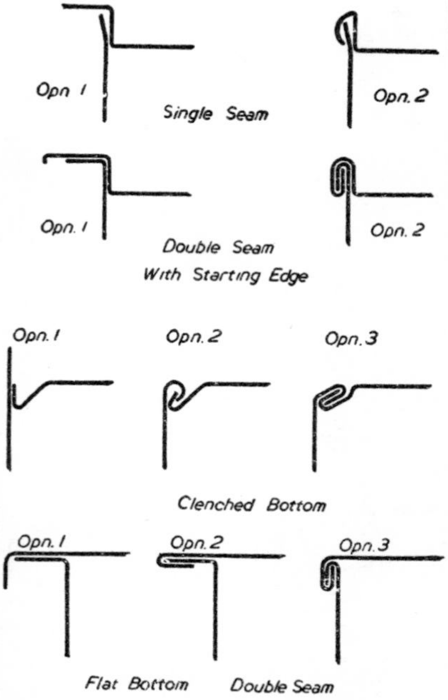

Opn. 1

Single Seam

Opn. 2

Opn. 1

Double Seam
With Starting Edge

Opn. 2

Opn. 1 Opn. 2 Opn. 3

Clenched Bottom

Opn. 1 Opn. 2 Opn. 3

Flat Bottom Double Seam

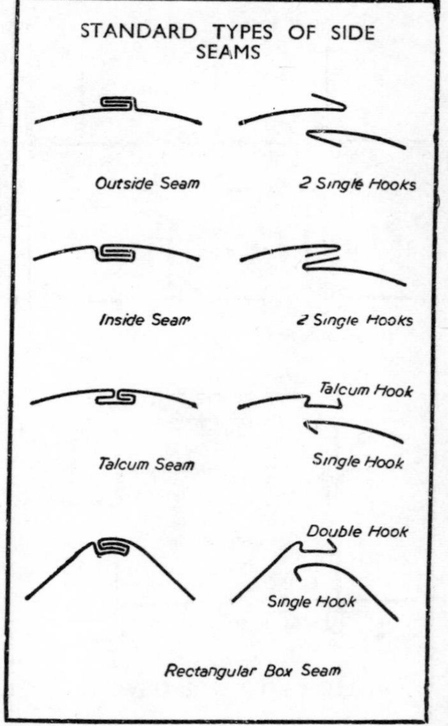
STANDARD TYPES OF SIDE SEAMS
Outside Seam
2 Single Hooks
Inside Seam
2 Single Hooks
Talcum Hook
Talcum Seam
Single Hook
Double Hook
Single Hook
Rectangular Box Seam

M

PRESS SELECTION

(Continued)

Horn presses or side wheel presses.—Used orginally for horning or grooving side seams, but when fitted with table (usually adjustable) suitable for light punching and raising and similar operations.

Table can be made to swing out so that press can be used for both purposes. Very useful for light work on large sheet metal articles requiring clearance on press front.

A further point in selecting the correct press type is that of deciding on the choice between a geared or ungeared press. Generalising again, direct-driven presses are used generally for work in which the pressure is required over a very small fraction of the working stroke, as in blanking, piercing, etc. If the work to be done is spread over an appreciable portion of the working stroke, then geared presses are to be preferred.

Most presses above 50 tons are better geared, and the gear ratio chosen to suit the conditions.

Correct press size.—Mathematics can aid experience here, but a very careful scrutiny is always required.

The sizes of press bed, ram face, etc., required are easy to establish. Then the "daylight" (distance bed to ram) and stroke have to be determined.

Bear very strongly in mind that the pressure given by the press declines considerably, away from the bottom of the stroke. The bottom of stroke tonnage usually given is far from an ideal press yard stick.

For blanking work only it is useful, but for drawing work the pressure required at the beginning of the draw and the depth of draw should be considered.

PRESS SELECTION
(Continued)

Then press capacities are better expressed in inch tons.

The bottom of the stroke pressure exerted by a press can be determined approximately from the formulæ :

Press capacity $= C\ d^2$ where d is the crankshaft diameter and C is a constant.

The value of the constant varies with a number of factors such as type of drive, stroke, press type, etc.

The following approximate rules might be given :

For double-crank, double-sided presses, stroke not exceeding crank diameter $C=4$.

Do. for very short strokes $C=4.5$ to 5.

For single crank open-fronted presses, short strokes $C=3.5$.

(Bear in mind that often in these presses, the press frame is the weaker member and care should be taken to see that it is adequate.)

For end wheel or horn presses $C=2.5$.

Drawing press sizes are easy to determine, as maker usually specifies maximum depth and diameter which may be drawn. Finally, the press maker should always be advised of the details of the maximum work the press has to perform. It is easier and better for both if this is done, instead of plain tonnage specifications.

ARITHMETICAL PROGRESSION

The term arithmetical progression refers to a series of numbers which increase or decrease by a constant difference. Thus: 3, 6, 9, 12, or 18, 15 12, 9, are arithmetical progressions, the constant difference being 3 in the first series and -3 in the second.

Let a = the first term of the series, z = the last term, n = the number of terms, d = the constant difference, S = the sum of all the terms.

$$a=z-d(n-1).\quad a=\frac{2S}{n}-z.\quad a=\frac{S}{n}-\frac{d}{2}(n-1).$$

$$z=a+d(n-1).$$

$$z=\frac{2S}{n}-a.\quad z=\frac{S}{n}+\frac{d}{2}(n-1).\quad n=\frac{z-a}{d}+1.$$

$$n=\frac{2S}{a+z}.$$

$$d=\frac{z-a}{n-1}.\quad d=\frac{(z+a)(z-a)}{2S-a-z}.\quad d=\frac{2(S-an)}{n(n-1)}$$

$$d=\frac{2(zn-S)}{n(n-1)}.$$

$$S=\frac{n(a+z)}{2}.\quad S=\frac{(a+z)(z+d-a)}{2d}.\quad S=n[a+\frac{d}{2}(n-1)]$$

$$S=n\left[z-\frac{d}{2}(n-1)\right].\quad a=\frac{d}{2}\pm\sqrt{(z+\frac{d}{2})^2-2dS}.$$

$$z=\frac{d}{2}\pm\sqrt{(a-\frac{d}{2})^2+2dS}.\quad n=\frac{1}{2}-\frac{a}{d}\pm\sqrt{(\frac{1}{2}-\frac{a}{d})^2+\frac{2S}{d}}$$

$$n=\frac{1}{2}+\frac{z}{d}\pm\sqrt{\left(\frac{1}{2}+\frac{z}{d}\right)^2-\frac{2S}{d}}$$

When the series is decreasing make the first term $=z$, and the last term $=a$. The Arithmetical Mean of two quantities, A and $B=\frac{A+B}{2}$

GEOMETRICAL PROGRESSION

A geometrical progression refers to a series of numbers which increase or decrease by a constant factor, or common ratio. For example, 3, 9, 27, 81, or 3, $-\frac{3}{4}$, $\frac{3}{16}$, $-\frac{3}{64}$, are Geometrical Progressions, the constant factor being 3 in the first series and $-\frac{1}{4}$ in the second. Let a = the first term, z = the last term, n = the number of terms, r = the constant factor, and S = the sum of the terms.

$a=\frac{z}{r^{n-1}}$. $a=S-r(S-z)$. $a=S\frac{r-1}{r^{n}-1}$. $z=ar^{n-1}$.

$z=S-\frac{S-a}{r}$. $z=S\left(\frac{r-1}{r^{n}-1}\right)r^{n-1}$. $r=\sqrt[n-1]{\frac{z}{a}}$.

$r=\frac{S-a}{S-z}$

$ar^{n}+S-rS-a=O$. $S=a\frac{(r^{n}-1)}{r-1}$. $S=a\frac{(1-r^{n})}{1-r}$.

$S=\frac{rz-a}{r-1}$.

$S=\frac{z\,(r^{n-1})}{(r-1)r^{n-1}}$ $n=1+\frac{\log z-\log a}{\log r}$.

$n=1+\frac{\log z-\log a}{\log(S+a)-\log(S-z)}$.

$n=\frac{\log[a+S(r-1)]-\log a}{\log r}$.

$n=1+\frac{\log z-\log[zr-S(r-1)]}{\log r}$.

$S=\frac{z^{n-1}\sqrt{z}-a^{n-1}\sqrt{a}}{\sqrt[n-1]{z}-\sqrt[n-1]{a}}$.

The Geometric Mean of two quantities, A and B $=\sqrt{AB}$

HARMONICAL PROGRESSION

Quantities are said to be in Harmonical Progression when, any three consecutive terms being taken, the first is to the third as the difference between the first and second is to the difference between the second and third. Thus, if x, y, z be the consecutive terms in a series, then, if $x : z :: x - y : y - z$, then x, y, z are in harmonical progression. If quantities are in harmonical progression, their reciprocals must also be in arithmetical progression. There is no simple method by which the sum of a harmonic series can be found.

The **Harmonic Mean of two quantities, A and B**

$$= \frac{2AB}{A+B}$$

EXTRACTING SQUARE ROOT

Mark off the number, the square root of which is to be found, into periods by marking a dot over every second figure commencing with the units place. Draw a vertical line to the left of the figure and a bracket on the right-hand side. Next, find the largest square in the left-hand period, and place this root behind the bracket. Next, the square of this root is subtracted from the first period, and the next period is brought down adjacent to the remainder and used as a dividend. Now, multiply the first root found by 2 and place this product to the left of the vertical line; then divide it into the left-hand figures of this new dividend, ignoring the right-hand figure. Attach the figure thus obtained to the root, and also to the divisor. Multiply this latest divisor by the figure of the root last obtained, finally subtracting the product from the dividend. Continue this operation until all periods have been brought down. If a decimal fraction is involved the periods for the decimal are marked off to the right of the decimal point.

The following examples will make the process clear. The first trial divisors are underlined in each case.

Example.

Find the square root of 1156:

```
  3 | 1156(34
    |  9
 64 |  256
    |  256
```

Find the square root of 54756:

```
  2 | 54756(234
    | 4
 43 | 147
    | 129
464 |  1856
    |  1856
```

EXTRACTING SQUARE ROOT—Contd.

In dealing with decimals, the periods relating to the decimal are marked off to the right as previously mentioned.

Find the square root of 39.476089 :

```
          .  . . .
     6 |39.476089(6.283
       |36
   122 | 347
       | 244
  1248 | 10360
       |  9984
 12563 |  37689
       |  37689
```

.....

EXTRACTING CUBE ROOT

Divide the number into periods by marking a dot over every third figure beginning at the units place as for square root. Find the greatest cube root in the figures in the left-hand period, and place this root on the left-hand side of the sum. Subtract the cube of this root from the left-hand period, and bringing down the next period to the remainder, using this as a dividend as before. Next, divide this dividend, omitting the last two figures, by three times the square of the root already found and attach this quotient to the root. Next, add together, thus obtaining the final divisor, firstly, the trial divisor with 00 attached, three times the product of the last root figure and the remainder of the root with one cipher attached, and the square of the last root figure. Multiply this final divisor by the figure of the root last obtained, and subtract the product from the dividend. The simplest method of extracting roots other than square or cube is by means of logarithms.

Example.

Find the cube root of 1728.

```
      .
   1 |1728(12
     |1
 364 | 728
     | 728
```

POWERS AND ROOTS OF π

n	$\frac{I}{n}$	n^2	n^3	$\sqrt{n}$	$\frac{I}{\sqrt{n}}$	$\sqrt[3]{n}$	$\frac{I}{\sqrt[3]{n}}$
$\pi = 3.142$	0.318	9.870	31.006	1.772	0.564	1.465	0.683
$2\pi = 6.283$	0.159	39.478	248.050	2.507	0.399	1.845	0.542
$\frac{\pi}{2} = 1.571$	0.637	2.467	3.878	1.253	0.798	1.162	0.860
$\frac{\pi}{3} = 1.047$	0.955	1.097	1.148	1.023	0.977	1.016	0.985
$\frac{4}{3}\pi = 4.189$	0.239	17.546	73.496	2.047	0.489	1.612	0.622
$\frac{\pi}{4} = 0.785$	1.274	0.617	0.484	0.886	1.128	0.923	1.084
$\frac{\pi}{6} = 0.524$	1.910	0.274	0.144	0.724	1.382	0.806	1.241
$\pi^2 = 9.870$	0.101	97.409	961.390	3.142	0.318	2.145	0.466
$\pi^3 = 31.066$	0.032	961.390	29.809.910	5.568	1.796	3.142	0.318
$\frac{\pi}{32} = 0.098$	10.186	0.0095	0.001	0.313	3.192	0.461	2.168

MENSURATION

A and a = area ; b = base ; C and c = circumference ; D and d = diameter ; h = height ; n° = number of degrees ; p = perpendicular ; R and r = radius ; S = span or chord ; V and v = volume and versed sine.

Square : $a = side^2$; side = $\sqrt{a}$; diagonal = side $\times \sqrt{2}$.

Rectangle or parallelogram : a = bp.

Trapezium (two sides parallel) : a = mean length parallel sides × distance between them.

Triangle : $a = \frac{1}{2}bp$.

Irregular figure : a = weight of template ÷ weight of square inch of similar material.

Side of square multiplied by 1.4142 equals diameter of its circumscribing circle.

A side multiplied by 4.443 equals circumference of its circumscribing circle.

A side multiplied by 1.128 equals diameter of a circle of equal area.

Area of an inscribed circle multiplied by 1.273 equals area of the square enclosing it.

MENSURATION *(continued)*

To find side of an equal square :

Multiply diameter by 0.8862 ; or divide diameter by 1.1284 ; or multiply circumference by 0.2821 ; or divide circumference by 3.545.

To find area of a circle :

Multiply circumference by $\frac{1}{4}$ of the diameter ; or multiply the square of diameter by 0.7854 ; or multiply the square of circumference by 0.07958 ; or multiply the square of $\frac{1}{2}$ diameter by 3.1416.

To find the surface of a sphere or globe:

Multiply the diameter by the circumference ; or multiply the square of diameter by 3.1416 ; or multiply 4 times the square of radius by 3.1416.

Cylinder.

To find the area of surface :

Multiply the diameter by $3\frac{1}{7} \times$ length.

Capacity $= 3\frac{1}{7} \times \text{radius}^2 \times \text{height}$.

Values and Powers of :

$\pi = 3.1415926536$, or 3.1416, or $\frac{22}{7}$ or $3\frac{1}{7}$;

$\pi^2 = 9.86965$; $\sqrt{\pi} = 1.772453$;

$\frac{1}{\pi} = 0.31831$; $\frac{\pi}{2} = 1.570796$;

$\frac{\pi}{3} = 1.047197$.

MENSURATION (*continued*)

Circle : $a = \pi r^2 = d^2\frac{\pi}{4} = 0.7854d^2 = 0.5$ cr. ; $c = 2\pi r = d\pi = 3.1416d = 3.54\sqrt{a} =$ (approximately) $\frac{22}{7}d$. Side of equal square $= 0.8862d$; side of inscribed square $= 0.7071d$; $d = .3183c$. A circle has the maximum area for a given perimeter.

Annulus of circle : $a = (D+d)(D-d)\frac{\pi}{4} = (D^2 - d^2)\frac{\pi}{4}$.

Chord of Circle $= \sin A \times D$, where A equals $\frac{1}{2}$ included angle at centre.

Segment of Circle : a = area of sector — area of triangle $= \frac{4v}{3}\sqrt{(0.625V)^2 + (\frac{1}{2}S)^2}$.

Length of Arc $= \pi\frac{n^\circ r}{180} = 0.0174533n^\circ r$; length of arc $= \frac{1}{3}\left(8\sqrt{\frac{S^2}{4} + v^2} - s\right)$; approximate length of arc $= \frac{1}{3}$ (8 times chord of $\frac{1}{2}$ arc — chord of whole arc).

$d = \frac{(\frac{1}{2}\text{ chord})}{v} + v$; radius of curve $= \frac{S^2}{SV} + \frac{V}{2}$.

Sector of circle : $a = 0.5r \times$ length arc ; $= n^\circ \times$ area circle $\div 360$.

MENSURATION (*continued*)

Ellipse : $a = \frac{\pi}{4}Dd = \pi Rr$; c (approx.) $= \sqrt{\frac{D^2 + d^2}{2}} \times \pi$; c (approx.) $= \pi\frac{Da}{2}$.

Parabola : $a = \frac{2}{3}bh$.

Cone or pyramid : surface $= \frac{\text{circ. of base} \times \text{slant length}}{2} + \text{base}$; contents = area of base × $\frac{1}{3}$vertical height.

Frustum of cone ; surface =

$(C + c) \times \frac{1}{2}$ slant height + ends ;

contents $= 0.2618h\,(D^2 + d^2 + Dd)$; $= \frac{1}{3}h\,(A + a + \sqrt{A \times a})$.

Wedge : volume $= \frac{1}{6}$ (length of edge + 2 length of back) bh.

Prism : volume = area base × height.

Sphere : surface $= d^2\pi = 4\pi r^2$; contents $= d^3\frac{\pi}{6} = \frac{4}{3}\pi r^3$.

Segment of sphere : r = rad. of base ; contents $= \frac{\pi}{6}h\ (3r^2 + h^2)$; r = rad. of sphere ; contents $= \frac{\pi}{3}h^2\ (3r - h)$.

Spherical zone : volume $= \frac{\pi}{2}h\ (\frac{1}{3}h^2 + R^2 + r^2)$; surface of convex part of segment or zone of sphere $= \pi d$ (of sph.) $h = 2\pi rh$.

MENSURATION (*continued*)

Mid sph. zone : volume $=(r+\frac{2}{3}h^2)\frac{\pi}{4}$.

Spheroid : volume $=$ revolving axis$^2 \times$ fixed axis $\times \frac{\pi}{6}$.

Cube or rectangular solid : volume $=$ length $\times$ breadth $\times$ thickness.

Prismoidal formula, volume $=$

$$\frac{\text{end areas} + 4 \times \text{mid area}}{6} \times \text{height}.$$

Solid of revolution : volume $=$ a of generating plane $\times$ c described by centroid of this plane during revolution. Areas of similar plane figures are as the squares of like sides. Volumes of similar solids are as the cubes of like sides. Rules relative to the circle, square, cylinder, etc. :

To find circumference of a circle :
Multiply diameter by 3.1416 ; or divide diameter by 0.3183.

To find diameter of a circle :
Multiply circumference by 0.3183 ; or divide circumference by 3.1416.

To find radius of a circle :
Multiply circumference by 0.15915 ; or divide circumference by 6.28318.

To find side of an inscribed square :
Multiply diameter by 0.7071 ; or multiply circumference by 0.2251 ; or divide circumference by 4.4428.

PLANE FIGURES

SQUARE
FOUR EQUAL SIDES
& FOUR RT ANGLES

RECTANGLE
OPPOSITE SIDES EQUAL
& FOUR RT ANGLES

RHOMBUS
OBLIQUE ANGLED FIGURE.
ALL SIDES & OPP ANGLES EQUAL

RHOMBOID
A RHOMBUS - BUT WITH
ONLY OPPOSITE SIDES
EQUAL

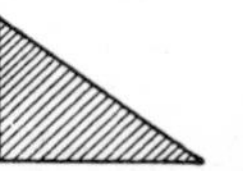

RIGHT-ANGLED TRIANGLE
HAVING ONE ANGLE
OF 90°

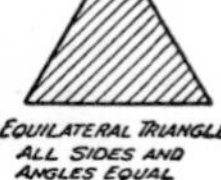

EQUILATERAL TRIANGLE
ALL SIDES AND
ANGLES EQUAL

ISOSCELES TRIANGLE
TWO SIDES EQUAL

SCALENE TRIANGLES
(HAVING NO TWO SIDES
EQUAL)

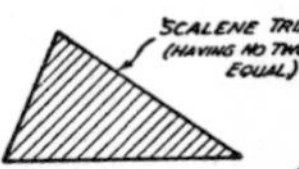

ACUTE-ANGLED TRIANGLE
ALL ANGLES LESS
THAN RT ANGLE

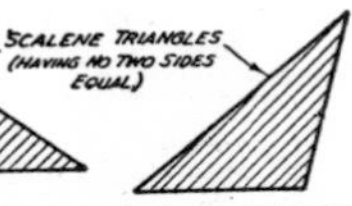

OBTUSE-ANGLED TRIANGLE
ONE ANGLE GREATER
THAN RT ANGLE

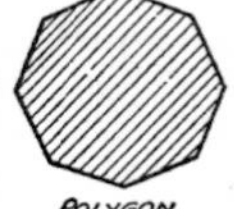

POLYGON
MULTI-SIDED FIGURE
WITH ALL SIDES EQUAL

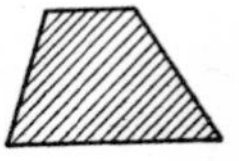

TRAPEZIUM
FOUR UNEQUAL SIDES.
TWO OF WHICH ARE
PARALLEL

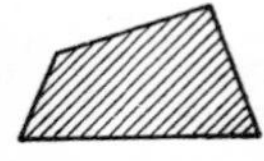

TRAPEZOID
ALL SIDES AND ANGLES
UNEQUAL

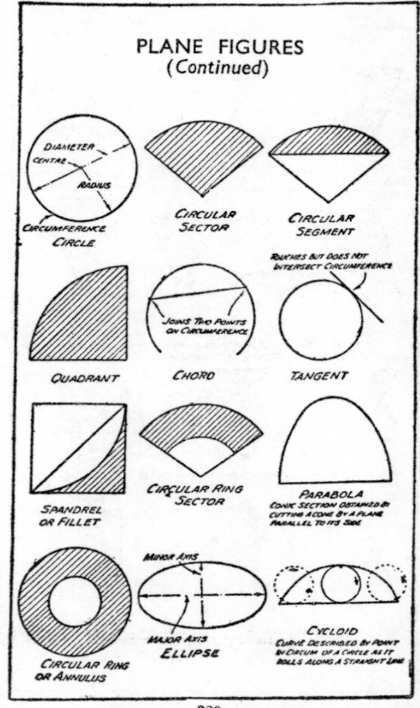
PLANE FIGURES
(Continued)
DIAMETER
CENTRE
RADIUS
CIRCUMFERENCE
CIRCLE
CIRCULAR SECTOR
CIRCULAR SEGMENT
QUADRANT
JOINS TWO POINTS ON CIRCUMFERENCE
CHORD
TOUCHES BUT DOES NOT INTERSECT CIRCUMFERENCE
TANGENT
SPANDREL OR FILLET
CIRCULAR RING SECTOR
PARABOLA
CONIC SECTION OBTAINED BY CUTTING A CONE BY A PLANE PARALLEL TO ITS SIDE
CIRCULAR RING OR ANNULUS
MINOR AXIS
MAJOR AXIS
ELLIPSE
CYCLOID
CURVE DESCRIBED BY POINT IN CIRCUM OF A CIRCLE AS IT ROLLS ALONG A STRAIGHT LINE

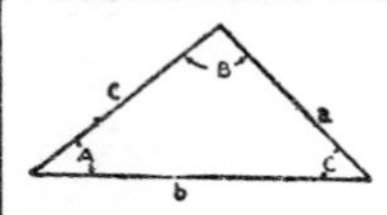

Fig. 1. Diagram for Table A.

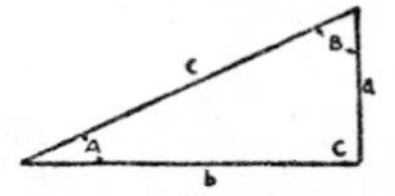

Fig. 2. Diagram for Table B.

TABLE A

See Fig. 1

Parts Given.	Parts to be Found.	Formulæ.
$a\ b\ c$	A	$\cos A = \frac{b^2+c^2-a^2}{2bc}$
$a\ b$ A	B	$\sin B = \frac{b \times \sin A}{a}$
$a\ b$ A	C	$C = 180° - (A+B)$
a A B	b	$b = \frac{a \times \sin B}{\sin A}$
a A B	c	$c = \frac{a \sin C}{\sin A}$ $= \frac{a \sin (180 - A - B)}{\sin A}$
$a\ b$ C	B	$B = 180° - (A+C)$

TABLE B

See Fig. 2.

Parts Given.	Parts to be Found.				
	A	B	a	b	c
a & c	$\sin A=\frac{a}{c}$	$\cos B=\frac{a}{c}$		$b=\sqrt{c^2-a^2}$	
a & b	$\tan A=\frac{a}{b}$	$\cot B=\frac{a}{b}$			$c=\sqrt{a^2+b^2}$
c & b	$\cos A=\frac{b}{c}$	$\sin B=\frac{b}{c}$	$a=\sqrt{c^2-b^2}$		
A & a		$B=90°-A$		$b=a\times\cot A$	$c=\frac{a}{\sin A}$
A & b		$B=90°-A$	$a=b\times\tan A$		$c=\frac{b}{\cos A}$
A & c		$B=90°-A$	$a=c\times\sin A$	$b=c\times\cos A$	

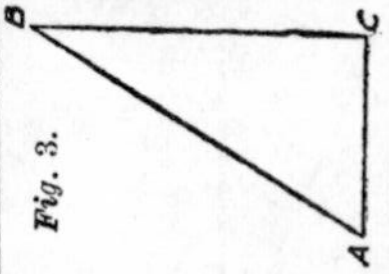

B C A

Fig. 3.

Fig. 3.—In any right-angled triangle:

$\tan A=\frac{BC}{AC}$, $\sin A=\frac{BC}{AB}$

$\cos A=\frac{AC}{AB}$, $\cot A=\frac{AC}{BC}$

$\sec A=\frac{AB}{AC}$, $\operatorname{cosec} A=\frac{AB}{BC}$

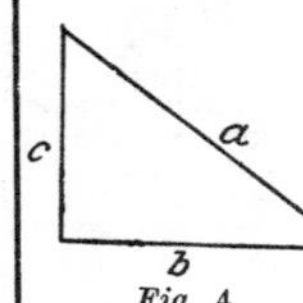

Fig. 4.

Fig. 4.—In any right-angled triangle :

$$a^2 = c^2 + b^2$$
$$c = \sqrt{a^2 - b^2}$$
$$b = \sqrt{a^2 - c^2}$$
$$a = \sqrt{b^2 + c^2}$$

Fig. 5.— $c + d : a + b :: b - a : d - c.$

$$d = \frac{c + d}{2} + \frac{d - c}{2}$$
$$x = \sqrt{b^2 - d^2}$$

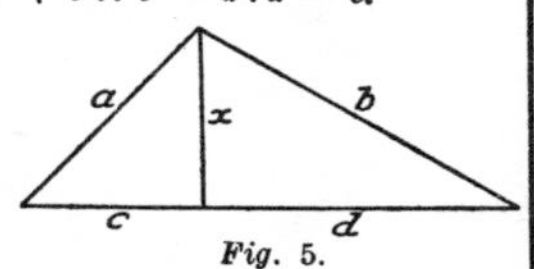

Fig. 5.

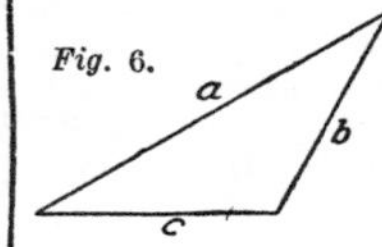

Fig. 6.

In Fig. 6, where the lengths of three sides only are known :
area =

$$\sqrt{s(s - a)\ (s - b)\ (s - c)}$$

where $s = \dfrac{a + b + c}{2}$

Fig. 7.—In this diagram :

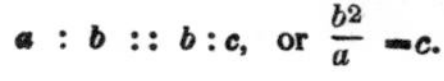

$a : b :: b : c$, **or** $\dfrac{b^2}{a} = c.$

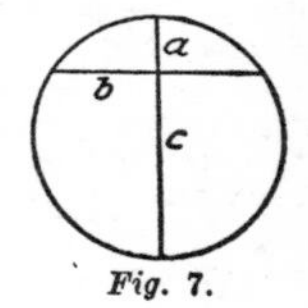

Fig. 7.

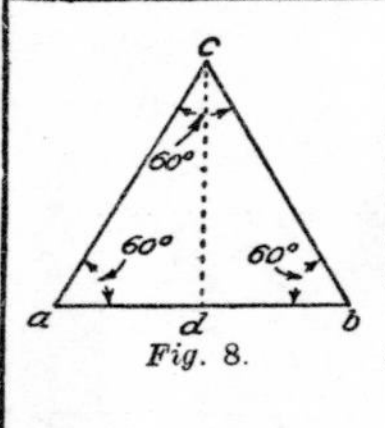

Fig. 8.

Fig. 8.—In an equilateral triangle $ab = 1$, then $cd = \sqrt{0.75} = 0.866$, and $ad = 0.5$; $ab = 2$, then $cd = \sqrt{3.0} = 1.732$, and $ad = 1$; $cd = 1$, then $ac = 1.155$ and $ad = 0.577$; $cd = 0.5$, then $ac = 0.577$ and $ad = 0.288$.

Fig. 9.—In a right-angled triangle with two equal acute angles, $bc = ac$. $bc = 1$, then $ab = \sqrt{2} = 1.414$; $ab = 1$, then $bc = \sqrt{0.5} = 0.707$.

Fig. 9.

Fig. 10.

Fig. 10 shows that parallelograms on the same base and between the same parallels are equal; thus ABCD = ADEF.

Fig. 11 demonstrates that triangles on the same base and between the same parallels are equal in area; thus ABC = ADC.

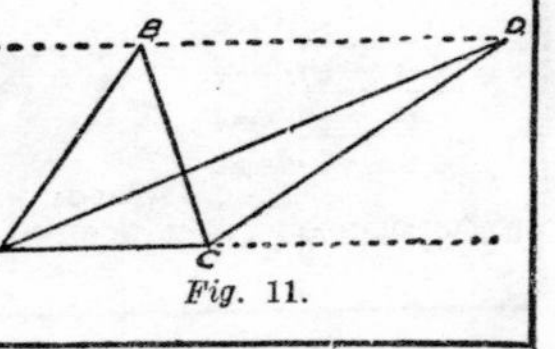

Fig. 11.

TRIGONOMETRICAL EQUIVALENTS

$\text{Sine} = \sqrt{1 - \text{Cos}^2}$.

$\text{Sine} = 1 \div \text{Cosec}$.

$\text{Sine} = \text{Cos} \div \text{Cotan}$.

$\text{Sine} = \text{Tan} \div \text{Sec}$.

$\text{Sin } 0^\circ = 0$.

$\text{Sin } 60^\circ = \text{Cos } 30^\circ$.

$\text{Cosine} = \sqrt{1 - \text{Sin}^2}$.

$\text{Cosine} = 1 \div \text{Sec}$.

$\text{Cosine} = \text{Sin} \times \text{Cotan}$.

$\text{Cosine} = \text{Sin} \div \text{Tan}$.

$\text{Cos } 0^\circ = 1$.

$\text{Secant} = 1 \div \text{Cos}$.

$\text{Secant} = \text{Tan} \div \text{Sin}$.

ANGLES BETWEEN 90° AND 180°

(when θ is between 90° and 180°)

$\text{Cot } \theta = -\text{Cot } (180 - \theta)$

$\text{Tan } \theta = -\text{Tan } (180 - \theta)$

$\text{Sin } \theta = \text{Sin } (180 - \theta)$

$\text{Sec } \theta = -\text{Sec } (180 - \theta)$

$\text{Cosec } \theta = \text{Cosec } (180 - \theta)$

$\text{Cos } \theta = -\text{Cos } (180 - \theta)$

COMPLEMENTARY ANGLES

$\text{Cos } \theta = \text{Sin } (90 - \theta)$

$\text{Sin } \theta = \text{Cos } (90 - \theta)$

$\text{Tan } \theta = \text{Cot } (90 - \theta)$

$\text{Sec } \theta = \text{Cosec } (90 - \theta)$

$\text{Cot } \theta = \text{Tan } (90 - \theta)$

$\text{Cosec } \theta = \text{Sec } (90 - \theta)$

$\text{Cosecant} = 1 \div \text{Sin}$.

$\text{Tangent} = 1 \div \text{Cotan}$.

$\text{Tangent} = \text{Sin} \div \text{Cos}$.

$\text{Cotangent} = 1 \div \text{Tan}$.

$\text{Cotangent} = \text{Cos} \div \text{Sin}$.

$\text{Versine} = 1 - \text{Cos}$.

$\text{Coversine} = 1 - \text{Sin}$.

$1 = \text{Tan} \times \text{Cotan}$.

$1 = \text{Sin}^2 + \text{Cos}^2$.

$\text{Secant}^2 = 1 + \text{Tan}^2$.

$\text{Cosec}^2 = 1 + \text{Cot}^2$.

EXAMPLES IN STEREOMETRY

(C = Solid Content)

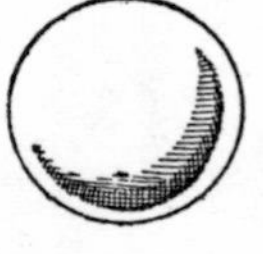

Sphere

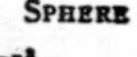

$$C = \frac{4\pi r^3}{3} = 4{\cdot}189r^3.$$

$$= \frac{\pi d^3}{6} = 0{\cdot}524d^3.$$

Torus

$$C = 2\pi^2 Rr^2$$

$$= 19{\cdot}74Rr^2.$$

$$= 2{\cdot}468Dd^2$$

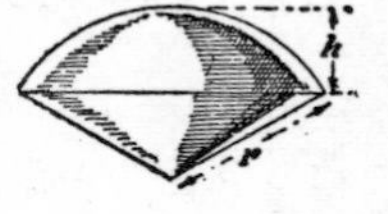

Sphere Sector

$$C = \frac{2}{3}\pi r^2 h = 2{\cdot}0944r^2h$$

$$= \frac{2}{3}\pi r^2(r \mp \sqrt{r^2 - \tfrac{1}{4}c^2}).$$

Circle Zone

$$C = \pi h^2(r - \frac{1}{3}h)$$

$$= \pi h^2\left(\frac{c^2 + 4h^2}{8h} - \frac{1}{3}h\right).$$

STEREOMETRY—(*Continued*)

CONE

$$C = \frac{\pi r^2 h}{3} = 1{\cdot}047r^2h$$
$$= 0{\cdot}2618d^2h.$$

CONIC FRUSTUM

$$C = \frac{1}{12}\pi h(D^2 + Dd + d^2)$$
$$= \frac{1}{3}\pi h(R^2 + Rr + r^2)$$
$$R = \frac{D}{2}.\ r = \frac{d}{2}.$$

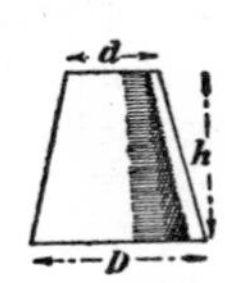

CYLINDER

$$C = \pi r^2 h = 0{\cdot}785d^2h$$
$$= \frac{p^2h}{4\pi} = 0{\cdot}0796p^2h.$$

p = length of periphery.

PROLATE SPHEROID

$$C = \frac{4}{3}\pi Rr^2 = 4{\cdot}189Rr^2$$
$$= \frac{1}{6}\pi Dd^2$$
$$= 0{\cdot}5236Dd^2.$$

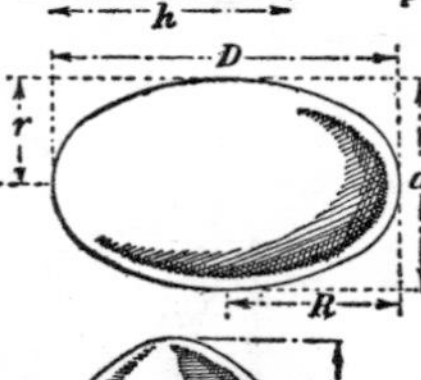

OBLATE SPHEROID

$$C = \frac{4}{3}\pi R^2r = 4{\cdot}189R^2$$
$$= \frac{1}{6}\pi D^2d$$
$$= 0{\cdot}5236D^2d.$$

PARABOLOID

$$C = \tfrac{1}{2}\pi r^2h = 1{\cdot}5708r^2h.$$

STEREOMETRY—(*Continued*)

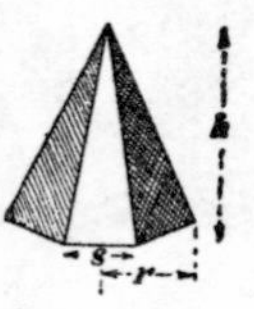

Pyramid

$C = \frac{1}{3}ah$

$= \frac{nsh}{6}\sqrt{r^2 - \frac{s^2}{4}}$

a = area of base.
n = number of sides.

Pyramidic Frustum

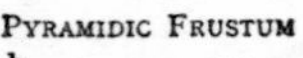

$C = \frac{h}{3}(A + a + \sqrt{Aa})$.

A and a = areas of the two ends.

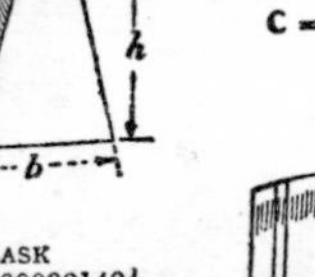

Wedge Frustum

$C = \frac{hs}{2}(a + b)$.

Cask

$C = 0{\cdot}000032149l\,(39D^2 + 26Dd + 25d^2)$ gallons; l, D, and d being in inches.

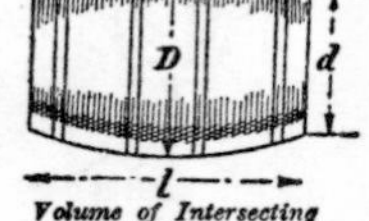

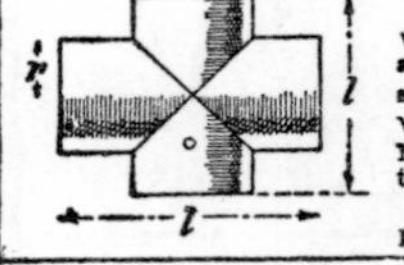

Volume of Intersecting Cylinders

For the general case in which the angle between the axes is ϕ, the volume of the solid is

$V = \pi r^2(l + l') - 5{\cdot}333r^3 \operatorname{cosec}\phi$.

Neither l nor l' must be less than $2r \operatorname{cosec}\phi$.

In the special case of perpendicular axes, $\operatorname{cosec}\phi = 1$.

LONGIMETRY

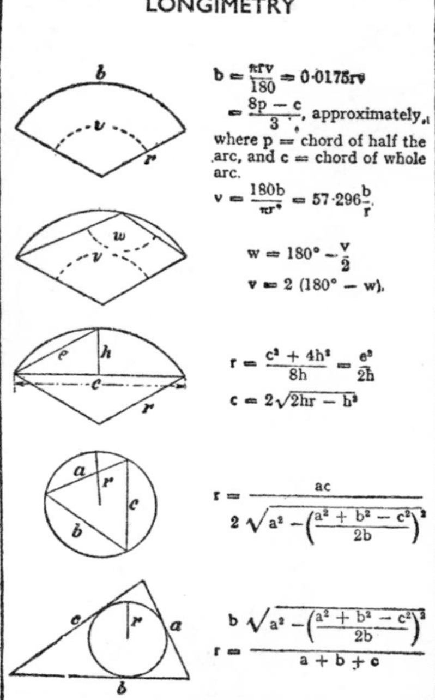

$$b = \frac{\pi r v}{180} = 0{\cdot}0175rv$$

$$= \frac{8p - c}{3}, \text{ approximately,}$$

where p = chord of half the arc, and c = chord of whole arc.

$$v = \frac{180b}{\pi r} = 57{\cdot}296\frac{b}{r}.$$

$$w = 180° - \frac{v}{2}$$

$$v = 2\,(180° - w).$$

$$r = \frac{c^2 + 4h^2}{8h} = \frac{e^2}{2h}$$

$$c = 2\sqrt{2hr - h^2}$$

$$r = \frac{ac}{2\sqrt{a^2 - \left(\frac{a^2 + b^2 - c^2}{2b}\right)^2}}$$

$$r = \frac{b\sqrt{a^2 - \left(\frac{a^2 + b^2 - c^2}{2b}\right)^2}}{a + b + c}$$

LONGIMETRY—(*Continued*)

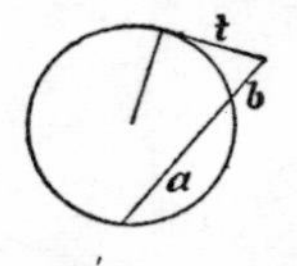

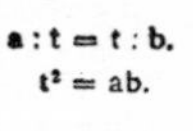

$$a : t = t : b.$$
$$t^2 = ab.$$

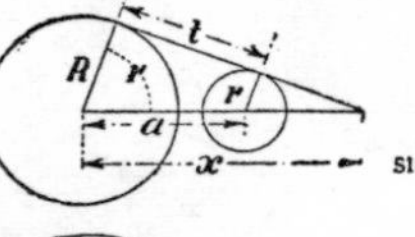

$$x = \frac{aR}{R - r}.$$
$$n = \sqrt{t^2 + (R - r)^2}.$$
$$t = \sqrt{a^2 - (R - r)^2}.$$
$$\sin v = \frac{t}{a}.$$

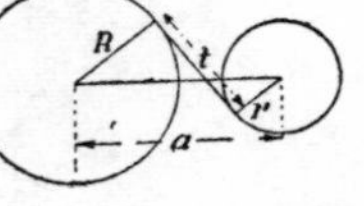

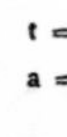

$$t = \sqrt{a^2 - (R + r)^2},$$
$$a = \sqrt{t^2 + (R + r)^2},$$

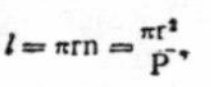

$$l = \pi rn = \frac{\pi r^2}{P},$$

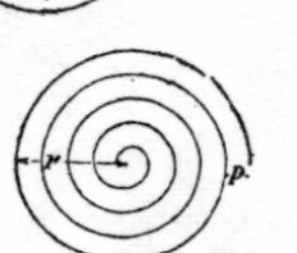

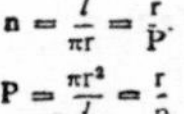

$$n = \frac{l}{\pi r} = \frac{r}{P}.$$
$$P = \frac{\pi r^2}{l} = \frac{r}{n}$$

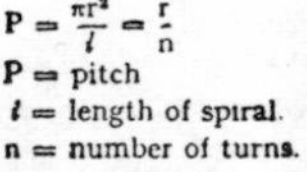

P = pitch

l = length of spiral.

n = number of turns.

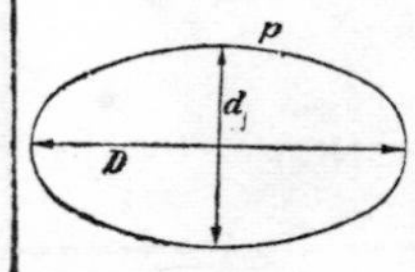

ELLIPSE

$$p = 2\sqrt{D^2 + 1{\cdot}4674d^2}.$$

p = length of periphery.

LONGIMETRY—(*Continued*)

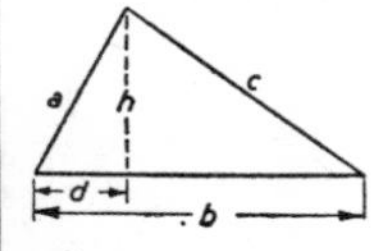

$$c^2 = a^2 + b^2 - 2bd$$
$$h = \sqrt{a^2 - d^2}.$$
$$d = \frac{a^2 + b^2 - c^2}{2b}.$$

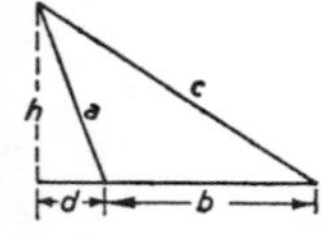

$$c^2 = a^2 + b^2 + 2bd\cdot$$
$$h = \sqrt{a^2 - d^2}.$$
$$d = \frac{c^2 - a^2 - b^2}{2b}.$$

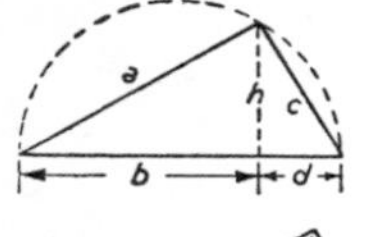

$$a : b :: c : h$$
$$\therefore \ ah = bc$$
$$h = \frac{bc}{a}$$
$$= \frac{ad}{c}$$

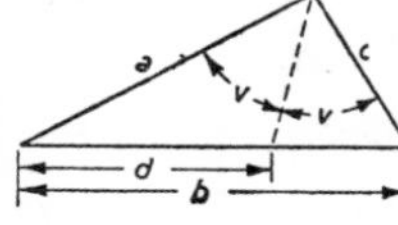

$$a : c = d : (b - d).$$
$$d = \frac{ab}{c + a}.$$
$$v = v.$$

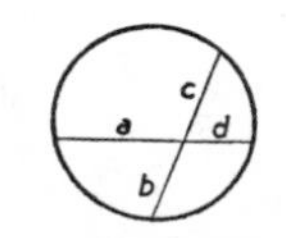

$$a : c = b : d.$$
$$ad = bc.$$

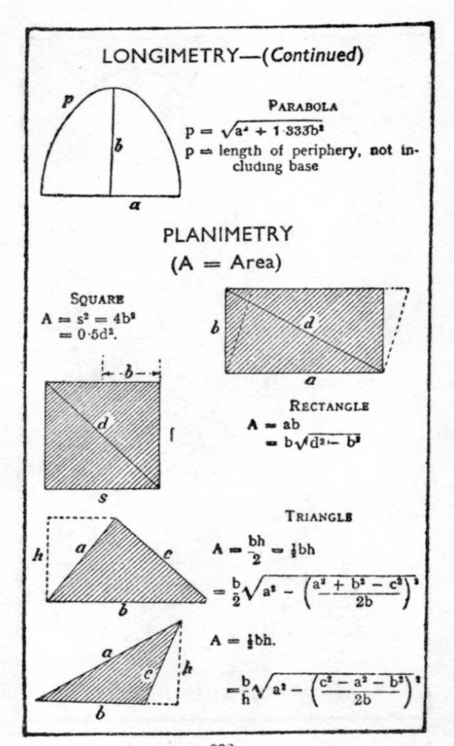
LONGIMETRY—(Continued)
PARABOLA
p = √(a² + 1·333b²)
p = length of periphery, not including base
p
b
a
PLANIMETRY
(A = Area)
SQUARE
A = s² = 4b²
= 0·5d².
b
d
s
RECTANGLE
A = ab
= b√(d² − b²)
b
d
a
TRIANGLE
A = bh/2 = ½bh
= (b/2)√(a² − ((a² + b² − c²)/2b)²)
h
a
c
b
A = ½bh.
= (b/h)√(a² − ((c² − a² − b²)/2b)²)
a
c
h
b

PLANIMETRY—(*Continued*)

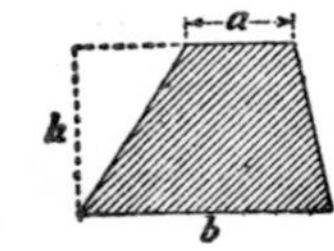

QUADRANGLE OR TRAPEZIUM

$A = \frac{1}{2}h\,(a + b).$

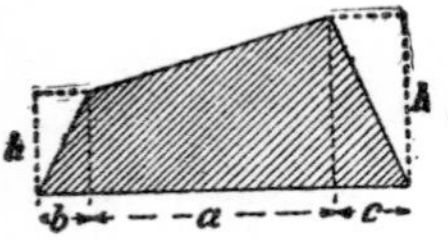

TRAPEZOID

$A = \frac{1}{2}\,(a[h + h'] + bh' + ch).$

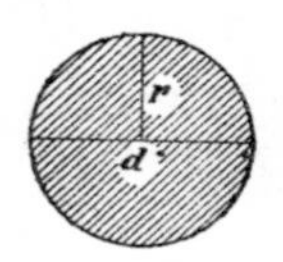

CIRCLE PLANE

$A = \pi r^2 = 0{\cdot}78d^2$

$= \frac{pr}{2} = {\cdot}00796p^2$

p = length of periphery.

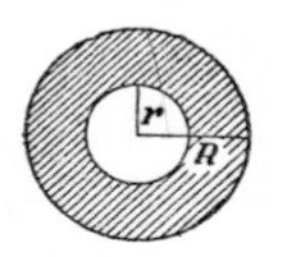

CIRCLE RING

$A = \pi\,(R^2 - r^2)$

$= \pi\,(R + r)\,(R - r)$

$= 0{\cdot}785\,(D^2 - d^2).$

$D = 2R. \quad d = 2r.$

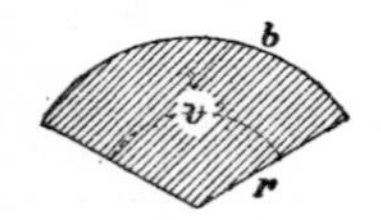

SECTOR

$A = \frac{1}{2}br$

$= \frac{\pi r^2 v}{360} = \frac{r^2 v}{114{\cdot}5}.$

PLANIMETRY—(*Continued*)

Segment

$$A = \tfrac{1}{2}[br - c(r - h)]$$

$$= \frac{h^3}{2c} + \frac{2}{3}ch, \text{ approx.}$$

Quadrant

$$A = 0{\cdot}785r^2 = 0{\cdot}3927c^2.$$

Corner Segment

$$A = 0{\cdot}215r^2 = 0{\cdot}1075c^2.$$

Ellipse

$$A = \pi Rr = 0{\cdot}785\ Dd$$

Parabola

$$A = \frac{2}{3}bd.$$

AREA OF IRREGULAR CURVILINEAR FIGURE

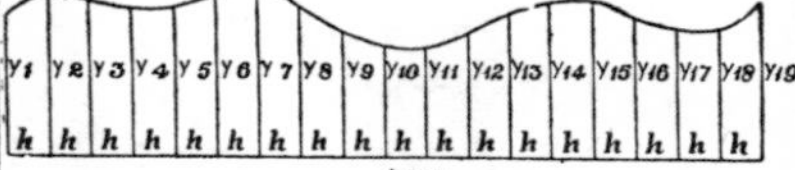

AREA

$$= \frac{3h}{10}\Big\{y_1 + y_3 + y_5 + y_7 + y_9 + y_{11} + y_{13} + y_{15} + y_{17} + y_{19} + 5\,(y_2 + y_4 + y_6 + y_8 + y_{10} + y_{12} + y_{14} + y_{16} + y_{18}) + y_4 + y_7 + y_{10} + y_{13} + y_{16}\Big\}$$

$$= \frac{3h}{10}\Big\{\text{sum of odd} + 5\ (\text{sum of even}) + \text{sum of every third excluding extremes}\Big\}$$

SURFACE AREA OF SOLIDS
(S = Surface)

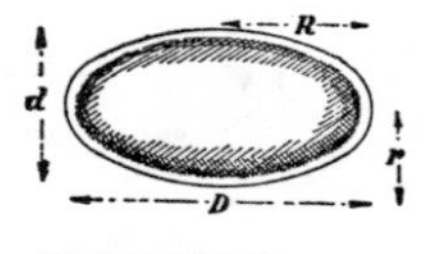

PROLATE SPHEROID

$$S = \frac{4\pi}{\sqrt{2}} \cdot r\sqrt{R^2 + r^2} = 8{\cdot}88r\sqrt{R^2 + r^2} = 2{\cdot}22d\sqrt{D^2 + d^2}.$$

OBLATE SPHEROID

$$S = \frac{4\pi}{\sqrt{2}} \cdot R\sqrt{R^2 + r^2}.$$

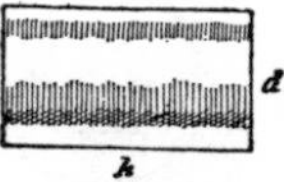

CYLINDER

$$S = 2\pi rh = \pi dh = 3{\cdot}1416dh.$$

$$h = \frac{S}{2\pi r} = \frac{S}{\pi d} = \frac{S}{3{\cdot}1416d}.$$

SPHERE

$$S = 4\pi r^2 = 12{\cdot}56r^2 = \pi d^2.$$

SURFACE AREA OF SOLIDS—(*Continued*)

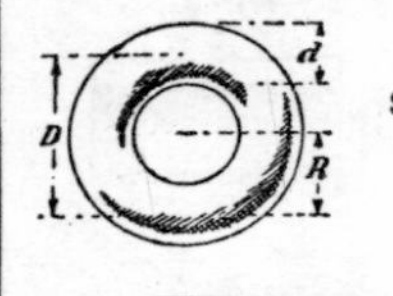

TORUS

$S = 4\pi^2 Rr = 39{\cdot}48Rr$
$= 9{\cdot}87Dd.$

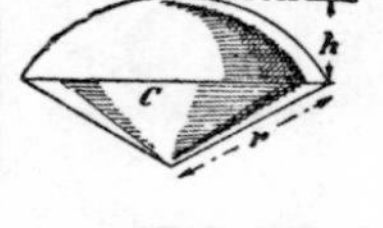

SPHERE SECTOR

$$S = \frac{\pi r}{2}(4h + c).$$

CIRCLE ZONE

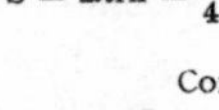

$$S = 2\pi rh = \frac{\pi}{4}(4h^2 + c^2).$$

CONE

$S = \pi Rs$

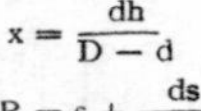

$= \pi R\sqrt{R^2 + h^2}.$

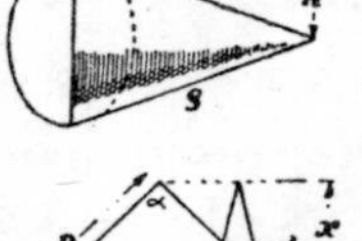

CONE

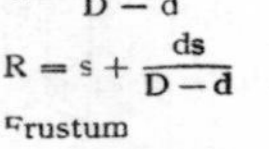

$$x = \frac{dh}{D - d}$$

$$R = s + \frac{ds}{D - d}$$

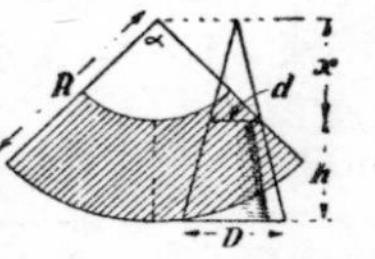

Frustum

$$= \frac{\pi s}{2}(D + d)$$

Apex angle α

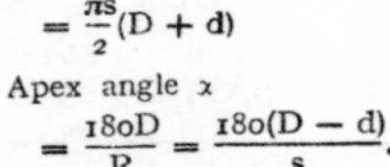

$$= \frac{180D}{R} = \frac{180(D - d)}{s}.$$

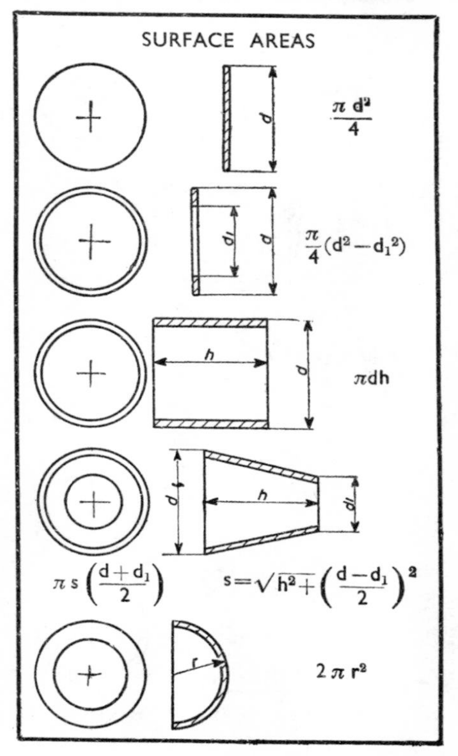
SURFACE AREAS
d
$\frac{\pi d^2}{4}$
d_1
d
$\frac{\pi}{4}(d^2 - d_1^2)$
h
d
πdh
d
h
d_1
$\pi s \left(\frac{d + d_1}{2}\right)$
$s = \sqrt{h^2 + \left(\frac{d - d_1}{2}\right)^2}$
r
$2 \pi r^2$

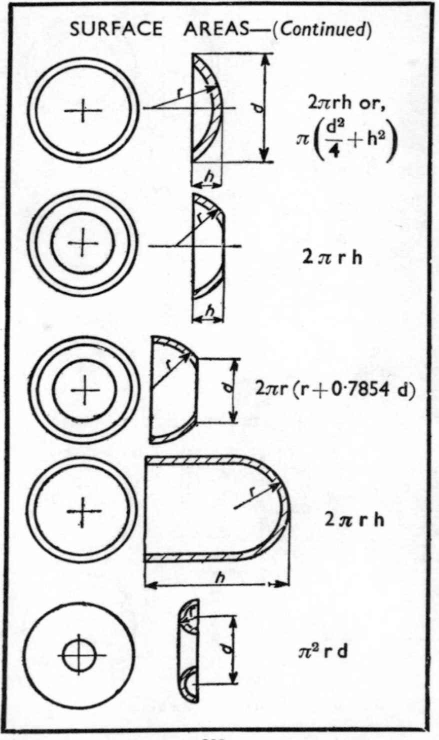
SURFACE AREAS—(Continued)
r
d
h
$2\pi rh$ or,
$\pi\left(\frac{d^2}{4}+h^2\right)$
$2\pi r h$
$2\pi r(r+0{\cdot}7854\ d)$
$2\pi r h$
$\pi^2 r d$

PULLEYS

Note.—$\frac{V1}{V}$ = Ratio of distances moved by **W** and **F**.

Single Fixed Pulley

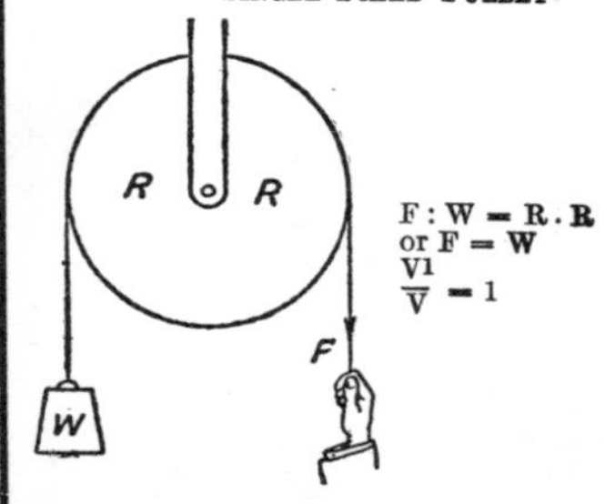

F : W = R . R
or F = W
$\frac{V1}{V} = 1$

Single Movable Pulley

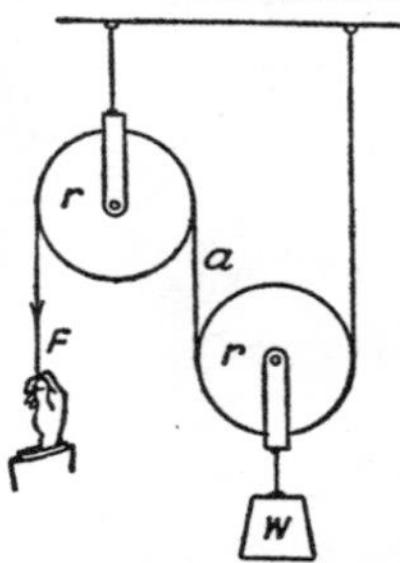

F : W = r : 2r
or F = ½ W
Note.—If the force is applied at "**a**" and acts upward, the result will be the same.
$\frac{V1}{V} = 2$

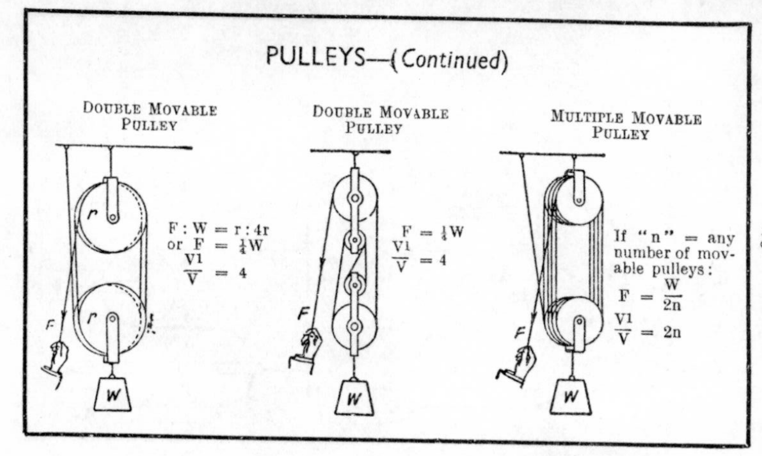
PULLEYS—(Continued)
DOUBLE MOVABLE PULLEY
r
r
F
W
F : W = r : 4r
or F = ¼W
V1/V = 4
DOUBLE MOVABLE PULLEY
F
W
F = ¼W
V1/V = 4
MULTIPLE MOVABLE PULLEY
F
W
If "n" = any number of movable pulleys:
F = W/2n
V1/V = 2n

PULLEYS
(Continued)

COMPOUND PULLEYS

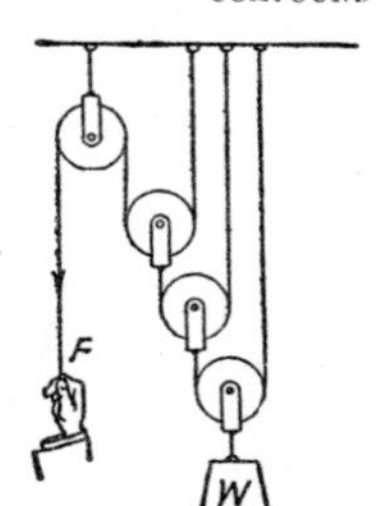

"n" = number of movable pulleys

$$F = \frac{W}{2^n}$$

$$W = 2^n F$$

$$\frac{V1}{V} = 2^n$$

OBLIQUE FIXED PULLEY

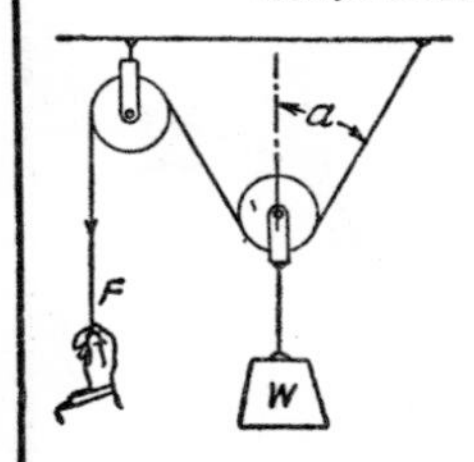

$$F : W = \text{Sec } a : 2$$

$$W = \frac{2F}{\text{Sec } a}$$

$$F = \frac{W \text{ Sec } a}{2}$$

PARALLELOGRAM OF FORCES

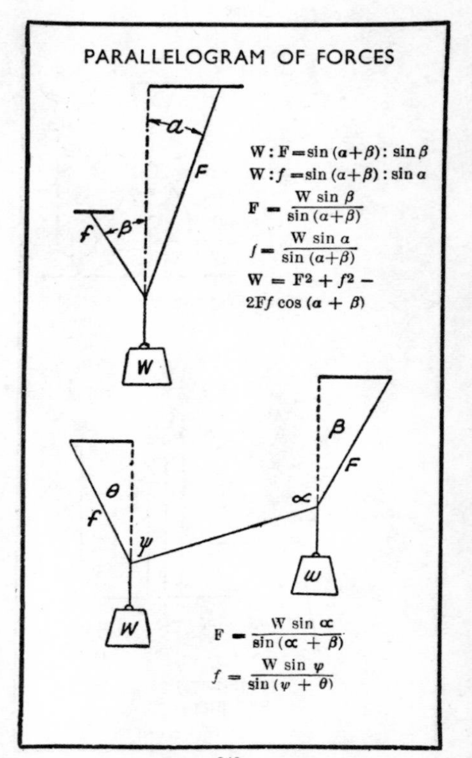

PARALLELOGRAM OF FORCES
(*Continued*)

INCLINED PLANE

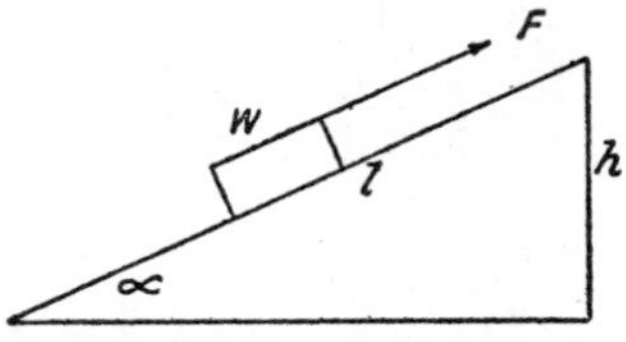

No Friction

$$F = \frac{Wh}{l} = W \sin \alpha$$

$$W = \frac{Fl}{h} = \frac{F}{\sin \alpha}$$

No Friction

$$F = W\frac{\sin \alpha}{\cos \beta} \quad W = F\frac{\cos \beta}{\sin \alpha}$$

With Friction

$$F = W\frac{\mu \cos \alpha + \sin \alpha}{\mu \sin \beta + \cos \beta}$$

where μ = coefficient of friction.

$$F = W\frac{\sin(\Phi + \alpha)}{\cos(\beta - \Phi)}$$

where Φ = limiting angle of resistance
$\mu = \tan \Phi$

WEDGE

$$F = \frac{Ra}{l} \qquad R = \frac{Fl}{a}$$

F = force required to drive the wedge

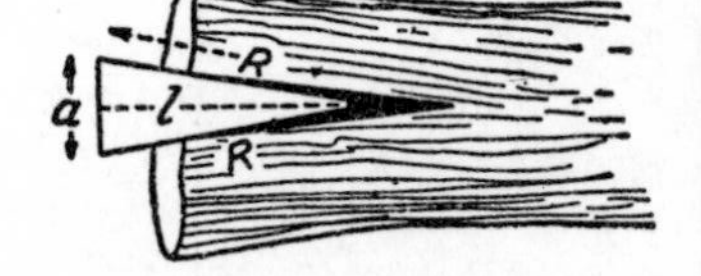

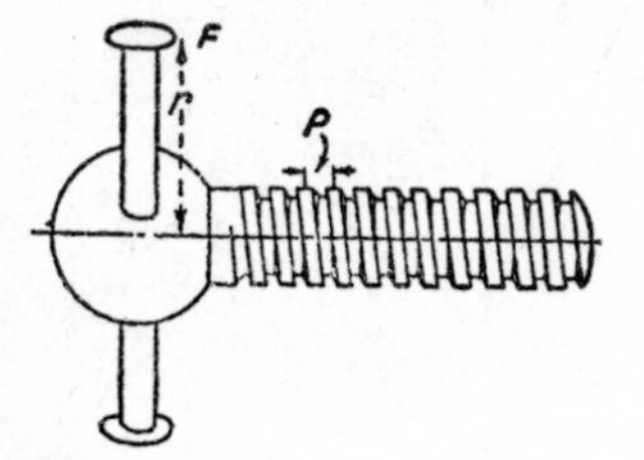

FORCE BY A SCREW

P = pitch of screw (distance between threads)

r = radius on which force F acts. $\pi = 3.1416$.

W = force exerted by screw at radius r.

$$F : W = P : 2\pi r. \quad F = \frac{WP}{2\pi r} \quad W = \frac{F2\pi r}{P}$$

Note. The above ignores friction forces.

CLAMPING FORCE EXERTED BY A NUT OR BOLT

The clamping force F exerted by a nut or bolt is given by the equation $F = 2T/\mu d$, where T is the applied torque, d is the threaded diameter of the bolt and μ is the coefficient of friction between the surfaces. Under normal conditions with some lubrication $\mu = 0{\cdot}4$ and the force F is given by the "5 rule", that is, $F = 5T/d$.

Example. When a nut on a 10 mm diameter bolt or stud is tightened to a torque of 5Nm, the force it will exert on the component which it is holding is

$$5 \times 5 \times \frac{1000}{10} = 2500 \text{ newtons}$$

If a super lubricant has been applied, μ can reduce to 0·2 when the 5 in the equation becomes 10; with no lubricant μ will increase to about 0·8 when the 5 becomes 2·5.

The variable effect due to the pitch angle for different threads is small and can usually be neglected.

FORCE ACTING AT AN ANGLE

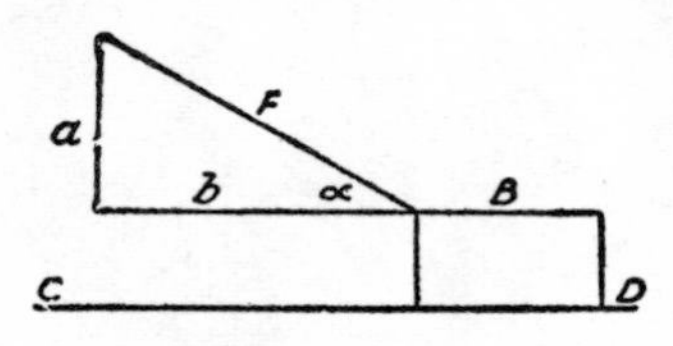

Let line F represent the magnitude and direction of a force acting at an angle α to move the body B on line CD. Then the line a represents a part of F which presses the body B against CD. The line b represents the magnitude of the force which actually moves the body B.

$b = \sqrt{F^2 - a^2}$

$b = F \cos a$

DEFINITION OF FORCE

A force has direction, magnitude and point of application. The direction of a force is that in which it tends to move the body upon which it is acting. The point of application is usually considered to be the centre of gravity of the body, and the magnitude is measured in kilogrammes.

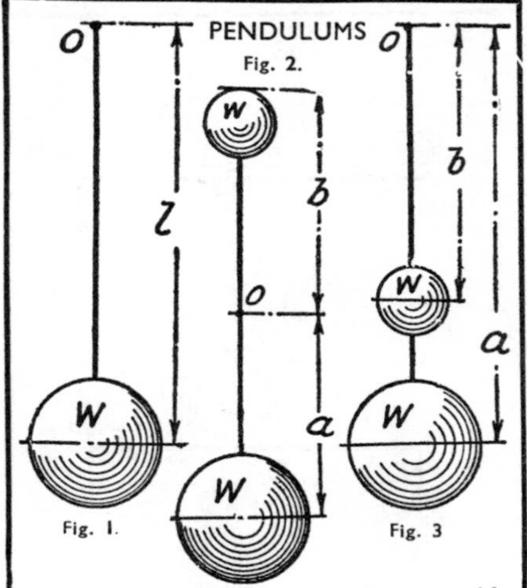

Fig. 1.—The Simple Pendulum, $t=\pi\sqrt{\frac{l}{g}}$;

l = length of pendulum: t = time in secs. of one oscillation; g = acceleration due to gravity.

Time for complete swing (two oscillations) $=2\pi\sqrt{\frac{l}{g}}$.

Fig. 2.—The Compound Pendulum in which o=centre of suspension and l=equivalent length of simple pendulum to give same time of oscillation, $l=\frac{a^2W+b^2w}{aW-bw}$.

Fig. 3.—Another form of Compound Pendulum, $l=\frac{a^2W+b^2w}{aW+bw}$.

FORMULÆ RELATIVE TO LEVERS

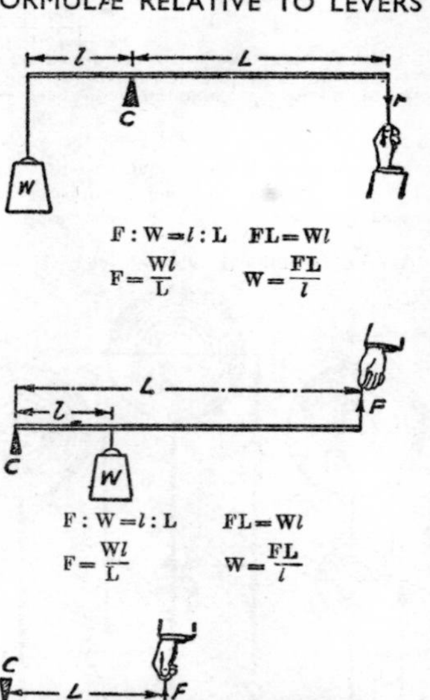

$F : W = l : L \qquad FL = Wl$

$F = \frac{Wl}{L} \qquad W = \frac{FL}{l}$

$F : W = l : L \qquad FL = Wl$

$F = \frac{Wl}{L} \qquad W = \frac{FL}{l}$

$F : W = l : L \qquad FL = Wl$

$F = \frac{Wl}{L} \qquad W = \frac{FL}{l}$

FORMULÆ RELATIVE TO LEVERS
(Continued)

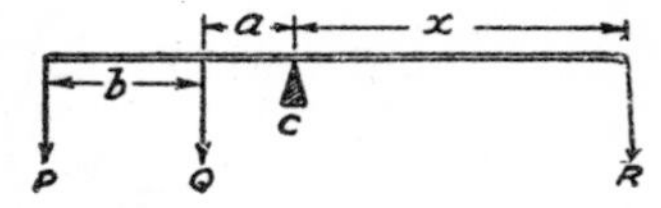

To find Fulcrum C when three forces act on one lever.

$$Rx = Qa + P(b+a)$$

$$x = \frac{Qa + P(b+a)}{R}$$

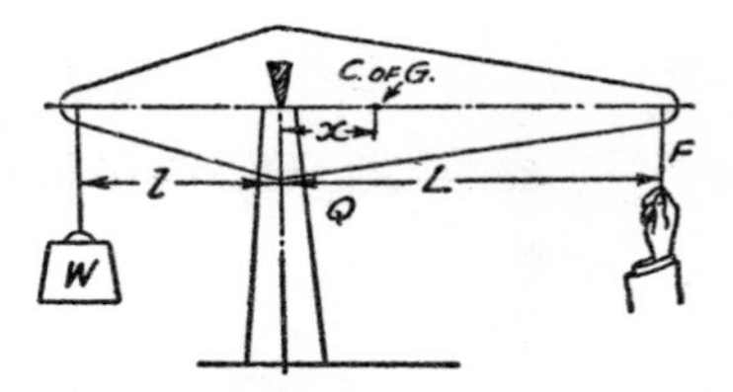

Q = Weight of the lever.

x = distance from centre of gravity of lever to fulcrum.

$$F = \frac{Wl - Qx}{L}$$

$$W = \frac{FL + Qx}{l}$$

FORMULÆ RELATIVE TO LEVERS

(Continued)

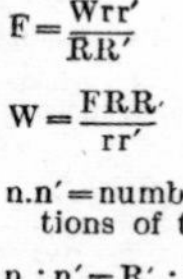

$F : W = r : R$

$FR = Wr$

$F = \frac{Wr}{R}$

$R = \frac{Wr}{F}$

$W = \frac{RF}{r}$

$r = \frac{RF}{W}$

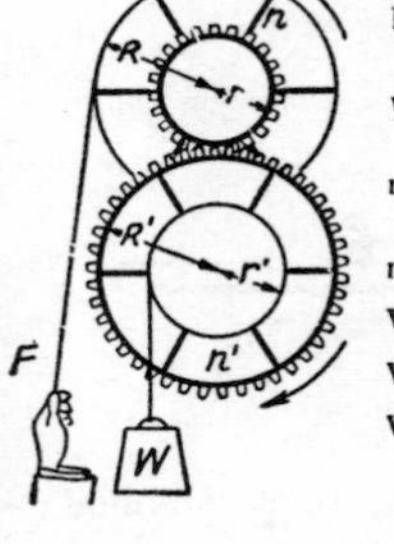

$F = \frac{Wrr'}{RR'}$

$W = \frac{FRR'}{rr'}$

n.n′ = number of revolutions of the wheels

$n : n' = R' : r$

$V : V' = RR' = rr'$

V = Velocity of F

V′ = Velocity of W

Formulæ Relating to Centrifugal Force, Tension and Governors

F = Centrifugal force in Newtons

M = Mass or weight of revolving body in kilogrammes.

V = Velocity of revol. body in metres per second.

R = Rad. of circle in which body revolves (metres)

n = Number of revolutions per minute.

g = Coefficient of terrestrial Acceleration = $9.81 m/s^2$

$\pi = 3.1416$.

M

R

$$F = \frac{Mv^2}{gR} = \frac{Mv^2}{9.81\,R}$$

$$M = \frac{FgR}{v^2} = \frac{892\,F}{Rn^2}$$

$$n = \sqrt{\frac{892\,F}{MR}}$$

$$F = \frac{4MR\pi^2n^2}{60^2g} = \frac{MRn^2}{892}$$

$$R = \frac{Mv^2}{Fg} = \frac{892\,F}{Mn^2}$$

$$v = \sqrt{\frac{FRg}{M}}$$

Formulæ Relating to Centrifugal Force, Tension and Governors

(Continued)

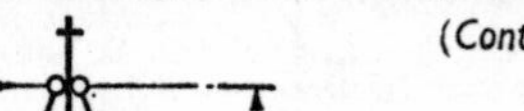

GOVERNOR

$$n = \frac{60}{2\pi}\sqrt{\frac{g}{h}} = \frac{30}{\sqrt{h}} = \frac{30}{\sqrt{l\cos a}}$$

$$h = \frac{892}{n^2} \quad l = \frac{892}{n^2 \cos a} = \frac{h}{\cos a}$$

$$\cos a = \frac{892}{n^2 l} = \frac{h}{l} \quad r = \sqrt{l^2 - h^2}$$

Formulæ Relating to Centrifugal Force, Tension and Governors

(Continued)

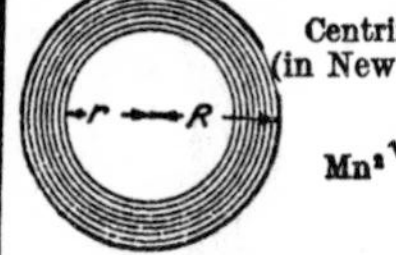

Centrifugal Tension (in Newtons) of a Ring.

$$Mn^2 \frac{\sqrt{R^2 + r^2}}{3004}$$

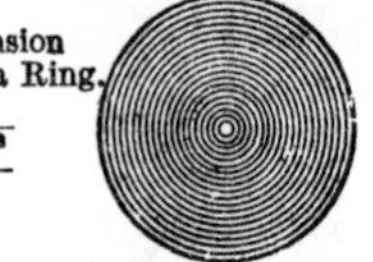

Centrifugal Tension of a Grindstone, Circle-plane, or Cylinder rotating round its centre.

$$= \frac{MRn^2}{3004}$$

Centrifugal tension of a Cylinder rotating round the diameter of its base.

$$\frac{Mn^2\sqrt{4l^2 + 3r^2}}{7408}$$

MOMENTS OF INERTIA, CONIC FRUSTUM AND FLYWHEEL FORMULÆ

CONIC FRUSTUM

$$x = \sqrt{\frac{h^2}{10}\left(\frac{R^2+3Rr+6r^2}{R^2+r^2+Rr}\right)+\frac{3}{20}\left(\frac{R^5-r^5}{R^3-r^3}\right)}$$

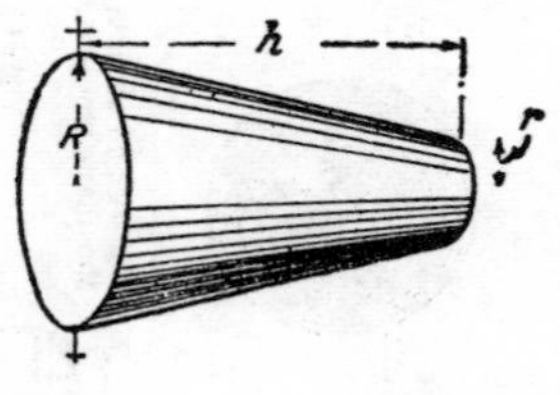

CYLINDER

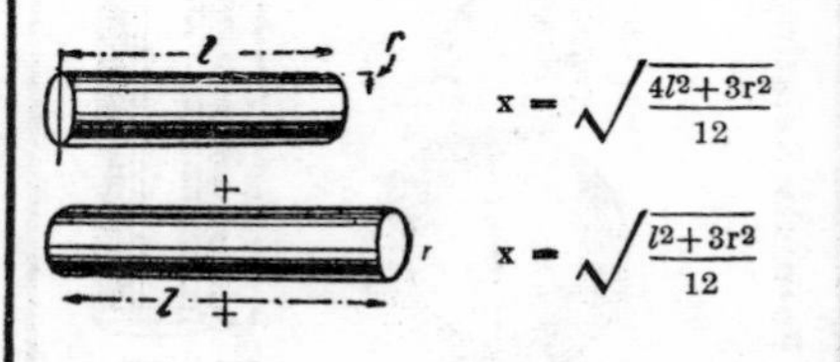

$$x = \sqrt{\frac{4l^2+3r^2}{12}}$$

$$x = \sqrt{\frac{l^2+3r^2}{12}}$$

MOMENTS OF INERTIA, CONIC FRUSTUM AND FLYWHEEL FORMULÆ
—Continued

Flywheel with Arms

$$x^2(W+w)=W\frac{R^2+r^2}{2}+w\frac{4r^2+b^2}{12}$$

$$=\sqrt{\frac{6W(R^2+r^2)+w(4r^2+b^2)}{12(W+w)}}$$

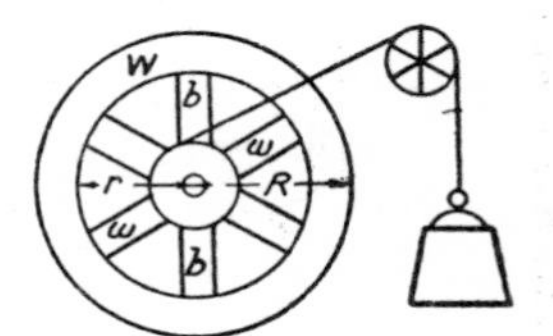

Parallelepiped

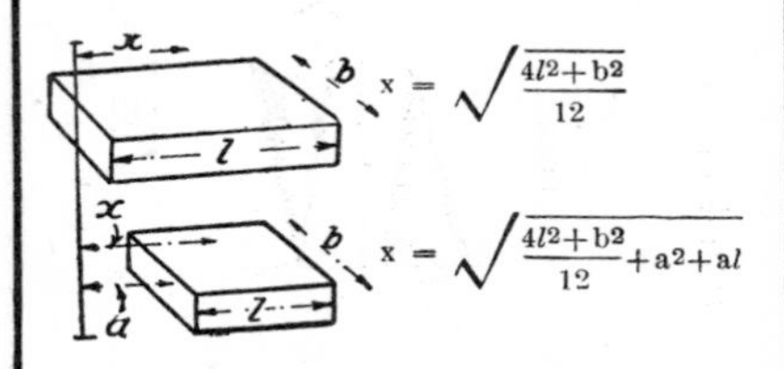

$$x=\sqrt{\frac{4l^2+b^2}{12}}$$

$$x=\sqrt{\frac{4l^2+b^2}{12}+a^2+al}$$

MOMENTS OF INERTIA, CONIC FRUSTUM AND FLYWHEEL FORMULÆ—*Continued*

CONE

$$x = \sqrt{\frac{12h^2+3R^2}{20}}$$

$$x = \sqrt{\frac{2h^2+3R^2}{20}}$$

FLYWHEEL

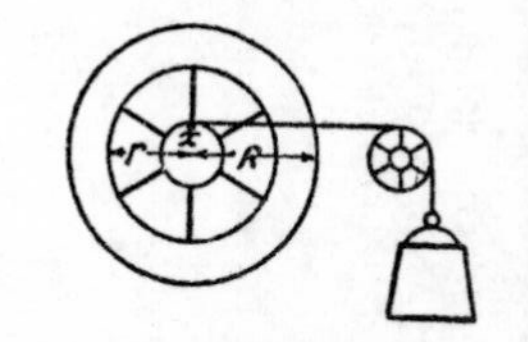

$$x = \sqrt{\frac{R^2+r^2}{2}}$$

MODULI OF SECTIONS AND MOMENTS OF INERTIA

	Area	Modulus of the Section $\frac{I}{\text{half depth}}$	Moment of Inertia I about the Axis shown in first column
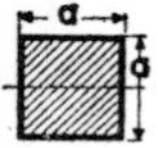 Square, side a	a^2	$\frac{a^3}{6}$	$\frac{a^4}{12}$
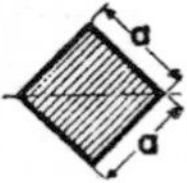 Square on diagonal, side a	a^2	$\frac{a^3\sqrt{2}}{12}$	$\frac{a^4}{12}$
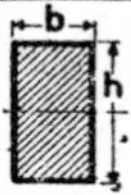 Rectangle, b × h	$b\,h$	$\frac{b\,h^2}{6}$	$\frac{b\,h^3}{12}$
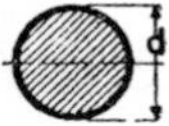 Circle, diameter d	$\frac{\pi\, d^2}{4}$	$\frac{\pi\, d^3}{32}$	$\frac{\pi\, d^4}{64}$
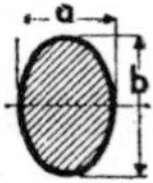 Ellipse, a × b	$\frac{\pi\, a\, b}{4}$	$\frac{\pi\, a\, b^2}{32}$	$\frac{\pi\, a\, b^3}{64}$

MODULI OF SECTIONS AND MOMENTS OF INERTIA

(*Continued*)

Section	Area	Modulus of the Section $\frac{I}{\text{half depth}}$	Moment of Inertia I about the Axis shown in first column
 B, h, b, H	$\frac{\pi}{4}(D^2-d^2)$	$\frac{\pi}{32}\frac{(D^4-d^4)}{D}$	$\frac{\pi}{64}(D^4-d^4)$
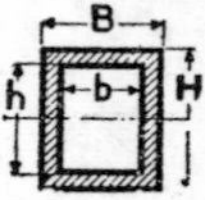 B, h, b, H	$BH-bh$	$\frac{BH^3-bh^3}{6H}$	$\frac{BH^3-bh^3}{12}$
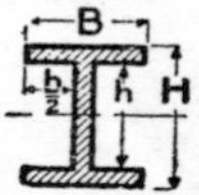 B, b/2, h, H	$BH-\frac{bh}{2}$	$\frac{BH^3-bh^3}{6H}$	$\frac{BH^3-bh^3}{12}$
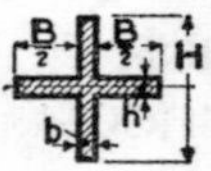 B/2, B/2, H, h, b	$bH+Bh$	$\frac{bH^3+Bh^3}{6H}$	$\frac{bH^3+Bh^3}{12}$

CHORDS OF CIRCLES (Method of finding length)

No. of Chords	Multiply Dia. by	No. of Chords	Multiply Dia. by	No. of Chords	Multiply Dia. by	No. of Chords	Multiply Dia. by	No. of Chords	Multiply Dia. by
3	.8660	23	.1362	43	.0730	63	.0499	83	.0378
4	.7071	24	.1305	44	.0713	64	.0491	84	.0374
5	.5878	25	.1253	45	.0698	65	.0483	85	.0370
6	.5000	26	.1205	46	.0682	66	.0476	86	.0365
7	.4339	27	.1161	47	.0668	67	.0469	87	.0361
8	.3827	28	.1120	48	.0654	68	.0462	88	.0357
9	.3420	29	.1081	49	.0641	69	.0455	89	.0353
10	.3090	30	.1045	50	.0628	70	.0449	90	.0349
11	.2817	31	.1012	51	.0616	71	.0442	91	.0345
12	.2588	32	.0980	52	.0604	72	.0436	92	.0341
13	.2393	33	.0951	53	.0592	73	.0430	93	.0338
14	.2225	34	.0923	54	.0581	74	.0424	94	.0334
15	.2079	35	.0896	55	.0571	75	.0419	95	.0331
16	.1951	36	.0872	56	.0561	76	.0413	96	.0327
17	.1838	37	.0848	57	.0551	77	.0408	97	.0324
18	.1736	38	.0826	58	.0541	78	.0403	98	.0321
19	.1646	39	.0805	59	.0532	79	.0398	99	.0317
20	.1564	40	.0785	60	.0523	80	.0393	100	.0314
21	.1490	41	.0765	61	.0515	81	.0388		
22	.1423	42	.0747	62	.0507	82	.0383		

CHORDS AND RADIANS

Degrees	Chords	Differences for 10′	Radians
0°	.0000	29	.0000
1°	.0175	29	.0175
2°	.0349	29	.0349
3°	.0524	29	.0524
4°	.0698	29	.0698
5°	.0872	29	.0873
6°	.1047	29	.1047
7°	.1221	29	.1222
8°	.1395	29	.1396
9°	.1569	29	.1571
10°	.1743	29	.1745
11°	.1917	29	.1920
12°	.2091	29	.2094
13°	.2264	29	.2269
14°	.2437	29	.2443
15°	.2611	29	.2618
16°	.2783	29	.2793
17°	.2956	29	.2967
18°	.3129	29	.3142
19°	.3301	29	.3316
20°	.3473	29	.3491
21°	.3645	29	.3665
22°	.3816	29	.3840
23°	.3987	28	.4014
24°	.4158	28	.4189
25°	.4329	28	.4363
26°	.4499	28	.4538
27°	.4669	28	.4712
28°	.4838	28	.4887
29°	.5008	28	.5061
30°	.5176	28	.5236
31°	.5345	28	.5411
32°	.5513	28	.5585
33°	.5680	28	.5760
34°	.5847	28	.5934
35°	.6014	28	.6109
36°	.6180	28	.6283
37°	.6346	28	.6458
38°	.6511	27	.6632
39°	.6676	27	.6807
40°	.6840	27	.6981
41°	.7004	27	.7156
42°	.7167	27	.7330
43°	.7330	27	.7505
44°	.7492	27	.7679
45°	.7654	27	.7854

Constant Differences	
1′	3″
2′	6″
3′	9″
4′	12″
5′	15″
6′	17″
7′	20″
8′	23″
9′	26″
10′	29″

Note. The table gives the lengths of chords, as fractions of the radius for specific angles.

∟ 1 right angle = $\frac{\pi}{2}$ radians = 1.5707963 radians

1 radian = 57.2958° = 57° 17′ 45″

To convert radians into seconds multiply by 206265.

CHORDS AND RADIANS (*Contd.*)

Degrees	Chords	Differences for 10′	Radians	Constant Differences
45°	.7654	27	.7854	1′ 3″
46°	.7815	27	.8029	2′ 6″
47°	.7975	27	.8203	3′ 9″
48°	.8135	27	.8378	4′ 12″
49°	.8294	26	.8552	5′ 15″
50°	.8452	26	.8727	6′ 17″
51°	.8610	26	.8901	7′ 20″
52°	.8767	26	.9076	8′ 23″
53°	.8924	26	.9250	9′ 26″
54°	.9080	26	.9425	10′ 29″
55°	.9235	26	.9599	—
56°	.9389	26	.9774	
57°	.9543	26	.9948	
58°	.9696	25	1.0123	
59°	.9848	25	1.0297	
60°	1.0000	25	1.0472	
61°	1.0151	25	1.0647	
62°	1.0301	25	1.0821	
63°	1.0450	25	1.0996	
64°	1.0598	25	1.1170	
65°	1.0746	24	1.1345	
66°	1.0893	24	1.1519	
67°	1.1039	24	1.1694	
68°	1.1184	24	1.1868	
69°	1.1328	24	1,2043	
70°	1.1472	24	1.2217	
71°	1.1614	24	1.2392	
72°	1.1756	23	1.2566	
73°	1.1896	23	1.2741	
74°	1.2036	23	1.2915	
75°	1.2175	23	1,3090	
76°	1.2313	23	1,3265	
77°	1.2450	23	1.3439	
78°	1.2586	23	1.3614	
79°	1.2722	22	1.3788	
80°	1.2856	22	1.3963	
81°	1.2989	22	1.4137	
82°	1.3121	22	1.4312	
83°	1.3252	22	1.4486	
84°	1.3383	22	1.4661	
85°	1.3512	21	1.4835	
86°	1.3640	21	1.5010	
87°	1.3767	21	1.5184	
88°	1.3893	21	1.5359	
89°	1.4018	21	1.5533	
90°	1.4142	—	1.5708	

B. Chord = diameter × sine of angle subtended at circumference.
= diameter × sine of semi-angle subtended at centre.

A radian is the angle subtended at the centre of a circle by an arc equal in length to the radius.

AREAS AND CIRCUMFERENCES OF CIRCLES.—1 to 68

Diameter	Area	Circumference	Diameter	Area	Circumference
1	0.7854	3.142	35	962.113	109.956
2	3.1416	6.283	36	1017.88	113.097
3	7.0686	9.425	37	1075.21	116.239
4	12.5664	12.566	38	1134.11	119.381
5	19.6350	15.708	39	1194.59	122.522
6	28.2743	18.850	40	1256.64	125.660
7	38.4845	21.991	41	1320.25	128.806
8	50.2655	25.133	42	1385.44	131 947
9	63.6173	28.274	43	1452.20	135.090
10	78.5398	31.416	44	1520.53	138.230
11	95.0332	34.558	45	1590.43	141.372
12	113.097	37.699	46	1661.90	144.514
13	132.732	40.841	47	1734.94	147.655
14	153.938	43.982	48	1809.56	150.800
15	176.715	47.124	49	1885.74	153.938
16	201.062	50.265	50	1963.50	157.080
17	226.980	53.407	51	2042.82	160.22
18	254.469	56.549	52	2123.72	163.36
19	283.529	59.690	53	2206.18	166.50
20	314.159	62.832	54	2290.22	169.65
21	346.361	65.973	55	2375.83	172.79
22	380.133	69.115	56	2463.01	175.93
23	415.476	72.257	57	2551.76	179.07
24	452.389	75.398	58	2642.08	182.21
25	490.874	78.540	59	2733.97	185.35
26	530.929	81.681	60	2827.43	188.50
27	572.555	84.823	61	2922.47	191.64
28	615.752	87.965	62	3019.07	194.78
29	660.520	91.106	63	3117.25	197.92
30	706.858	94.248	64	3216.99	201.06
31	754.768	97.389	65	3318.31	204.20
32	804.248	100.531	66	3421.19	207.35
33	855.299	103.673	67	3525.65	210.49
34	907.920	106.814	68	3631.68	213.63

AREAS AND CIRCUMFERENCES OF CIRCLES.—69-136

Diameter	Area	Circumference	Diameter	Area	Circumference
69	3739.28	216.77	103	8332.29	323.58
70	3848.45	219.91	104	8494.87	326.73
71	3959.19	223.05	105	8659.01	329.87
72	4071.50	226.19	106	8824.73	333.01
73	4185.39	229.34	107	8992.02	336.15
74	4300.84	232.48	108	9160.88	339.29
75	4417.86	235.62	109	9331.32	342.43
76	4536.46	238.76	110	9503.32	345.58
77	4656.63	241.90	111	9676.89	348.72
78	4778.36	245.04	112	9852.03	351.86
79	4901.67	248.19	113	10028.7	355.00
80	5026.55	251.33	114	10207.0	358.14
81	5153.00	254.47	115	10386.9	361.28
82	5281.02	257.61	116	10568.3	364.42
83	5410.61	260.75	117	10751.3	367.57
84	5541.77	263.89	118	10935.9	370.71
85	5674.50	267.04	119	11122.0	373.85
86	5808.80	270.18	120	11309.7	376.99
87	5944.68	273.32	121	11499.0	380.13
88	6082.12	276.46	122	11689.9	383.27
89	6221.14	279.60	123	11882.3	386.42
90	6361.73	282.74	124	12076.3	389.56
91	6503.88	285.88	125	12271.8	392.70
92	6647.61	289.03	126	12469.0	395.84
93	6792.91	292.17	127	12667.7	398.98
94	6939.78	295.31	128	12868.0	402.12
95	7088.22	298.45	129	13069.8	405.27
96	7238.23	301.59	130	13273.2	408.41
97	7389.81	304.73	131	13478.2	411.55
98	7542.96	307.88	132	13684.8	414.69
99	7697.69	311.02	133	13892.9	417.83
100	7853.98	314.16	134	14102.6	420.97
101	8011.85	317.30	135	14313.9	424.12
102	8171.28	320.44	136	14526.7	427.26

AREAS AND CIRCUMFERENCES OF CIRCLES.—137-204

Diameter	Area	Circumference	Diameter	Area	Circumference
137	14741.1	430.40	171	22965.8	537.21
138	14957.1	433.54	172	23235.2	540.35
139	15174.7	436.68	173	23506.2	543.50
140	15393.8	439.82	174	23778.7	546.64
141	15614.5	442.96	175	24052.8	549.78
142	15836.8	446.11	176	24328.5	552.92
143	16060.6	449.25	177	24605.7	556.06
144	16286.0	452.39	178	24884.6	559.20
145	16513.0	455.53	179	25164.9	562.35
146	16741.5	458.67	180	25446.9	565.49
147	16971.7	461.81	181	25730.4	568.63
148	17293.4	464.96	182	26015.5	571.77
149	17436.6	468.10	183	26302.2	574.91
150	17671.5	471.24	184	26590.4	578.05
151	17907.9	474.38	185	26880.3	581.19
152	18145.8	477.52	186	27171.6	584.34
153	18385.4	480.66	187	27464.6	587.48
154	18626.5	483.81	188	27759.1	590.62
155	18869.2	486.95	189	28055.2	593.76
156	19113.4	490.09	190	28352.9	596.90
157	19359.3	493.23	191	28652.1	600.04
158	19606.7	496.37	192	28952.9	603.19
159	19855.7	499.51	193	29255.3	606.33
160	20106.2	502.65	194	29559.2	609.47
161	20358.3	505.80	195	29864.8	612.61
162	20612.0	508.94	196	30171.9	615.75
163	20867.2	512.08	197	30480.5	618.89
164	21124.1	515.22	198	30790.7	622.04
165	21382.5	518.36	199	31102.6	625.18
166	21642.4	521.50	200	31415.9	628.32
167	21904.0	524.65	201	31730.9	631.46
168	22167.1	527.79	202	32047.4	634.60
169	22431.8	530.93	203	32365.5	637.74
170	22698.0	534.07	204	32685.1	640.89

AREAS AND CIRCUMFERENCES OF CIRCLES.—205 to 272

Diameter	Area	Circumference	Diameter	Area	Circumference
205	33006.4	644.03	239	44862.7	750.84
206	33329.2	647.17	240	45238.9	753.98
207	33653.5	650.31	241	45616.7	757.12
208	33979.5	653.45	242	45996.1	760.27
209	34307.0	656.59	243	46377.0	763.41
210	34636.1	659.73	244	46759.5	766.55
211	34966.7	662.88	245	47143.5	769.69
212	35298.9	666.02	246	47529.2	772.83
213	35632.7	669.16	247	47916.4	775.97
214	35968.1	672.30	248	48305.1	779.12
215	36305.0	675.44	249	48695.5	782.26
216	36643.5	678.58	250	49087.4	785.40
217	36983.6	681.73	251	49480.9	788.54
218	37325.3	684.87	252	49875.9	791.68
219	37668.5	688.01	253	50272.6	794.82
220	38013.3	691.15	254	50670.7	797.96
221	38359.6	694.29	255	51070.5	801.11
222	38707.6	697.43	256	51471.9	804.25
223	39057.1	700.58	257	51874.8	807.39
224	39408.1	703.72	258	52279.2	810.53
225	39760.8	706.86	259	52685.3	813.67
226	40115.0	710.00	260	53092.9	816.81
227	40470.8	713.14	261	53502.1	819.96
228	40828.1	716.28	262	53912.9	823.10
229	41187.1	719.42	263	54325.2	826.24
230	41547.6	722.57	264	54739.1	829.38
231	41909.6	725.71	265	55154.6	832.52
232	42273.3	728.85	266	55571.6	835.66
233	42638.5	731.99	267	55990.3	838.81
234	43005.3	735.13	268	56410.4	841.95
235	43373.6	738.27	269	56832.2	845.09
236	43743.5	741.42	270	57255.5	848.23
237	44115.0	744.56	271	57680.4	851.37
238	44488.1	747.70	272	58106.9	854.51

AREAS AND CIRCUMFERENCES OF CIRCLES.—273-340

Diameter	Area	Circumference	Diameter	Area	Circumference
273	58534.9	857.66	307	74023.0	964.47
274	58964.6	860.80	308	74506.0	967.61
275	59395.7	863.94	309	74990.6	970.75
276	59828.5	867.08	310	75476.8	973.89
277	60262.8	870.22	311	75964.5	977.04
278	60698.7	873.36	312	76453.8	980.18
279	61136.2	876.50	313	76944.7	983.32
280	61575.2	879.65	314	77437.1	986.46
281	62015.8	882.79	315	77931.1	989.60
282	62458.0	885.93	316	78426.7	992.74
283	62901.8	889.07	317	78923.9	995.88
284	63347.1	892.21	318	79422.6	999.03
285	63794.0	895.35	319	79922.9	1002.2
286	64242.4	898.50	320	80424.8	1005.3
287	64692.5	901.64	321	80928.2	1008.5
288	65144.1	904.78	322	81433.2	1011.6
289	65597.2	907.92	323	81939.8	1014.7
290	66052.0	911.06	324	82448.0	1017.9
291	66508.3	914.20	325	82957.7	1021.0
292	66966.2	917.35	326	83469.0	1024.2
293	67425.6	920.49	327	83981.8	1027.3
294	67886.7	923.63	328	84496.3	1030.4
295	68349.3	926.77	329	85012.3	1033.6
296	68813.5	929.91	330	85529.9	1036.7
297	69279.2	933.05	331	86049.0	1039.9
298	69746.5	936.19	332	86569.7	1043.0
299	70215.4	939.34	333	87092.0	1046.2
300	70685.8	942.48	334	87615.9	1049.3
301	71157.9	945.62	335	88141.3	1052.4
302	71631.5	948.76	336	88668.3	1055.6
303	72106.6	951.90	337	89196.9	1058.7
304	72583.4	955.04	338	89727.0	1061.9
305	73061.7	958.19	339	90258.7	1065.0
306	73541.5	961.33	340	90792.0	1068.1

AREAS AND CIRCUMFERENCES OF CIRCLES.—341 to 410

Diameter	Area	Circumference	Diameter	Area	Circumference
341	91326.9	1071.3	376	111036	1181.2
342	91863.3	1074.4	377	111628	1184.4
343	92401.3	1077.6	378	112221	1187.5
344	92940.9	1080.7	379	112815	1190.7
345	93482.0	1083.8	380	113411	1193.8
346	94024.7	1087.0	381	114009	1196.9
347	94569.0	1090.1	382	114608	1200.1
348	95114.9	1093.3	383	115209	1203.2
349	95662.3	1096.4	384	115812	1206.4
350	96211.3	1099.6	385	116416	1209.5
351	96761.8	1102.7	386	117021	1212.7
352	97314.0	1105.8	387	117628	1215.8
353	97867.7	1109.0	388	118237	1218.9
354	98423.0	1112.1	389	118847	1222.1
355	98979.8	1115.3	390	119459	1225.2
356	99538.2	1118.4	391	120072	1228.4
357	100098	1121.5	392	120687	1231.5
358	100660	1124.7	393	121304	1234.6
359	101223	1127.8	394	121922	1237.8
360	101788	1131.0	395	122542	1240.9
361	102354	1134.1	396	123163	1244.1
362	102922	1137.3	397	123786	1247.2
363	103491	1140.4	398	124410	1250.4
364	104062	1143.5	399	125036	1253.5
365	104635	1146.7	400	125664	1256.6
366	105209	1149.8	401	126293	1259.8
367	105785	1153.0	402	126923	1262.9
368	106362	1156.1	403	127556	1266.1
369	106941	1159.2	404	128190	1269.2
370	107521	1162.4	405	128825	1272.3
371	108103	1165.5	406	129462	1275.5
372	108687	1168.7	407	130100	1278.6
373	109272	1171.8	408	130741	1281.8
374	109858	1175.0	409	131382	1284.9
375	110447	1178.1	410	132025	1288.1

AREAS AND CIRCUMFERENCES OF CIRCLES.—411 to 478

Diameter	Area	Circumference	Diameter	Area	Circumference
411	132670	1291.2	445	155528	1398.0
412	133317	1294.3	446	156228	1401.2
413	133965	1297.5	447	156930	1404.3
414	134614	1300.6	448	157633	1407.4
415	135265	1303.8	449	158337	1410.6
416	135918	1306.9	450	159043	1413.7
417	136572	1310.0	451	159751	1416.9
418	137228	1313.2	452	160460	1420.0
419	137885	1316.3	453	161171	1423.1
420	138544	1319.5	454	161883	1426.3
421	139205	1322.6	455	162597	1429.4
422	139867	1325.8	456	163313	1432.6
423	140531	1328.9	457	164030	1435.7
424	141196	1332.0	458	164748	1438.9
425	141863	1335.2	459	165468	1442.0
426	142531	1338.3	460	166190	1445.1
427	143201	1341.5	461	166914	1448.3
428	143872	1344.6	462	167639	1451.4
429	144545	1347.7	463	168365	1454.6
430	145220	1350.9	464	169093	1457.7
431	145896	1354.0	465	169823	1460.8
432	146574	1357.2	466	170554	1464.0
433	147254	1360.3	467	171287	1467.1
434	147934	1363.5	468	172021	1470.3
435	148617	1366.6	469	172757	1473.4
436	149301	1369.7	470	173494	1476.5
437	149987	1372.9	471	174234	1479.7
438	150674	1376.0	472	174974	1482.8
439	151363	1379.2	473	175716	1486.0
440	152053	1382.3	474	176460	1489.1
441	152745	1385.4	475	177205	1492.3
442	153439	1388.6	476	177952	1495.4
443	154134	1391.7	477	178701	1498.5
444	154830	1394.9	478	179451	1501.7

AREAS AND CIRCUMFERENCES OF CIRCLES.—479 to 546

Diameter	Area	Circumference	Diameter	Area	Circumference
479	180203	1504.8	513	206692	1611.6
480	180956	1508.0	514	207499	1614.8
481	181711	1511.1	515	208307	1617.9
482	182467	1514.3	516	209117	1621.1
483	183225	1517.4	517	209928	1624.2
484	183984	1520.5	518	210741	1627.3
485	184745	1523.7	519	211556	1630.5
486	185508	1526.8	520	212372	1633.6
487	186272	1530.0	521	213189	1636.8
488	187038	1533.1	522	214008	1639.9
489	187805	1536.2	523	214829	1643.1
490	188574	1539.4	524	215651	1646.2
491	189345	1542.5	525	216475	1649.3
492	190117	1545.7	526	217301	1652.5
493	190890	1548.8	527	218128	1655.6
494	191665	1551.9	528	218956	1658.8
495	192442	1555.1	529	219787	1661.9
496	193221	1558.2	530	220618	1665.0
497	194000	1561.4	531	221452	1668.2
498	194782	1564.5	532	222287	1671.3
499	195565	1567.7	533	223123	1674.5
500	196350	1570.8	534	223961	1677.6
501	197136	1573.9	535	224801	1680.8
502	197923	1577.1	536	225642	1683.9
503	198713	1580.2	537	226484	1687.0
504	199504	1583.4	538	227329	1690.2
505	200296	1586.5	539	228175	1693.3
506	201090	1589.7	540	229022	1696.5
507	201886	1592.8	541	229871	1699.6
508	202683	1595.9	542	230722	1702.7
509	203482	1599.1	543	231574	1705.9
510	204282	1602.2	544	232428	1709.0
511	205084	1605.4	545	233283	1712.2
512	205887	1608.5	546	234140	1715.3

AREAS AND CIRCUMFERENCES OF CIRCLES.—547-614

Diameter	Area	Circumference	Diameter	Area	Circumference
547	234998	1718.5	581	265120	1825.3
548	235858	1721.6	582	266033	1828.4
549	236720	1724.7	583	266948	1831.6
550	237583	1727.9	584	267865	1834.7
			585	268783	1837.8
551	238448	1731.0			
552	239314	1734.2	586	269703	1841.0
553	240182	1737.3	587	270624	1844.1
554	241051	1740.4	588	271547	1847.3
555	241922	1743.6	589	272471	1850.4
			590	273397	1853.5
556	242795	1746.7			
557	243669	1749.9	591	274325	1856.7
558	244545	1753.0	592	275254	1859.8
559	245422	1756.2	593	276184	1863.0
560	246301	1759.3	594	277117	1866.1
			595	278051	1869.3
561	247181	1762.4			
562	248063	1765.6	596	278986	1872.4
563	248947	1768.7	597	279923	1875.5
564	249832	1771.9	598	280862	1878.7
565	250719	1775.0	599	281802	1881.8
			600	282743	1885.0
566	251607	1778.1			
567	252497	1781.3	601	283687	1881.1
568	253388	1784.4	602	284631	1891.2
569	254281	1787.6	603	285578	1894.4
570	255176	1790.7	604	286526	1897.5
			605	287475	1900.7
571	256072	1793.9			
572	256970	1797.0	606	288426	1903.8
573	257869	1800.1	607	289379	1907.0
574	258770	1803.3	608	290333	1910.1
575	259672	1806.4	609	291289	1913.2
			610	292247	1916.4
576	260576	1809.6			
577	261482	1812.7	611	293206	1919.5
578	262389	1815.8	612	294166	1922.7
579	263298	1819.0	613	295128	1925.8
580	264208	1822.1	614	296092	1928.9

AREAS AND CIRCUMFERENCES OF CIRCLES.—615 to 682

Diameter	Area	Circumference	Diameter	Area	Circumference
615	297057	1932.1	649	330810	2038.9
616	298024	1935.2	650	331831	2042.0
617	298992	1938.4	651	332853	2045.2
618	299962	1941.5	652	333876	2048.3
619	300934	1944.7	653	334901	2051.5
620	301907	1947.8	654	335927	2054.6
621	302882	1950.9	655	336955	2057.7
622	303858	1954.1	656	337985	2060.9
623	304836	1957.2	657	339016	2064.0
624	305815	1960.4	658	340049	2067.2
625	306796	1963.5	659	341083	2070.3
626	307779	1966.6	660	342119	2073.5
627	308763	1969.8	661	343157	2076.6
628	309748	1972.9	662	344196	2079.7
629	310736	1976.1	663	345237	2082.9
630	311725	1979.2	664	346279	2086.0
631	312715	1982.4	665	347323	2089.2
632	313707	1985.5	666	348368	2092.3
633	314700	1988.6	667	349415	2095.4
634	315696	1991.8	668	350464	2098.6
635	316692	1994.9	669	351514	2101.7
636	317690	1998.1	670	352565	2104.9
637	318690	2001.2	671	353618	2108.0
638	319692	2004.3	672	354673	2111.2
639	320695	2007.5	673	355730	2114.3
640	321699	2010.6	674	356788	2117.4
641	322705	2013.8	675	357847	2120.6
642	323713	2016.9	676	358908	2123.7
643	324722	2020.0	677	359971	2126.9
644	325733	2023.2	678	361035	2130.0
645	326745	2026.3	679	362101	2133.1
646	327759	2029.5	680	363168	2136.3
647	328775	2032.6	681	364237	2139.4
648	329792	2035.8	682	365308	2142.6

AREAS AND CIRCUMFERENCES OF CIRCLES.—683 to 750

Diameter	Area	Circumference	Diameter	Area	Circumference
683	366380	2145.7	717	403765	2252.5
684	367453	2148.9	718	404892	2255.7
685	368528	2152.0	719	406020	2258.8
686	369605	2155.1	720	407150	2261.9
687	370684	2158.3	721	408282	2265.1
688	371764	2161.4	722	409416	2268.2
689	372845	2164.6	723	410550	2271.4
690	373928	2167.7	724	411687	2274.5
691	375013	2170.8	725	412825	2277.7
692	376099	2174.0	726	413965	2280.8
693	377187	2177.1	727	415106	2283.9
694	378276	2180.3	728	416248	2287.1
695	379367	2183.4	729	417393	2290.2
696	380459	2186.6	730	418539	2293.4
697	381554	2189.7	731	419686	2296.5
698	382649	2192.8	732	420835	2299.7
699	383746	2196.0	733	421986	2302.8
700	384845	2199.1	734	423138	2305.9
701	385945	2202.3	735	424292	2309.1
702	387047	2205.4	736	425447	2312.2
703	388151	2208.5	737	426604	2315.4
704	389256	2211.7	738	427762	2318.5
705	390363	2214.8	739	428922	2321.6
706	391471	2218.0	740	430084	2324.8
707	392580	2221.1	741	431247	2327.9
708	393692	2224.3	742	432412	2331.1
709	394805	2227.4	743	433578	2334.2
710	395919	2230.5	744	434746	2337.3
711	397035	2233.7	745	435916	2340.5
712	398153	2236.8	746	437087	2343.6
713	399272	2240.0	747	438259	2346.8
714	400393	2243.1	748	439433	2349.9
715	401515	2246.2	749	440609	2353.1
716	402629	2249.4	750	441786	2356.2

AREAS AND CIRCUMFERENCES OF CIRCLES.—751 TO 818

Diameter	*Area*	*Circumference*	*Diameter*	*Area*	*Circumference*
751	442965	2359·3	785	483982	2466·2
752	444146	2362·5	786	485216	2469·3
753	445328	2365·6	787	486451	2472·4
754	446511	2368·8	788	487688	2475·6
755	447697	2371·9	789	488927	2478·7
756	448883	2375·0	790	490167	2481·9
757	450072	2378·2	791	491409	2485·0
758	451262	2381·3	792	492652	2488·1
759	452453	2384·5	793	493897	2491·3
760	453646	2387·6	794	495143	2494·4
761	454841	2390·8	795	496391	2497·6
762	456037	2393·9	796	497641	2500·7
763	457234	2397·0	797	498892	2503·8
764	458434	2400·2	798	500145	2507·0
765	459635	2403·3	799	501399	2510·1
766	460837	2406·5	800	502655	2513·3
767	462042	2409·6	801	503912	2516·4
768	463247	2412·7	802	505171	2519·6
769	464454	2415·9	803	506432	2522·7
770	465663	2419·0	804	507694	2525·8
771	466873	2422·2	805	508958	2529·0
772	468085	2425·3	806	510223	2532·1
773	469298	2428·5	807	511490	2535·3
774	470513	2431·6	808	512758	2538·4
775	471730	2434·7	809	514028	2541·5
776	472948	2437·9	810	515300	2544·7
777	474168	2441·0	811	516573	2547·8
778	475389	2444·2	812	517848	2551·0
779	476612	2447·3	813	519124	2554·1
780	477836	2450·4	814	520402	2557·3
781	479062	2453·6	815	521681	2560·4
782	480290	2456·7	816	522962	2563·5
783	481519	2459·9	817	524245	2566·7
784	482750	2463·0	818	525529	2569·8

AREAS AND CIRCUMFERENCES OF CIRCLES.—819 TO 886

Diameter	Area	Circumference	Diameter	Area	Circumference
819	526814	2573·0	853	571463	2679·8
820	528102	2576·1	854	572803	2682·9
821	529391	2579·2	855	574146	2686·1
822	530681	2582·4	856	575490	2689·2
823	531973	2585·5	857	576835	2692·3
824	533267	2588·7	858	578182	2695·5
825	534562	2591·8	859	579530	2698·6
826	535858	2595·0	860	580880	2701·8
827	537157	2598·1	861	582232	2704·9
828	538456	2601·2	862	583585	2708·1
829	539758	2604·4	863	584940	2711·2
830	541061	2607·5	864	586297	2714·3
831	542365	2610·7	865	587655	2717·5
832	543671	2613·8	866	589014	2720·6
833	544979	2616·9	867	590375	2723·8
834	546288	2620·1	868	591738	2726·9
835	547599	2623·2	869	593102	2730·0
836	548912	2626·4	870	594468	2733·2
837	550226	2629·5	871	595835	2736·3
838	551541	2632·7	872	597204	2739·5
839	552858	2635·8	873	598575	2742·6
840	554177	2638·9	874	599947	2745·8
841	555497	2642·1	875	601320	2748·9
842	556819	2645·2	876	602696	2752·0
843	558142	2648·4	877	604073	2755·2
844	559467	2651·5	878	605451	2758·3
845	560794	2654·6	879	606831	2761·5
846	562122	2657·8	880	608212	2764·6
847	563452	2660·9	881	609595	2767·7
848	564783	2664·1	882	610980	2770·9
849	566116	2667·2	883	612366	2774·0
850	567450	2670·4	884	613754	2777·2
851	568786	2673·5	885	615143	2780·3
852	570124	2676·6	886	616534	2783·5

AREAS AND CIRCUMFERENCES OF CIRCLES.—887 TO 954

Diameter	*Area*	*Circumference*	*Diameter*	*Area*	*Circumference*
887	617927	2786·6	921	666207	2893·4
888	619321	2789·7	922	667654	2896·5
889	620717	2792·9	923	669103	2899·7
890	622114	2796·0	924	670554	2902·8
891	623513	2799·2	925	672006	2906·0
892	624913	2802·3	926	673460	2909·1
893	626315	2805·4	927	674915	2912·3
894	627718	2808·6	928	676372	2915·4
895	629124	2811·7	929	677831	2918·5
896	630530	2814·9	930	679291	2921·7
897	631938	2818·0	931	680752	2924·8
898	633348	2821·2	932	682216	2928·0
899	634760	2824·3	933	683680	2931·1
900	636173	2827·4	934	685147	2934·2
901	637587	2830·6	935	686615	2937·4
902	639003	2833·7	936	688084	2940·5
903	640421	2836·9	937	689555	2943·7
904	641840	2840·0	938	691028	2946·8
905	643261	2843·1	939	692502	2950·0
906	644683	2846·3	940	693978	2953·1
907	646107	2849·4	941	695455	2956·2
908	647533	2852·6	942	696934	2959·4
909	648960	2855·7	943	698415	2962·5
910	650388	2858·8	944	699897	2965·7
911	651818	2862·0	945	701380	2968·8
912	653250	2865·1	946	702865	2971·9
913	654684	2868·3	947	704352	2975·1
914	656118	2871·4	948	705840	2978·2
915	657555	2874·6	949	707330	2981·4
916	658993	2877·7	950	708822	2984·5
917	660433	2880·8	951	710315	2987·7
918	661874	2884·0	952	711809	2990·8
919	663317	2887·1	953	713306	2993·9
920	664761	2890·3	954	714803	2997·1

AREAS AND CIRCUMFERENCES OF CIRCLES.—955 TO 999

Diameter	*Area*	*Circumference*	*Diameter*	*Area*	*Circumference*
955	716303	3000·2	978	751221	3072·5
956	717804	3003·4	979	752758	3075·6
957	719306	3006·5	980	754296	3078·8
958	720810	3009·6	981	755837	3081·9
959	722316	3012·8	982	757378	3085·0
960	723823	3015·9	983	758922	3088·2
961	725332	3019·1	984	760466	3091·3
962	726842	3022·2	985	762013	3094·5
963	728354	3025·4	986	763561	3097·6
964	729867	3028·5	987	765111	3100·8
965	731382	3031·6	988	766662	3103·9
966	732899	3034·8	989	768214	3107·0
967	734417	3037·9	990	769769	3110·2
968	735937	3041·1	991	771325	3113·3
969	737458	3044·2	992	772882	3116·5
970	738981	3047·3	993	774441	3119·6
971	740506	3050·5	994	776002	3122·7
972	742032	3053·6	995	777564	3125·9
973	743559	3056·8	996	779128	3129·0
974	745088	3059·9	997	780693	3132·2
975	746619	3063·1	998	782260	3135·2
976	748151	3066·2	999	783828	3138·5
977	749685	3069·3			

POWERS AND ROOTS

No.	Squares.	Cubes.	Square Roots.	Cube Roots.
1	1	1	1.000	1.000
2	4	8	1.414	1.260
3	9	27	1.732	1.442
4	16	64	2.000	1.587
5	25	125	2.236	1.710
6	36	216	2.449	1.817
7	49	343	2.646	1.913
8	64	512	2.828	2.000
9	81	729	3.000	2.080
10	100	1 000	3.162	2.154
11	121	1 331	3.317	2.224
12	144	1 728	3.464	2.289
13	169	2 197	3.606	2.351
14	196	2 744	3.742	2.410
15	225	3 375	3.873	2.466
16	256	4 096	4.000	2.520
17	289	4 913	4.123	2.571
18	324	5 832	4.243	2.621
19	361	6 859	4.359	2.668
20	400	8 000	4.472	2.714
21	441	9 261	4.583	2.759
22	484	10 648	4.690	2.802
23	529	12 167	4.796	2.844
24	576	13 824	4.899	2.884
25	625	15 625	5.000	2.924
26	676	17 576	5.099	2.962
27	729	19 683	5.196	3.000
28	784	21 952	5.292	3.037
29	841	24 389	5.385	3.072
30	900	27 000	5.477	3.107
31	961	29 791	5.568	3.141
32	1 024	32 768	5.657	3.175
33	1 089	35 937	5.745	3.208
34	1 156	39 304	5.831	3.240
35	1 225	42 875	5.916	3.271

POWERS AND ROOTS (*continued*)

No.	Squares.	Cubes	Square Roots.	Cube Roots.
36	1 296	46 656	6.000	3.302
37	1 369	50 653	6.083	3.332
38	1 444	54 872	6.164	3.362
39	1 521	59 319	6.245	3.391
40	1 600	64 000	6.325	3.420
41	1 681	68 921	6.403	3.448
42	1 764	74 088	6.481	3.476
43	1 849	79 507	6.557	3.503
44	1 936	85 184	6.633	3.530
45	2 025	91 125	6.708	3.557
46	2 116	97 336	6.782	3.583
47	2 209	103 823	6.856	3.609
48	2 304	110 592	6.928	3.634
49	2 401	117 649	7.000	3.659
50	2 500	125 000	7.071	3.684
51	2 601	132 651	7.141	3.708
52	2 704	140 608	7.211	3.733
53	2 809	148 877	7.280	3.756
54	2 916	157 464	7.348	3.780
55	3 025	166 375	7.416	3.803
56	3 136	175 616	7.483	3.826
57	3 249	185 193	7.550	3.849
58	3 364	195 112	7.616	3.871
59	3 481	205 379	7.681	3.893
60	3 600	216 000	7.746	3.915
61	3 721	226 981	7.810	3.936
62	3 844	238 328	7.874	3.958
63	3 969	250 047	7.937	3.979
64	4 096	262 144	8.000	4.000
65	4 225	274 625	8.062	4.021
66	4 356	287 496	8.124	4.041
67	4 489	300 763	8.185	4.062
68	4 624	314 432	8.246	4.082
69	4 761	328 509	8.307	4.102

POWERS AND ROOTS (*continued*)

No.	Squares.	Cubes.	Square Roots.	Cube Roots.
70	4 900	343 000	8.367	4.121
71	5 041	357 911	8.426	4.141
72	5 184	373 248	8.485	4.160
73	5 329	389 017	8.544	4.179
74	5 476	405 224	8.602	4.198
75	5 625	421 875	8.660	4.217
76	5 776	438 976	8.718	4.236
77	5 929	456 533	8.775	4.254
78	6 084	474 552	8.832	4.273
79	6 241	493 039	8.888	4.291
80	6 400	512 000	8.944	4.309
81	6 561	531 441	9.000	4.327
82	6 724	551 368	9.055	4.344
83	6 889	571 787	9.110	4.362
84	7 056	592 704	9.165	4.380
85	7 225	614 125	9.220	4.397
86	7 396	636 056	9.274	4.414
87	7 569	658 503	9.327	4.431
88	7 744	681 472	9.381	4.448
89	7 921	704 969	9.434	4.465
90	8 100	729 000	9.487	4.481
91	8 281	753 571	9.539	4.498
92	8 464	778 688	9.592	4.514
93	8 649	804 357	9.644	4.531
94	8 836	830 584	9.695	4.547
95	9 025	857 375	9.747	4.563
96	9 216	884 736	9.798	4.579
97	9 409	912 673	9.849	4.595
98	9 604	941 192	9.899	4.610
99	9 801	970 299	9.950	4.626
100	10 000	1 000 000	10.000	4.642

SQUARES, CUBES AND RECIPROCALS

Nos.	Squares	Cubes	Reciprocals	Nos.	Squares	Cubes	Reciprocals
1	1	1	1.000000000	17	289	4913	.058823529
2	4	8	.500000000	18	324	5832	.055555556
3	9	27	.333333333	19	361	6859	.052631579
4	16	64	.250000000	20	400	8000	050000000
5	25	125	.200000000	21	441	9261	.047619048
6	36	216	.166666667	22	484	10648	.045454545
7	49	343	.142857143	23	529	12167	.043478261
8	64	512	.125000000	24	576	13824	.041666667
9	81	729	.111111111	25	625	15625	.040000000
10	100	1000	.100000000	26	676	17576	.038461538
11	121	1331	.090909091	27	729	19683	.037037037
12	144	1728	.083333333	28	784	21952	.035714286
13	169	2197	.076923077	29	841	24389	.034482759
14	196	2744	.071428571	30	900	27000	.033333333
15	225	3375	.066666667	31	961	29791	.032258065
16	256	4096	.062500000	32	1024	32768	.031250000

SQUARES, CUBES AND RECIPROCALS (Continued)

Nos.	Squares	Cubes	Reciprocals	Nos.	Squares	Cubes	Reciprocals
33	1089	35937	.030303030	50	2500	125000	.020000000
34	1156	39304	.029411765	51	2601	132651	.019607843
35	1225	42875	.028571429	52	2704	140608	.019230769
36	1296	46656	.027777778	53	2809	148877	.018867925
37	1369	50653	.027027027	54	2916	157464	.018518519
38	1444	54872	.026315789	55	3025	166375	.018181818
39	1521	59319	.025641026	56	3136	175616	.017857143
40	1600	64000	.025000000	57	3249	185193	.017543860
41	1681	68921	.024390244	58	3364	195112	.017241379
42	1764	74088	.023809524	59	3481	205379	.016949153
43	1849	79507	.023255814	60	3600	216000	.016666667
44	1936	85184	.022727273	61	3721	226981	.016393443
45	2025	91125	.022222222	62	3844	238328	.016129032
46	2116	97336	.021739130	63	3969	250047	.015873016
47	2209	103823	.021276596	64	4096	262144	.015625000
48	2304	110592	.020833333	65	4225	274625	.015384615
49	2401	117649	.020408163	66	4356	287496	.015151515

SQUARES, CUBES AND RECIPROCALS (Continued)

Nos.	Squares	Cubes	Reciprocals	Nos.	Squares	Cubes	Reciprocals
67	4489	300763	.014925373	84	7056	592704	.011900762
68	4624	314432	.014705882	85	7225	614125	.011764746
69	4761	328509	.014492754	86	7396	636056	.011627907
70	4900	343000	.014285714	87	7569	658503	.011494253
71	5041	357911	.014084507	88	7744	681472	.011363636
72	5184	373248	.013888889	89	7921	704969	.011235955
73	5329	389017	.013698630	90	8100	729000	.011111111
74	5476	405224	.013513514	91	8281	753571	.010989011
75	5625	421875	.013333333	92	8464	778688	.010869565
76	5776	438976	.013157895	93	8649	804357	.010752688
77	5929	456533	.012987013	94	8836	830584	.010638298
78	6084	474552	.012820513	95	9025	857375	.010526316
79	6241	493039	.012658228	96	9216	884736	.010416667
80	6400	512000	.012500000	97	9409	912673	.010309278
81	6561	531441	.012345679	98	9604	941192	.010204082
82	6724	551368	.012195122	99	9801	970299	.010101010
83	6889	571787	.012048193	100	10000	1000000	.010000000

FACTORS OF NUMBERS UP TO 9999

The tables in the following pages give the lowest and highest factors of every odd number (excluding those ending in 5) up to 9999.

To determine the factors of any number in this range,

(1) Divide it by 2 as many times as are necessary to make the quotient an odd number.

(2) Divide the odd quotient by 5 as many times as possible without leaving a remainder.

(3) Use the table to find the lowest and highest factors of the last quotient.

(4) Use the table to find the lowest and highest factors of the highest factor found by (3).

(5) Repeat this procedure until a prime number (denoted by p) is reached. This, and the previous factors form the complete list of factors of the original number.

Example 1. To determine the factors of 9044

$$\frac{9044}{2}=4522 \qquad \frac{4522}{2}=2261$$

From the table, the lowest and highest factors of 2261 are 7 and 323.

From the table, the lowest and highest factors of 323 are 17 and 19.

From the table, 19 is seen to be a prime number.

Hence, $9044=2\times2\times7\times17\times19$.

Example 2. To determine the factors of 8925

$$\frac{8925}{5}=1785 \qquad \frac{1785}{5}=357$$

From the table, $357=3\times119$.

From the table, $119=7\times17$, and 17 is a prime number.

Hence, $8925=5\times5\times3\times7\times17$.

FACTORS OF NUMBERS UP TO 9999—(*Continued*)

From to	0 100	100 200	200 300	300 400	400 500	500 600	600 700	700 800	800 900	900 1000	1000 1100	1100 1200
1	p	p	3 67	7 43	p	3 167	p	p	3 267	17 53	7 143	3 367
3	p	p	7 29	3 101	13 31	p	3 201	19 37	11 73	3 301	17 59	p
7	p	p	3 69	p	11 37	3 169	p	7 101	3 269	p	19 53	3 369
9	3 3	p	11 19	3 103	p	p	3 203	p	p	3 303	p	p
11	p	3 37	p	p	3 137	7. 73	13 47	3 237	p	p	3 337	11 101
13	p	p	3 71	p	7 59	3 171	p	23 31	3 271	11 83	p	3 371
17	p	3 39	7 31	p	3 139	11 47	p	3 239	19 43	7 131	3 339	p
19	p	7 17	3 73	11 29	p	3 173	p	p	3 273	p	p	3 373
21	3 7	11 11	13 17	3 107	p	p	3 207	7 103	p	3 307	p	19 59
23	p	3 41	p	17 19	3 141	p	7 89	3 241	p	13 71	3 341	p
27	3 9	p	p	3 109	7 61	17 31	3 209	p	p	3 309	13 79	7 161
29	p	3 43	p	7 47	3 143	23 23	17 37	3 243	p	p	3 343	p
31	p	p	3 77	p	p	3 177	p	17 43	3 277	7 133	p	3 377
33	3 11	7 19	p	3 111	p	13 41	3 211	p	7 119	3 311	p	11 103
37	p	p	3 79	p	19 23	3 179	7 91	11 67	3 279	p	17 61	3 379
39	3 13	p	p	3 113	p	7 77	3 213	p	p	3 313	p	17 67
41	p	3 47	p	11 31	3 147	p	p	3 247	29 29	p	3 347	7 163
43	p	11 13	3 81	7 49	p	3 181	p	p	3 281	23 41	7 149	3 381
47	p	3 49	13 19	p	3 149	p	p	3 249	7 121	p	3 349	31 37
49	7 7	p	3 83	p	p	3 183	11 59	7 107	3 283	13 73	p	3 383

FACTORS OF NUMBERS UP TO 9999—(*Continued*)

From to	0 100	100 200	200 300	300 400	400 500	500 600	600 700	700 800	800 900	900 1000	1000 1100	1100 1200
51	3 17	p	p	3 117	11 41	19 29	3 217	p	23 37	3 317	p	p
53	p	3 51	11 23	p	3 151	7 79	p	3 251	p	p	3 351	p
57	3 19	p	p	3 119	p	p	3 219	p	p	3 319	7 151	13 89
59	p	3 53	7 37	p	3 153	13 43	p	3 253	p	7 137	3 353	19 61
61	p	7 23	3 87	19 19	p	3 187	p	p	3 287	31 31	p	3 387
63	3 21	p	p	3 121	p	p	3 221	7 109	p	3 321	p	p
67	p	p	3 89	p	p	3 189	23 29	13 59	3 289	p	11 97	3 389
69	3 23	13 13	p	3 123	7 67	p	3 223	p	11 79	3 323	p	7 167
71	p	3 57	p	7 53	3 157	p	11 61	3 257	13 67	p	3 357	p
73	p	p	3 91	p	11 43	3 191	p	p	3 291	7 139	29 37	3 391
77	7 11	3 59	p	13 29	3 159	p	p	3 259	p	p	3 359	11 107
79	p	p	3 93	p	p	3 193	7 97	19 41	3 293	11 89	13 83	3 393
81	3 27	p	p	3 127	13 37	7 83	3 227	11 71	p	3 327	23 47	p
83	p	3 61	p	p	3 161	11 53	p	3 261	p	p	3 361	7 169
87	3 29	11 17	7 41	3 129	p	p	3 229	p	p	3 329	p	p
89	p	3 63	17 17	p	3 163	19 31	13 53	3 263	7 127	23 43	3 363	29 41
91	7 13	p	3 97	17 23	p	3 197	p	7 113	3 297	p	p	3 397
93	3 31	p	p	3 131	17 29	p	3 231	13 61	19 47	3 331	p	p
97	p	p	3 99	p	7 71	3 199	17 41	p	3 299	p	p	3 399
99	3 33	p	13 23	3 133	p	p	3 233	17 47	29 31	3 333	7 157	11 109

FACTORS OF NUMBERS UP TO 9999—(*Continued*)

From to	1200 1300	1300 1400	1400 1500	1500 1600	1600 1700	1700 1800	1800 1900	1900 2000	2000 2100	2100 2200	2200 2300	2300 2400
1	p	p	3 467	19 79	p	3 567	p	p	3 667	11 191	31 71	3 767
3	3 401	p	23 61	3 501	7 229	13 131	3 601	11 173	p	3 701	p	7 329
7	17 71	p	3 469	11 137	p	3 569	13 139	p	3 669	7 301	p	3 769
9	3 403	7 187	p	3 503	p	p	3 603	23 83	7 287	3 703	47 47	p
11	7 173	3 437	17 83	p	3 537	29 59	p	3 637	p	p	3 737	p
13	p	13 101	3 471	17 89	p	3 571	7 259	p	3 671	p	p	3 771
17	p	3 439	13 109	37 41	3 539	17 101	23 79	3 639	p	29 73	3 739	7 331
19	23 53	p	3 473	7 217	p	3 573	17 107	19 101	3 673	13 163	7 317	3 773
21	3 407	p	7 203	3 507	p	p	3 607	17 113	43 47	3 707	p	11 211
23	p	3 441	p	p	3 541	p	p	3 641	7 289	11 193	3 741	23 101
27	3 409	p	p	3 509	p	11 157	3 609	41 47	p	3 709	17 131	13 179
29	p	3 443	p	11 139	3 543	7 247	31 59	3 643	p	p	3 743	17 137
31	p	11 121	3 477	p	7 233	3 577	p	p	3 677	p	23 97	3 777
33	3 411	31 43	p	3 511	23 71	p	3 611	p	19 107	3 711	7 319	p
37	p	7 191	3 479	29 53	p	3 579	11 167	13 149	3 679	p	p	3 779
39	3 413	13 103	p	3 513	11 149	37 47	3 613	7 277	p	3 713	p	p
41	17 73	3 447	11 131	23 67	3 547	p	7 263	3 647	13 157	p	3 747	p
43	11 113	17 79	3 481	p	31 53	3 581	19 97	29 67	3 681	p	p	3 781
47	29 43	3 449	p	7 221	3 549	p	p	3 649	23 89	19 113	3 749	p
49	p	19 71	3 483	p	17 97	3 583	43 43	p	3 683	7 307	13 173	3 783

FACTORS OF NUMBERS UP TO 9999—(*Continued*)

From to	1200 1300	1300 1400	1400 1500	1500 1600	1600 1700	1700 1800	1800 1900	1900 2000	2000 2100	2100 2200	2200 2300	2300 2400
51	3 417	7 193	p	3 517	13 127	17 103	3 617	p	7 293	3 717	p	p
53	7 179	3 451	p	p	3 551	p	17 109	3 651	p	p	3 751	13 181
57	3 419	23 59	31 47	3 519	p	7 251	3 619	19 103	11 187	3 719	37 61	p
59	p	3 453	p	p	3 553	p	11 169	3 653	29 71	17 127	3 753	7 337
61	13 97	p	3 487	7 223	11 151	3 587	p	37 53	3 687	p	7 323	3 787
63	3 421	29 47	7 209	3 521	p	41 43	3 621	13 151	p	3 721	31 73	17 139
67	7 181	p	3 489	p	p	3 589	p	7 281	3 689	11 197	p	3 789
69	3 423	37 37	13 113	3 523	p	29 61	3 623	11 179	p	3 723	p	23 103
71	31 41	3 457	p	p	3 557	7 253	p	3 657	19 109	13 167	3 757	p
73	19 67	p	3 491	11 143	7 239	3 591	p	p	3 691	41 53	p	3 791
77	p	3 459	7 211	19 83	3 559	p	p	3 659	31 67	7 311	3 759	p
79	p	7 197	3 493	p	23 73	3 593	p	p	3 693	p	43 53	3 793
81	3 427	p	p	3 527	41 41	13 137	3 627	7 283	p	3 727	p	p
83	p	3 461	p	p	3 561	p	7 269	3 661	p	37 59	3 761	p
87	3 429	19 73	p	3 529	7 241	p	3 629	p	p	3 729	p	7 341
89	p	3 463	p	7 227	3 563	p	p	3 663	p	11 199	3 763	p
91	p	13 107	3 497	37 43	19 89	3 597	31 61	11 181	3 697	7 313	29 79	3 797
93	3 431	7 199	p	3 531	p	11 163	3 631	p	7 299	3 731	p	p
97	p	11 127	3 499	p	p	3 599	7 271	p	3 699	13 169	p	3 799
99	3 433	p	p	3 533	p	7 257	3 633	p	p	3 733	11 209	p

FACTORS OF NUMBERS UP TO 9999—(*Continued*)

From to	2400 2500	2500 2600	2600 2700	2700 2800	2800 2900	2900 3000	3000 3100	3100 3200	3200 3300	3300 3400	3400 3500	3500 3600
1	7 343	41 61	3 867	37 73	p	3 967	p	7 443	3 1067	p	19 179	3 1167
3	3 801	p	19 137	3 901	p	p	3 1001	29 107	p	3 1101	41 83	31 113
7	29 83	23 109	3 869	p	7 401	3 969	31 97	13 239	3 1069	p	p	3 1169
9	3 803	13 193	p	3 903	53 53	p	3 1003	p	p	3 1103	7 487	11 319
11	p	3 837	7 373	p	3 937	41 71	p	3 1037	13 247	7 473	3 1137	p
13	19 127	7 359	3 871	p	29 97	3 971	23 131	11 283	3 1071	p	p	3 1171
17	p	3 839	p	11 247	3 939	p	7 431	3 1039	p	31 107	3 1139	p
19	41 59	11 229	3 873	p	p	3 973	p	p	3 1073	p	13 263	3 1173
21	3 807	p	p	3 907	7 403	23 127	3 1007	p	p	3 1107	11 311	7 503
23	p	3 841	43 61	7 389	3 941	37 79	p	3 1041	11 293	p	3 1141	13 271
27	3 809	7 361	37 71	3 909	11 257	p	3 1009	53 59	7 461	3 1109	23 149	p
29	7 347	3 843	11 239	p	3 943	29 101	13 233	3 1043	p	p	3 1143	p
31	11 221	p	3 877	p	19 149	3 977	7 433	31 101	3 1077	p	47 73	3 1177
33	3 811	17 149	p	3 911	p	7 419	3 1011	13 241	53 61	3 1111	p	p
37	p	43 59	3 879	7 391	p	3 979	p	p	3 1079	47 71	7 491	3 1179
39	3 813	p	7 377	3 913	17 167	p	3 1013	43 73	41 79	3 1113	19 181	p
41	p	3 847	19 139	p	3 947	17 173	p	3 1047	7 463	13 257	3 1147	p
43	7 349	p	3 881	13 211	p	3 981	17 179	7 449	3 1081	p	11 313	3 1181
47	p	3 849	p	41 67	3 949	7 421	11 277	3 1049	17 191	p	3 1149	p
49	31 79	p	3 883	p	7 407	3 983	p	47 67	3 1083	17 197	p	3 1183

FACTORS OF NUMBERS UP TO 9999—(*Continued*)

From to	2400 2500	2500 2600	2600 2700	2700 2800	2800 2900	2900 3000	3000 3100	3100 3200	3200 3300	3300 3400	3400 3500	3500 3600
51	3 817	p	11 241	3 917	p	13 227	3 1017	23 137	p	3 1117	7 493	53 67
53	11 223	3 851	7 379	p	3 951	p	43 71	3 1051	p	7 479	3 1151	11 323
57	3 819	p	p	3 919	p	p	3 1019	7 451	p	3 1119	p	p
59	p	3 853	p	31 89	3 953	11 269	7 437	3 1053	p	p	3 1153	p
61	23 107	13 197	3 887	11 251	p	3 987	p	29 109	3 1087	p	p	3 1187
63	3 821	11 233	p	3 921	7 409	p	3 1021	p	13 251	3 1121	p	7 509
67	p	17 151	3 889	p	47 61	3 989	p	p	3 1089	7 481	p	3 1189
69	3 823	7 367	17 157	3 923	19 151	p	3 1023	p	7 467	3 1123	p	43 83
71	7 353	3 857	p	17 163	3 957	p	37 83	3 1057	p	p	3 1157	p
73	p	31 83	3 891	47 59	13 221	3 991	7 439	19 167	3 1091	p	23 151	3 1191
77	p	3 859	p	p	3 959	13 229	17 181	3 1059	29 113	11 307	3 1159	7 511
79	37 67	p	3 893	7 397	p	3 993	p	11 289	3 1093	31 109	7 497	3 1193
81	3 827	29 89	7 383	3 927	43 67	11 271	3 1027	p	17 193	3 1127	59 59	p
83	13 191	3 861	p	11 253	3 961	19 157	p	3 1061	7 469	17 199	3 1161	p
87	3 829	13 199	p	3 929	p	29 103	3 1029	p	19 173	3 1129	11 317	17 211
89	19 131	3 863	p	p	3 963	7 427	p	3 1063	11 299	p	3 1163	37 97
91	47 53	p	3 897	p	7 413	3 997	11 281	p	3 1097	p	p	3 1197
93	3 831	p	p	3 931	11 263	41 73	3 1031	31 103	37 89	3 1131	7 499	p
97	11 227	7 371	3 899	p	p	3 999	19 163	23 139	3 1099	43 79	13 269	3 1199
99	3 833	23 113	p	3 933	13 223	p	3 1033	7 457	p	3 1133	p	59 61

FACTORS OF NUMBERS UP TO 9999—(*Continued*)

From to	3600 3700	3700 3800	3800 3900	3900 4000	4000 4100	4100 4200	4200 4300	4300 4400	4400 4500	4500 4600	4600 4700	4700 4800
1	13 277	p	3 1267	47 83	p	3 1367	p	11 391	3 1467	7 643	43 107	3 1567
3	3 1201	7 529	p	3 1301	p	11 373	3 1401	13 331	7 629	3 1501	p	p
7	p	11 337	3 1269	p	p	3 1369	7 601	59 73	3 1469	p	17 271	3 1569
9	3 1203	p	13 293	3 1303	19 211	7 587	3 1403	31 139	p	3 1503	11 419	17 277
11	23 157	3 1237	37 103	p	3 1337	p	p	3 1437	11 401	13 347	3 1537	7 673
13	p	47 79	3 1271	7 559	p	3 1371	11 383	19 227	3 1471	p	7 659	3 1571
17	p	3 1239	11 347	p	3 1339	23 179	p	3 1439	7 631	p	3 1539	53 89
19	7 517	p	3 1273	p	p	3 1373	p	7 617	3 1473	p	31 149	3 1573
21	3 1207	61 61	p	3 1307	p	13 317	3 1407	29 149	p	3 1507	p	p
23	p	3 1241	p	p	3 1341	7 589	41 103	3 1441	p	p	3 1541	p
27	3 1209	p	43 89	3 1309	p	p	3 1409	p	19 233	3 1509	7 661	29 163
29	19 191	3 1243	7 547	p	3 1343	p	p	3 1443	43 103	7 647	3 1543	p
31	p	7 533	3 1277	p	29 139	3 1377	p	61 71	3 1477	23 197	11 421	3 1577
33	3 1211	p	p	3 1311	37 109	p	3 1411	7 619	11 403	3 1511	41 113	p
37	p	37 101	3 1279	31 127	11 367	3 1379	19 223	p	3 1479	13 349	p	3 1579
39	3 1213	p	11 349	3 1313	7 577	p	3 1413	p	23 193	3 1513	p	7 677
41	11 331	3 1247	23 167	7 563	3 1347	41 101	p	3 1447	p	19 239	3 1547	11 431
43	p	19 197	3 1281	p	13 311	3 1381	p	43 101	3 1481	7 649	p	3 1581
47	7 521	3 1249	p	p	3 1349	11 377	31 137	3 1449	p	p	3 1549	47 101
49	41 89	23 163	3 1283	11 359	p	3 1383	7 607	p	3 1483	p	p	3 1583

FACTORS OF NUMBERS UP TO 9999—(*Continued*)

From to	3600 3700	3700 3800	3800 3900	3900 4000	4000 4100	4100 4200	4200 4300	4300 4400	4400 4500	4500 4600	4600 4700	4700 4800
51	3 1217	11 341	p	3 1317	p	7 593	3 1417	19 229	p	3 1517	p	p
53	13 281	3 1251	p	59 67	3 1351	p	p	3 1451	61 73	29 157	3 1551	7 679
57	3 1219	13 289	7 551	3 1319	p	p	3 1419	p	p	3 1519	p	67 71
59	p	3 1253	17 227	37 107	3 1353	p	p	3 1453	7 637	47 97	3 1553	p
61	7 523	p	3 1287	17 233	31 131	3 1387	p	7 623	3 1487	p	59 79	3 1587
63	3 1221	53 71	p	3 1321	17 239	23 181	3 1421	p	p	3 1521	p	11 433
67	19 193	p	3 1289	p	7 581	3 1389	17 251	11 397	3 1489	p	13 359	3 1589
69	3 1223	p	53 73	3 1323	13 313	11 379	3 1423	17 257	41 109	3 1523	7 667	19 251
71	p	3 1257	7 553	11 361	3 1357	43 97	p	3 1457	17 263	7 653	3 1557	13 367
73	p	7 539	3 1291	29 137	p	3 1391	p	p	3 1491	17 269	p	3 1591
77	p	3 1259	p	41 97	3 1359	p	7 611	3 1459	11 407	23 199	3 1559	17 281
79	13 283	p	3 1293	23 173	p	3 1393	11 389	29 151	3 1493	19 241	p	3 1593
81	3 1227	19 199	p	3 1327	7 583	37 113	3 1427	13 337	p	3 1527	31 151	7 683
83	29 127	3 1261	11 353	7 569	3 1361	47 89	p	3 1461	p	p	3 1561	p
87	3 1229	7 541	13 299	3 1329	61 67	53 79	3 1429	41 107	7 641	3 1529	43 109	p
89	7 527	3 1263	p	p	3 1363	59 71	p	3 1463	67 67	13 353	3 1563	p
91	p	17 223	3 1297	13 307	p	3 1397	7 613	p	3 1497	p	p	3 1597
93	3 1231	p	17 229	3 1331	p	7 599	3 1431	23 191	p	3 1531	13 361	p
97	p	p	3 1299	7 571	17 241	3 1399	p	p	3 1499	p	7 671	3 1599
99	3 1233	29 131	7 557	3 1333	p	13 323	3 1433	53 83	11 409	3 1533	37 127	p

FACTORS OF NUMBERS UP TO 9999—(*Continued*)

From to	4800 4900	4900 5000	5000 5100	5100 5200	5200 5300	5300 5400	5400 5500	5500 5600	5600 5700	5700 5800	5800 5900	5900 6000
1	p	13 377	3 1667	p	7 743	3 1767	11 491	p	3 1867	p	p	3 1967
3	3 1601	p	p	3 1701	11 473	p	3 1801	p	13 431	3 1901	7 829	p
7	11 437	7 701	3 1669	p	41 127	3 1769	p	p	3 1869	13 439	p	3 1969
9	3 1603	p	p	3 1703	p	p	3 1803	7 787	71 79	3 1903	37 157	19 311
11	17 283	3 1637	p	19 269	3 1737	47 113	7 773	3 1837	31 181	p	3 1937	23 257
13	p	17 289	3 1671	p	13 401	3 1771	p	37 149	3 1871	29 197	p	3 1971
17	p	3 1639	29 173	7 731	3 1739	13 409	p	3 1839	41 137	p	3 1939	61 97
19	61 79	p	3 1673	p	17 307	3 1773	p	p	3 1873	7 817	11 529	3 1973
21	3 1607	7 703	p	3 1707	23 227	17 313	3 1807	p	7 803	3 1907	p	31 191
23	7 689	3 1641	p	47 109	3 1741	p	11 493	3 1841	p	59 97	3 1941	p
27	3 1609	13 379	11 457	3 1709	p	7 761	3 1809	p	17 331	3 1909	p	p
29	11 439	3 1643	47 107	23 223	3 1743	73 73	61 89	3 1843	13 433	17 337	3 1943	7 847
31	p	p	3 1677	7 733	p	3 1777	p	p	3 1877	11 521	7 833	3 1977
33	3 1611	p	7 719	3 1711	p	p	3 1811	11 503	43 131	3 1911	19 307	17 349
37	7 691	p	3 1679	11 467	p	3 1779	p	7 791	3 1879	p	13 449	3 1979
39	3 1613	11 449	p	3 1713	13 403	19 281	3 1813	29 191	p	3 1913	p	p
41	47 103	3 1647	71 71	53 97	3 1747	7 763	p	3 1847	p	p	3 1947	13 457
43	29 167	p	3 1681	37 139	7 749	3 1781	p	23 241	3 1881	p	p	3 1981
47	37 131	3 1649	7 721	p	3 1749	p	13 419	3 1849	p	7 821	3 1949	19 313
49	13 373	7 707	3 1683	19 271	29 181	3 1783	p	31 179	3 1883	p	p	3 1983

FACTORS OF NUMBERS UP TO 9999—(*Continued*)

From to	4800 4900	4900 5000	5000 5100	5100 5200	5200 5300	5300 5400	5400 5500	5500 5600	5600 5700	5700 5800	5800 5900	5900 6000
51	3 1617	p	p	3 1717	59 89	p	3 1817	7 793	p	3 1917	p	11 541
53	23 211	3 1651	31 163	p	3 1751	53 101	7 779	3 1851	p	11 523	3 1951	p
57	3 1619	p	13 389	3 1719	7 751	11 487	3 1819	p	p	3 1919	p	7 851
59	43 113	3 1653	p	7 737	3 1753	23 233	53 103	3 1853	p	13 443	3 1953	59 101
61	p	11 451	3 1687	13 397	p	3 1787	43 127	67 83	3 1887	7 823	p	3 1987
63	3 1621	7 709	61 83	3 1721	19 277	31 173	3 1821	p	7 809	3 1921	11 533	67 89
67	31 157	p	3 1689	p	23 229	3 1789	7 781	19 293	3 1889	73 79	p	3 1989
69	3 1623	p	37 137	3 1723	11 479	7 767	3 1823	p	p	3 1923	p	47 127
71	p	3 1657	11 461	p	3 1757	41 131	p	3 1857	53 107	29 199	3 1957	7 853
73	11 443	p	3 1691	7 739	p	3 1791	13 421	p	3 1891	23 251	7 839	3 1991
77	p	3 1659	p	31 167	3 1759	19 283	p	3 1859	7 811	53 109	3 1959	43 139
79	7 697	13 383	3 1693	p	p	3 1793	p	7 797	3 1893	p	p	3 1993
81	3 1627	17 293	p	3 1727	p	p	3 1827	p	13 437	3 1927	p	p
83	19 257	3 1661	13 391	71 73	3 1761	7 769	p	3 1861	p	p	3 1961	31 193
87	3 1629	p	p	3 1729	17 311	p	3 1829	37 151	11 517	3 1929	7 841	p
89	p	3 1663	7 727	p	3 1763	17 317	11 499	3 1863	p	7 327	3 1963	53 113
91	67 73	7 713	3 1697	29 179	11 481	3 1797	17 323	p	3 1897	p	43 137	3 1997
93	3 1631	p	11 463	3 1731	67 79	p	3 1831	7 799	p	3 1931	71 83	13 461
97	59 83	19 263	3 1699	p	p	3 1799	23 239	29 193	3 1899	11 527	p	3 1999
99	3 1633	p	p	3 1733	7 757	p	3 1833	11 509	41 139	3 1933	17 347	7 857

FACTORS OF NUMBERS UP TO 9999—(*Continued*)

From to	6000 6100	6100 6200	6200 6300	6300 6400	6400 6500	6500 6600	6600 6700	6700 6800	6800 6900	6900 7000	7000 7100	7100 7200
1	17 353	p	3 2067	p	37 173	3 2167	7 943	p	3 2267	67 103	p	3 2367
3	3 2001	17 359	p	3 2101	19 337	7 929	3 2201	p	p	3 2301	47 149	p
7	p	31 197	3 2069	7 901	43 149	3 2169	p	19 353	3 2269	p	7 1001	3 2369
9	3 2003	41 149	7 887	3 2103	13 493	23 283	3 2203	p	11 619	3 2303	43 163	p
11	p	3 2037	p	p	3 2137	17 383	11 601	3 2237	7 973	p	3 2337	13 547
13	7 859	p	3 2071	59 107	11 583	3 2171	17 389	7 959	3 2271	31 223	p	3 2371
17	11 547	3 2039	p	p	3 2139	7 931	13 509	3 2239	17 401	p	3 2339	11 647
19	13 463	29 211	3 2073	71 89	7 917	3 2173	p	p	3 2273	11 629	p	3 2373
21	3 2007	p	p	3 2107	p	p	3 2207	11 611	19 359	3 2307	7 1003	p
23	19 317	3 2041	7 889	p	3 2141	11 593	37 179	3 2241	p	7 989	3 2341	17 419
27	3 2009	11 557	13 479	3 2109	p	61 107	3 2209	7 961	p	3 2309	p	p
29	p	3 2043	p	p	3 2143	p	7 947	3 2243	p	13 533	3 2343	p
31	37 163	p	3 2077	13 487	59 109	3 2177	19 349	53 127	3 2277	29 239	79 89	3 2377
33	3 2011	p	23 271	3 2111	7 919	47 139	3 2211	p	p	3 2311	13 541	7 1019
37	p	17 361	3 2079	p	41 157	3 2179	p	p	3 2279	7 991	31 227	3 2379
39	3 2013	7 877	17 367	3 2113	47 137	13 503	3 2213	23 293	7 977	3 2313	p	11 649
41	7 863	3 2047	79 79	17 373	3 2147	31 211	29 229	3 2247	p	11 631	3 2347	37 193
43	p	p	3 2081	p	17 379	3 2181	7 949	11 613	3 2281	53 131	p	3 2381
47	p	3 2049	p	11 577	3 2149	p	17 391	3 2249	41 167	p	3 2349	7 1021
49	23 263	11 559	3 2083	7 907	p	3 2183	61 109	17 397	3 2283	p	7 1007	3 2383

FACTORS OF NUMBERS UP TO 9999—(*Continued*)

From to	6000 6100	6100 6200	6200 6300	6300 6400	6400 6500	6500 6600	6600 6700	6700 6800	6800 6900	6900 7000	7000 7100	7100 7200
51	3 2017	p	7 893	3 2117	p	p	3 2217	43 157	13 527	3 2317	11 641	p
53	p	3 2051	13 481	p	3 2151	p	p	3 2251	7 979	17 409	3 2351	23 311
57	3 2019	47 131	p	3 2119	11 587	79 83	3 2219	29 233	p	3 2319	p	17 421
59	73 83	3 2053	11 569	p	3 2153	7 937	p	3 2253	19 361	p	3 2353	p
61	11 551	61 101	3 2087	p	7 923	3 2187	p	p	3 2287	p	23 307	3 2387
63	3 2021	p	p	3 2121	23 281	p	3 2221	p	p	3 2321	7 1009	13 551
67	p	7 881	3 2089	p	29 223	3 2189	59 113	67 101	3 2289	p	37 191	3 2389
69	3 2023	31 199	p	3 2123	p	p	3 2223	7 967	p	3 2323	p	67 107
71	13 467	3 2057	p	23 277	3 2157	p	7 953	3 2257	p	p	3 2357	71 101
73	p	p	3 2091	p	p	3 2191	p	13 521	3 2291	19 367	11 643	3 2391
77	59 103	3 2059	p	7 911	3 2159	p	11 607	3 2259	13 529	p	3 2359	p
79	p	37 167	3 2093	p	11 589	3 2193	p	p	3 2293	7 997	p	3 2393
81	3 2027	7 883	11 571	3 2127	p	p	3 2227	p	7 983	3 2327	73 97	43 167
83	7 869	3 2061	61 103	13 491	3 2161	29 227	41 163	3 2261	p	p	3 2361	11 653
87	3 2029	23 269	p	3 2129	13 499	7 941	3 2229	11 617	71 97	3 2329	19 373	p
89	p	3 2063	19 331	p	3 2163	11 599	p	3 2263	83 83	29 241	3 2363	7 1027
91	p	41 151	3 2097	7 913	p	3 2197	p	p	3 2297	p	7 1013	3 2397
93	3 2031	11 563	7 899	3 2131	43 151	19 347	3 2231	p	61 113	3 2331	41 173	p
97	7 871	p	3 2099	p	73 89	3 2199	37 181	7 971	3 2299	p	47 151	3 2399
99	3 2033	p	p	3 2133	67 97	p	3 2233	13 523	p	3 2333	31 229	23 313

FACTORS OF NUMBERS UP TO 9999—(*Continued*)

From to	7200 7300	7300 7400	7400 7500	7500 7600	7600 7700	7700 7800	7800 7900	7900 8000	8000 8100	8100 8200	8200 8300	8300 8400
1	19 379	7 1043	3 2467	13 577	11 691	3 2567	29 269	p	3 2667	p	59 139	3 2767
3	3 2401	67 109	11 673	3 2501	p	p	3 2601	7 1129	53 151	3 2701	13 631	19 437
7	p	p	3 2469	p	p	3 2569	37 211	p	3 2669	11 737	29 283	3 2769
9	3 2403	p	31 239	3 2503	7 1087	13 593	3 2603	11 719	p	3 2703	p	7 1187
11	p	3 2437	p	7 1073	3 2537	11 701	73 107	3 2637	p	p	3 2737	p
13	p	71 103	3 2471	11 683	23 331	3 2571	13 601	41 193	3 2671	7 1159	43 191	3 2771
17	7 1031	3 2439	p	p	3 2539	p	p	3 2639	p	p	3 2739	p
19	p	13 563	3 2473	73 103	19 401	3 2573	7 1117	p	3 2673	23 353	p	3 2773
21	3 2407	p	41 181	3 2507	p	7 1103	3 2607	89 89	13 617	3 2707	p	53 157
23	31 233	3 2441	13 571	p	3 2541	p	p	3 2641	71 113	p	3 2741	7 1189
27	3 2409	17 431	7 1061	3 2509	29 263	p	3 2609	p	23 349	3 2709	19 433	11 757
29	p	3 2443	17 437	p	3 2543	59 131	p	3 2643	7 1147	11 739	3 2743	p
31	7 1033	p	3 2477	17 443	13 587	3 2577	41 191	7 1133	3 2677	47 173	p	3 2777
33	3 2411	p	p	3 2511	17 449	11 703	3 2611	p	29 277	3 2711	p	13 641
37	p	11 667	3 2479	p	7 1091	3 2579	17 461	p	3 2679	79 103	p	3 2779
39	3 2413	41 179	43 173	3 2513	p	71 109	3 2613	17 467	p	3 2713	7 1177	31 269
41	13 557	3 2447	7 1063	p	3 2547	p	p	3 2647	11 731	7 1163	3 2747	19 439
43	p	7 1049	3 2481	19 397	p	3 2581	11 713	13 611	3 2681	17 479	p	3 2781
47	p	3 2449	11 677	p	3 2549	61 127	7 1121	3 2649	13 619	p	3 2749	17 491
49	11 659	p	3 2483	p	p	3 2583	47 167	p	3 2683	29 281	73 113	3 2783

FACTORS OF NUMBERS UP TO 9999—(*Continued*)

From to	7200 7300	7300 7400	7400 7500	7500 7600	7600 7700	7700 7800	7800 7900	7900 8000	8000 8100	8100 8200	8200 8300	8300 8400
51	3 2417	p	p	3 2517	7 1093	23 337	3 2617	p	83 97	3 2717	37 223	7 1193
53	p	3 2451	29 257	7 1079	3 2551	p	p	3 2651	p	31 263	3 2751	p
57	3 2419	7 1051	p	3 2519	13 589	p	3 2619	73 109	7 1151	3 2719	23 359	61 137
59	7 1037	3 2453	p	p	3 2553	p	29 271	3 2653	p	41 199	3 2753	13 643
61	53 137	17 433	3 2487	p	47 163	3 2587	7 1123	19 419	3 2687	p	11 751	3 2787
63	3 2421	37 199	17 439	3 2521	79 97	7 1109	3 2621	p	11 733	3 2721	p	p
67	13 559	53 139	3 2489	7 1081	11 697	3 2589	p	31 257	3 2689	p	7 1181	3 2789
69	3 2423	p	7 1067	3 2523	p	17 457	3 2623	13 613	p	3 2723	p	p
71	11 661	3 2457	31 241	67 113	3 2557	19 409	17 463	3 2657	7 1153	p	3 2757	1 761
73	7 1039	73 101	3 2491	p	p	3 2591	p	7 1139	3 2691	11 743	p	3 2791
77	19 383	3 2459	p	p	3 2559	7 1111	p	3 2659	41 197	13 629	3 2759	p
79	29 251	47 157	3 2493	11 689	7 1097	3 2593	p	79 101	3 2693	p	17 487	3 2793
81	3 2427	11 671	p	3 2527	p	31 251	3 2627	23 347	p	3 2727	7 1183	17 493
83	p	3 2461	7 1069	p	3 2561	43 181	p	3 2661	59 137	7 1169	3 2761	83 101
87	3 2429	83 89	p	3 2529	p	13 599	3 2629	7 1141	p	3 2729	p	p
89	37 197	3 2463	p	p	3 2563	p	7 1127	3 2663	p	19 431	3 2763	p
91	23 317	19 389	3 2497	p	p	3 2597	13 607	61 131	3 2697	p	p	3 2797
93	3 2431	p	59 127	3 2531	7 1099	p	3 2631	p	p	3 2731	p	7 1199
97	p	13 569	3 2499	71 107	43 179	3 2599	53 149	11 727	3 2699	7 1171	p	3 2799
99	3 2433	7 1057	p	3 2533	p	11 709	3 2633	19 421	7 1157	3 2733	43 193	37 227

FACTORS OF NUMBERS UP TO 9999—(*Continued*)

From to	8400 8500	8500 8600	8600 8700	8700 8800	8800 8900	8900 9000	9000 9100	9100 9200	9200 9300	9300 9400	9400 9500	9500 9600
1	31 271	p	3 2867	7 1243	13 677	3 2967	p	19 479	3 3067	71 131	7 1343	3 3167
3	3 2801	11 773	7 1229	3 2901	p	29 307	3 3001	p	p	3 3101	p	13 731
7	7 1201	47 181	3 2869	p	p	3 2969	p	7 1301	3 3069	41 227	23 409	3 3169
9	3 2803	67 127	p	3 2903	23 383	59 151	3 3003	p	p	3 3103	97 97	37 257
11	13 647	3 2837	79 109	31 281	3 2937	7 1273	p	3 3037	61 151	p	3 3137	p
13	47 179	p	3 2871	p	7 1259	3 2971	p	13 701	3 3071	67 139	p	3 3171
17	19 443	3 2839	7 1231	23 379	3 2939	37 241	71 127	3 3039	13 709	7 1331	3 3139	31 307
19	p	7 1217	3 2873	p	p	3 2973	29 311	11 829	3 3073	p	p	3 3173
21	3 2807	p	37 233	3 2907	p	11 811	3 3007	7 1303	p	3 3107	p	p
23	p	3 2841	p	11 793	3 2941	p	7 1289	3 3041	23 401	p	3 3141	89 107
27	3 2809	p	p	3 2909	7 1261	79 113	3 3009	p	p	3 3109	11 857	7 1361
29	p	3 2843	p	7 1247	3 2943	p	p	3 3043	11 839	19 491	3 3143	13 733
31	p	19 449	3 2877	p	p	3 2977	11 821	23 397	3 3077	7 1333	p	3 3177
33	3 2811	7 1219	89 97	3 2911	11 803	p	3 3011	p	7 1319	3 3111	p	p
37	11 767	p	3 2879	p	p	3 2979	7 1291	p	3 3079	p	p	3 3179
39	3 2813	p	53 163	3 2913	p	7 1277	3 3013	13 703	p	3 3113	p	p
41	23 367	3 2847	p	p	3 2947	p	p	3 3047	p	p	3 3147	7 1363
43	p	p	3 2881	7 1249	37 239	3 2981	p	41 223	3 3081	p	7 1349	3 3181
47	p	3 2849	p	p	3 2949	23 389	83 109	3 3049	7 1321	13 719	3 3149	p
49	7 1207	83 103	3 2883	13 673	p	3 2983	p	7 1307	3 3083	p	11 859	3 3183

FACTORS OF NUMBERS UP TO 9999—(*Continued*)

From to	8400 8500	8500 8600	8600 8700	8700 8800	8800 8900	8900 9000	9000 9100	9100 9200	9200 9300	9300 9400	9400 9500	9500 9600
51	3 2817	17 503	41 211	3 2917	53 167	p	3 3017	p	11 841	3 3117	13 727	p
53	79 107	3 2851	17 509	p	3 2951	7 1279	11 823	3 3051	19 487	47 199	3 3151	41 233
57	3 2819	43 199	11 787	3 2919	17 521	13 689	3 3019	p	p	3 3119	7 1351	19 503
59	11 769	3 2853	7 1237	19 461	3 2953	17 527	p	3 3053	47 197	7 1337	3 3153	11 869
61	p	7 1223	3 2887	p	p	3 2987	13 697	p	3 3087	11 851	p	3 3187
63	3 2821	p	p	3 2921	p	p	3 3021	7 1309	59 157	3 3121	p	73 131
67	p	13 659	3 2889	11 797	p	3 2989	p	89 103	3 3089	17 551	p	3 3189
69	3 2823	11 779	p	3 2923	7 1267	p	3 3023	53 173	13 713	3 3123	17 557	7 1367
71	43 197	3 2857	13 667	7 1253	3 2957	p	47 193	3 3057	73 127	p	3 3157	17 563
73	37 229	p	3 2891	31 283	19 467	3 2991	43 211	p	3 3091	7 1339	p	3 3191
77	7 1211	3 2859	p	67 131	3 2959	47 191	29 313	3 3059	p	p	3 3159	61 157
79	61 139	23 373	3 2893	p	13 683	3 2993	7 1297	67 137	3 3093	83 113	p	3 3193
81	3 2827	p	p	3 2927	83 107	7 1283	3 3027	p	p	3 3127	19 499	11 871
83	17 499	3 2861	19 457	p	3 2961	13 691	31 293	3 3061	p	11 853	3 3161	7 1369
87	3 2829	31 277	7 1241	3 2929	p	11 817	3 3029	p	37 251	3 3129	53 179	p
89	13 653	3 2863	p	11 799	3 2963	89 101	61 149	3 3063	7 1327	41 229	3 3163	43 223
91	7 1213	11 781	3 2897	59 149	17 523	3 2997	p	7 1313	3 3097	p	p	3 3197
93	3 2831	13 661	p	3 2931	p	17 529	3 3031	29 317	p	3 3131	11 863	53 181
97	29 293	p	3 2899	19 463	7 1271	3 2999	11 827	17 541	3 3099	p	p	3 3199
99	3 2833	p	p	3 2933	11 809	p	3 3033	p	17 547	3 3133	7 1357	29 331

FACTORS OF NUMBERS UP TO 9999—(*Continued*)

From to	9600 9700	9700 9800	9800 9900	9900 10000	From to	9600 9700	9700 9800	9800 9900	9900 9999
1	p	89 109	3 3267	p	51	3 3217	7 1393	p	3 3317
3	3 3201	31 313	p	3 3301	53	7 1379	3 3251	59 167	37 269
7	13 739	17 571	3 3269	p	57	3 3219	11 887	p	3 3319
9	3 3203	7 1387	17 577	3 3303	59	13 743	3 3253	p	23 433
11	7 1373	3 3237	p	11 901	61	p	43 227	3 3287	7 1423
13	p	11 883	3 3271	23 431	63	3 3221	13 751	7 1409	3 3321
17	59 163	3 3239	p	47 211	67	7 1381	p	3 3289	p
19	p	p	3 3273	7 1417	69	3 3223	p	71 139	3 3323
21	3 3207	p	7 1403	3 3307	71	19 509	3 3257	p	13 767
23	p	3 3241	11 893	p	73	17 569	29 337	3 3291	p
27	3 3209	71 137	31 317	3 3309	77	p	3 3259	7 1411	11 907
29	p	3 3243	p	p	79	p	7 1397	3 3293	17 587
31	p	37 263	3 3277	p	81	3 3227	p	41 241	3 3327
33	3 3211	p	p	3 3311	83	23 421	3 3261	p	67 149
37	23 419	7 1391	3 3279	19 523	87	3 3229	p	p	3 3329
39	3 3213	p	p	3 3313	89	p	3 3263	11 899	7 1427
41	31 311	3 3247	13 757	p	91	11 881	p	3 3297	97 103
43	p	p	3 3281	61 163	93	3 3231	7 1399	13 761	3 3331
47	11 877	3 3249	43 229	7 1421	97	p	97 101	3 3299	13 769
49	p	p	3 3283	p	99	3 3233	41 239	19 521	3 3333

NATURAL SINES

Degrees	0′	6′	12′	18′	24′	30′	36′	42′	48′	54′	Mean Differences				
											1′	2′	3′	4′	5′
0	.0000	0017	0035	0052	0070	0087	0105	0122	0140	0157	3	6	9	12	15
1	.0175	0192	0209	0227	0244	0262	0279	0297	0314	0332	3	6	9	12	15
2	.0349	0366	0384	0401	0419	0436	0454	0471	0488	0506	3	6	9	12	15
3	.0523	0541	0558	0576	0593	0610	0628	0645	0663	0680	3	6	9	12	15
4	.0698	0715	0732	0750	0767	0785	0802	0819	0837	0854	3	6	9	12	14
5	.0872	0889	0906	0924	0941	0958	0976	0993	1011	1028	3	6	9	12	14
6	.1045	1063	1080	1097	1115	1132	1149	1167	1184	1201	3	6	9	12	14
7	.1219	1236	1253	1271	1288	1305	1323	1340	1357	1374	3	6	9	12	14
8	.1392	1409	1426	1444	1461	1478	1495	1513	1530	1547	3	6	9	12	14
9	.1564	1582	1599	1616	1633	1650	1668	1685	1702	1719	3	6	9	12	14
10	.1736	1754	1771	1788	1805	1822	1840	1857	1874	1891	3	6	9	11	14
11	.1908	1925	1942	1959	1977	1994	2011	2028	2045	2062	3	6	9	11	14
12	.2079	2096	2113	2130	2147	2164	2181	2198	2215	2233	3	6	9	11	14
13	.2250	2267	2284	2300	2317	2334	2351	2368	2385	2402	3	6	8	11	14
14	.2419	2436	2453	2470	2487	2504	2521	2538	2554	2571	3	6	8	11	14
15	.2588	2605	2622	2639	2656	2672	2689	2706	2723	2740	3	6	8	11	14
16	.2756	2773	2790	2807	2823	2840	2857	2874	2890	2907	3	6	8	11	14
17	.2924	2940	2957	2974	2990	3007	3024	3040	3057	3074	3	6	8	11	14

NATURAL SINES (Continued)

Degrees	0′	6′	12′	18′	24′	30′	36′	42′	48′	54′	Mean Differences 1′	2′	3′	4′	5′
18	.3090	3107	3123	3140	3156	3173	3190	3206	3223	3239	3	6	8	11	14
19	.3256	3272	3289	3305	3322	3338	3355	3371	3387	3404	3	5	8	11	14
20	.3420	3437	3453	3469	3486	3502	3518	3535	3551	3567	3	5	8	11	14
21	.3584	3600	3616	3633	3649	3665	3681	3697	3714	3730	3	5	8	11	14
22	.3746	3762	3778	3795	3811	3827	3843	3859	3875	3891	3	5	8	11	14
23	.3907	3923	3939	3955	3971	3987	4003	4019	4035	4051	3	5	8	11	14
24	.4067	4083	4099	4115	4131	4147	4163	4179	4195	4210	3	5	8	11	13
25	.4226	4242	4258	4274	4289	4305	4321	4337	4352	4368	3	5	8	11	13
26	.4384	4399	4415	4431	4446	4462	4478	4493	4509	4524	3	5	8	10	13
27	.4540	4555	4571	4586	4602	4617	4633	4648	4664	4679	3	5	8	10	13
28	.4695	4710	4726	4741	4756	4772	4787	4802	4818	4833	3	5	8	10	13
29	.4848	4863	4879	4894	4909	4924	4939	4955	4970	4985	3	5	8	10	13
30	.5000	5015	5030	5045	5060	5075	5090	5105	5120	5135	3	5	8	10	13
31	.5150	5165	5180	5195	5210	5225	5240	5255	5270	5284	2	5	7	10	12
32	.5299	5314	5329	5344	5358	5373	5388	5402	5417	5432	2	5	7	10	12
33	.5446	5461	5476	5490	5505	5519	5534	5548	5563	5577	2	5	7	10	12
34	.5592	5606	5621	5635	5650	5664	5678	5693	5707	5721	2	5	7	10	12
35	.5736	5750	5764	5779	5793	5807	5821	5835	5850	5864	2	5	7	9	12

NATURAL SINES (Continued)

Degrees	0′	6′	12′	18′	24′	30′	36′	42′	48′	54′	Mean Differences 1′	2′	3′	4′	5′
36	.5878	5892	5906	5920	5934	5948	5962	5976	5990	6004	2	5	7	9	12
37	.6018	6032	6046	6060	6074	6088	6101	6115	6129	6143	2	5	7	9	12
38	.6157	6170	6184	6198	6211	6225	6239	6252	6266	6280	2	5	7	9	11
39	.6293	6307	6320	6334	6347	6361	6374	6388	6401	6414	2	4	7	9	11
40	.6428	6441	6455	6468	6481	6494	6508	6521	6534	6547	2	4	7	9	11
41	.6561	6574	6587	6600	6613	6626	6639	6652	6665	6678	2	4	7	9	11
42	.6691	6704	6717	6730	6743	6756	6769	6782	6794	6807	2	4	6	9	11
43	.6820	6833	6845	6858	6871	6884	6896	6909	6921	6934	2	4	6	8	11
44	.6947	6959	6972	6984	6997	7009	7022	7034	7046	7059	2	4	6	8	10
45	.7071	7083	7096	7108	7120	7133	7145	7157	7169	7181	2	4	6	8	10
46	.7193	7206	7218	7230	7242	7254	7266	7278	7290	7302	2	4	6	8	10
47	.7314	7325	7337	7349	7361	7373	7385	7396	7408	7420	2	4	6	8	10
48	.7431	7443	7455	7466	7478	7490	7501	7513	7524	7536	2	4	6	8	10
49	.7547	7559	7570	7581	7593	7604	7615	7627	7638	7649	2	4	6	8	9
50	.7660	7672	7683	7694	7705	7716	7727	7738	7749	7760	2	4	6	7	9
51	.7771	7782	7793	7804	7815	7826	7837	7848	7859	7869	2	4	5	7	9
52	.7880	7891	7902	7912	7923	7934	7944	7955	7965	7976	2	4	5	7	9
53	.7986	7997	8007	8018	8028	8039	8049	8059	8070	8080	2	3	5	7	9

NATURAL SINES (Continued)

Degrees	0′	6′	12′	18′	24′	30′	36′	42′	48′	54′	Mean Differences 1′	2′	3′	4′	5′
54	.8090	8100	8111	8121	8131	8141	8151	8161	8171	8181	2	3	5	7	8
55	.8192	8202	8211	8221	8231	8241	8251	8261	8271	8281	2	3	5	7	8
56	.8290	8300	8310	8320	8329	8339	8348	8358	8368	8377	2	3	5	6	8
57	.8387	8396	8406	8415	8425	8434	8443	8453	8462	8471	2	3	5	6	8
58	.8480	8490	8499	8508	8517	8526	8536	8545	8554	8563	2	3	5	6	8
59	.8572	8581	8590	8599	8607	8616	8625	8634	8643	8652	1	3	4	6	7
60	.8660	8669	8678	8686	8695	8704	8712	8721	8729	8738	1	3	4	6	7
61	.8746	8755	8763	8771	8780	8788	8796	8805	8813	8821	1	3	4	6	7
62	.8829	8838	8846	8854	8862	8870	8878	8886	8894	8902	1	3	4	5	7
63	.8910	8918	8926	8934	8942	8949	8957	8965	8973	8980	1	3	4	5	6
64	.8988	8996	9003	9011	9018	9026	9033	9041	9048	9056	1	3	4	5	6
65	.9063	9070	9078	9085	9092	9100	9107	9114	9121	9128	1	2	4	5	6
66	.9135	9143	9150	9157	9164	9171	9178	9184	9191	9198	1	2	3	5	6
67	.9205	9212	9219	9225	9232	9239	9245	9252	9259	9265	1	2	3	4	6
68	.9272	9278	9285	9291	9298	9304	9311	9317	9323	9330	1	2	3	4	5
69	.9336	9342	9348	9354	9361	9367	9373	9379	9385	9391	1	2	3	4	5
70	.9397	9403	9409	9415	9421	9426	9432	9438	9444	9449	1	2	3	4	5
71	.9455	9461	9466	9472	9478	9483	9489	9494	9500	9505	1	2	3	4	5

NATURAL SINES (Continued)

Degrees	0′	6′	12′	18′	24′	30′	36′	42′	48′	54′	Mean Differences 1′	2′	3′	4′	5′
72	.9511	9516	9521	9527	9532	9537	9542	9548	9553	9558	1	2	3	3	4
73	.9563	9568	9573	9578	9583	9588	9593	9598	9603	9608	1	2	2	3	4
74	.9613	9617	9622	9627	9632	9636	9641	9646	9650	9655	1	2	2	3	4
75	.9659	9664	9668	9673	9677	9681	9686	9690	9694	9699	1	1	2	3	4
76	.9703	9707	9711	9715	9720	9724	9728	9732	9736	9740	1	1	2	3	3
77	.9744	9748	9751	9755	9759	9763	9767	9770	9774	9778	1	1	2	3	3
78	.9781	9785	9789	9792	9796	9799	9803	9806	9810	9813	1	1	2	2	3
79	.9816	9820	9823	9826	9829	9833	9836	9839	9842	9845	1	1	2	2	3
80	.9848	9851	9854	9857	9860	9863	9866	9869	9871	9874	0	1	1	2	2
81	.9877	9880	9882	9885	9888	9890	9893	9895	9898	9900	0	1	1	2	2
82	.9903	9905	9907	9910	9912	9914	9917	9919	9921	9923	0	1	1	2	2
83	.9925	9928	9930	9932	9934	9936	9938	9940	9942	9943	0	1	1	1	2
84	.9945	9947	9949	9951	9952	9954	9956	9957	9959	9960	0	1	1	1	2
85	.9962	9963	9965	9966	9968	9969	9971	9972	9973	9974	0	0	1	1	1
86	.9976	9977	9978	9979	9980	9981	9982	9983	9984	9985	0	0	1	1	1
87	.9986	9987	9988	9989	9990	9990	9991	9992	9993	9993	0	0	0	1	1
88	.9994	9995	9995	9996	9996	9997	9997	9997	9998	9998	0	0	0	0	0
89	.9998	9999	9999	9999	9999	1.000	1.000	1.000	1.000	1.000	0	0	0	0	0

LOGARITHMIC SINES

Degrees	0′	6′	12′	18′	24′	30′	36′	42′	48′	54′	Mean Differences 1′	2′	3′	4′	5′
0	—∞	7.2419	5429	7190	8439	9408	0200	0870	1450	1961					
1	8.2419	2832	3210	3558	3880	4179	4459	4723	4971	5206					
2	8.5428	5640	5842	6035	6220	6397	6567	6731	6889	7041					
3	8.7188	7330	7468	7602	7731	7857	7979	8098	8213	8326					
4	8.8436	8543	8647	8749	8849	8946	9042	9135	9226	9315	16	32	48	64	80
5	8.9403	9489	9573	9655	9736	9816	9894	9970	0046	0120	13	26	39	52	65
6	9.0192	0264	0334	0403	0472	0539	0605	0670	0734	0797	11	22	33	44	55
7	9.0859	0920	0981	1040	1099	1157	1214	1271	1326	1381	10	19	29	38	48
8	9.1436	1489	1542	1594	1646	1697	1747	1797	1847	1895	8	17	25	34	42
9	9.1943	1991	2038	2085	2131	2176	2221	2266	2310	2353	8	15	23	30	38
10	9.2397	2439	2482	2524	2565	2606	2647	2687	2727	2767	7	14	20	27	34
11	9.2806	2845	2883	2921	2959	2997	3034	3070	3107	3143	6	12	19	25	31
12	9.3179	3214	3250	3284	3319	3353	3387	3421	3455	3488	6	11	17	23	28
13	9.3521	3554	3586	3618	3650	3682	3713	3745	3775	3806	5	11	16	21	26
14	9.3837	3867	3897	3927	3957	3986	4015	4044	4073	4102	5	10	15	20	24
15	9.4130	4158	4186	4214	4242	4269	4296	4323	4350	4377	5	9	14	18	23
16	9.4403	4430	4456	4482	4508	4533	4559	4584	4609	4634	4	9	13	17	21
17	9.4659	4684	4709	4733	4757	4781	4805	4829	4853	4876	4	8	12	16	20

LOGARITHMIC SINES (Continued)

Degrees	0′	6′	12′	18′	24′	30′	36′	42′	48′	54′	Mean Differences 1′	2′	3′	4′	5′
18	9.4900	4923	4946	4969	4992	5015	5037	5060	5082	5104	4	8	11	15	19
19	9.5126	5148	5170	5192	5213	5235	5256	5278	5299	5320	4	7	11	14	18
20	9.5341	5361	5382	5402	5423	5443	5463	5484	5504	5523	3	7	10	14	17
21	9.5543	5563	5583	5602	5621	5641	5660	5679	5698	5717	3	6	10	13	16
22	9.5736	5754	5773	5792	5810	5828	5847	5865	5883	5901	3	6	9	12	15
23	9.5919	5937	5954	5972	5990	6007	6024	6042	6059	6076	3	6	9	12	15
24	9.6093	6110	6127	6144	6161	6177	6194	6210	6227	6243	3	6	8	11	14
25	9.6259	6276	6292	6308	6324	6340	6356	6371	6387	6403	3	5	8	11	13
26	9.6418	6434	6449	6465	6480	6495	6510	6526	6541	6556	3	5	8	10	13
27	9.6570	6585	6600	6615	6629	6644	6659	6673	6687	6702	2	5	7	10	12
28	9.6716	6730	6744	6759	6773	6787	6801	6814	6828	6842	2	5	7	9	12
29	9.6856	6869	6883	6896	6910	6923	6937	6950	6963	6977	2	4	7	9	11
30	9.6990	7003	7016	7029	7042	7055	7068	7080	7093	7106	2	4	6	9	11
31	9.7118	7131	7144	7156	7168	7181	7193	7205	7218	7230	2	4	6	8	10
32	9.7242	7254	7266	7278	7290	7302	7314	7326	7338	7349	2	4	6	8	10
33	9.7361	7373	7384	7396	7407	7419	7430	7442	7453	7464	2	4	6	8	10
34	9.7476	7487	7498	7509	7520	7531	7542	7553	7564	7575	2	4	6	7	9
35	9.7586	7597	7607	7618	7629	7640	7650	7661	7671	7682	2	4	5	7	9

LOGARITHMIC SINES (Continued)

Degrees	0′	6′	12′	18′	24′	30′	36′	42′	48′	54′	Mean Differences 1′	2′	3′	4′	5′
36	9.7692	7703	7713	7723	7734	7744	7754	7764	7774	7785	2	3	5	7	9
37	9.7795	7805	7815	7825	7835	7844	7854	7864	7874	7884	2	3	5	7	8
38	9.7893	7903	7913	7922	7932	7941	7951	7960	7970	7979	2	3	5	7	8
39	9.7989	7998	8007	8017	8026	8035	8044	8053	8063	8072	2	3	5	6	8
40	9.8081	8090	8099	8108	8117	8125	8134	8143	8152	8161	1	3	4	6	7
41	9.8169	8178	8187	8195	8204	8213	8221	8230	8238	8247	1	3	4	6	7
42	9.8255	8264	8272	8280	8289	8297	8305	8313	8322	8330	1	3	4	6	7
43	9.8338	8346	8354	8362	8370	8378	8386	8394	8402	8410	1	3	4	5	7
44	9.8418	8426	8433	8441	8449	8457	8464	8472	8480	8487	1	3	4	5	6
45	9.8495	8502	8510	8517	8525	8532	8540	8547	8555	8562	1	2	4	5	6
46	9.8569	8577	8584	8591	8598	8606	8613	8620	8627	8634	1	2	4	5	6
47	9.8641	8648	8655	8662	8669	8676	8683	8690	8697	8704	1	2	3	5	6
48	9.8711	8718	8724	8731	8738	8745	8751	8758	8765	8771	1	2	3	4	6
49	9.8778	8784	8791	8797	8804	8810	8817	8823	8830	8836	1	2	3	4	5
50	9.8843	8849	8855	8862	8868	8874	8880	8887	8893	8899	1	2	3	4	5
51	9.8905	8911	8917	8923	8929	8935	8941	8947	8953	8959	1	2	3	4	5
52	9.8965	8971	8977	8983	8989	8995	9000	9006	9012	9018	1	2	3	4	5
53	9.9023	9029	9035	9041	9046	9052	9057	9063	9069	9074	1	2	3	4	5

LOGARITHMIC SINES (Continued)

Degrees	0′	6′	12′	18′	24′	30′	36′	42′	48′	54′	Mean Differences 1′	2′	3′	4′	5′
54	9.9080	9085	9091	9096	9101	9107	9112	9118	9123	9128	1	2	3	4	5
55	9.9134	9139	9144	9149	9155	9160	9165	9170	9175	9181	1	2	3	3	4
56	9.9186	9191	9196	9201	9206	9211	9216	9221	9226	9231	1	2	3	3	4
57	9.9236	9241	9246	9251	9255	9260	9265	9270	9275	9279	1	2	2	3	4
58	9.9284	9289	9294	9298	9303	9308	9312	9317	9322	9326	1	2	2	3	4
59	9.9331	9335	9340	9344	9349	9351	9358	9362	9367	9371	1	1	2	3	4
60	9.9375	9380	9384	9388	9393	9397	9401	9406	9410	9414	1	1	2	3	4
61	9.9418	9422	9427	9431	9435	9439	9443	9447	9451	9455	1	1	2	3	3
62	9.9459	9463	9467	9471	9475	9479	9483	9487	9491	9495	1	1	2	3	3
63	9.9499	9503	9507	9510	9514	9518	9522	9525	9529	9533	1	1	2	3	3
64	9.9537	9540	9544	9548	9551	9555	9558	9562	9566	9569	1	1	2	2	3
65	9.9573	9576	9580	9583	9587	9590	9594	9597	9601	9604	1	1	2	2	3
66	9.9607	9611	9614	9617	9621	9624	9627	9631	9634	9637	1	1	2	2	3
67	9.9640	9643	9647	9650	9653	9656	9659	9662	9666	9669	1	1	2	2	3
68	9.9672	9675	9678	9681	9684	9687	9690	9693	9696	9699	0	1	1	2	2
69	9.9702	9704	9707	9710	9713	9716	9719	9722	9724	9727	0	1	1	2	2
70	9.9730	9733	9735	9738	9741	9743	9746	9749	9751	9754	0	1	1	2	2
71	9.9757	9759	9762	9764	9767	9770	9772	9775	9777	9780	0	1	1	2	2

LOGARITHMIC SINES (Continued)

Degrees	0′	6′	12′	18′	24′	30′	36′	42′	48′	54′	Mean Differences 1′	2′	3′	4′	5′
72	9.9782	9785	9787	9789	9792	9794	9797	9799	9801	9804	0	1	1	2	2
73	9.9806	9808	9811	9813	9815	9817	9820	9822	9824	9826	0	1	1	2	2
74	9.9828	9831	9833	9835	9837	9839	9841	9843	9845	9847	0	1	1	1	2
75	9.9849	9851	9853	9855	9857	9859	9861	9863	9865	9867	0	1	1	1	2
76	9.9869	9871	9873	9875	9876	9878	9880	9882	9884	9885	0	1	1	1	2
77	9.9837	9889	9891	9892	9894	9896	9897	9899	9901	9902	0	1	1	1	1
78	9.9904	9906	9907	9909	9910	9912	9913	9915	9916	9918	0	1	1	1	1
79	9.9919	9921	9922	9924	9925	9927	9928	9929	9931	9932	0	0	1	1	1
80	9.9934	9935	9936	9937	9939	9940	9941	9943	9944	9945	0	0	1	1	1
81	9.9946	9947	9949	9950	9951	9952	9953	9954	9955	9956	0	0	1	1	1
82	9.9958	9959	9960	9961	9962	9963	9964	9965	9966	9967	0	0	1	1	1
83	9.9968	9968	9969	9970	9971	9972	9973	9974	9975	9975	0	0	0	0	1
84	9.9976	9977	9978	9978	9979	9980	9981	9981	9982	9983	0	0	0	0	1
85	9.9983	9984	9985	9985	9986	9987	9987	9988	9988	9989	0	0	0	0	0
86	9.9989	9990	9990	9991	9991	9992	9992	9993	9993	9994	0	0	0	0	0
87	9.9994	9994	9995	9995	9996	9996	9996	9996	9997	9997	0	0	0	0	0
88	9.9997	9998	9998	9998	9998	9999	9999	9999	9999	9999	0	0	0	0	0
89	9.9999	9999	10.00	10.00	10.00	10.00	10.00	10.00	10.00	10.00	0	0	0	0	0

NATURAL COSINES

Degrees	0′	6′	12′	18′	24′	30′	36′	42′	48′	54′	Mean Differences				
											1′	2′	3′	4′	5′
0	1.000	1.000	1.000	1.000	1.000	1.000	.9999	9999	9999	9999	0	0	0	0	0
1	.9998	9998	9998	9997	9997	9997	9996	9996	9995	9995	0	0	0	0	0
2	.9994	9993	9993	9992	9991	9990	9990	9989	9988	9987	0	0	0	1	1
3	.9986	9985	9984	9983	9982	9981	9980	9979	9978	9977	0	0	1	1	1
4	.9976	9974	9973	9972	9971	9969	9968	9966	9965	9963	0	0	1	1	1
5	.9962	9960	9959	9957	9956	9954	9952	9951	9949	9947	0	1	1	1	2
6	.9945	9943	9942	9940	9938	9936	9934	9932	9930	9928	0	1	1	1	2
7	.9925	9923	9921	9919	9917	9914	9912	9910	9907	9905	0	1	1	2	2
8	.9903	9900	9898	9895	9893	9890	9888	9885	9882	9880	0	1	1	2	2
9	.9877	9874	9871	9869	9866	9863	9860	9857	9854	9851	0	1	1	2	2
10	.9848	9845	9842	9839	9836	9833	9829	9826	9823	9820	1	1	2	2	3
11	.9816	9813	9810	9806	9803	9799	9796	9792	9789	9785	1	1	2	2	3
12	.9781	9778	9774	9770	9767	9763	9759	9755	9751	9748	1	1	2	3	3
13	.9744	9740	9736	9732	9728	9724	9720	9715	9711	9707	1	1	2	3	3
14	.9703	9699	9694	9690	9686	9681	9677	9673	9668	9664	1	1	2	3	4
15	.9659	9655	9650	9646	9641	9636	9632	9627	9622	9617	1	2	2	3	4
16	.9613	9608	9603	9598	9593	9588	9583	9578	9573	9568	1	2	2	3	4
17	.9563	9558	9553	9548	9542	9537	9532	9527	9521	9516	1	2	3	3	4

N.B.—Subtract Mean Differences.

NATURAL COSINES (Continued)

Degrees	0′	6′	12′	18′	24′	30′	36′	42′	48′	54′	Mean Differences 1′	2′	3′	4′	5′
18	.9511	9505	9500	9494	9489	9483	9478	9472	9466	9461	1	2	3	4	5
19	.9455	9449	9444	9438	9432	9426	9421	9415	9409	9403	1	2	3	4	5
20	.9397	9391	9385	9379	9373	9367	9361	9354	9348	9342	1	2	3	4	5
21	.9336	9330	9323	9317	9311	9304	9298	9291	9285	9278	1	2	3	4	5
22	.9272	9265	9259	9252	9245	9239	9232	9225	9219	9212	1	2	3	4	6
23	.9205	9198	9191	9184	9178	9171	9164	9157	9150	9143	1	2	3	5	6
24	.9135	9128	9121	9114	9107	9100	9092	9085	9078	9070	1	2	4	5	6
25	.9063	9056	9048	9041	9033	9026	9018	9011	9003	8996	1	3	4	5	6
26	.8988	8980	8973	8965	8957	8949	8942	8934	8926	8918	1	3	4	5	6
27	.8910	8902	8894	8886	8878	8870	8862	8854	8846	8838	1	3	4	5	7
28	.8829	8821	8813	8805	8796	8788	8780	8771	8763	8755	1	3	4	6	7
29	.8746	8738	8729	8721	8712	8704	8695	8686	8678	8669	1	3	4	6	7
30	.8660	8652	8643	8634	8625	8616	8607	8599	8590	8581	1	3	4	6	7
31	.8572	8563	8554	8545	8536	8526	8517	8508	8499	8490	2	3	5	6	8
32	.8480	8471	8462	8453	8443	8434	8425	8415	8406	8396	2	3	5	6	8
33	.8387	8377	8368	8358	8348	8339	8329	8320	8310	8300	2	3	5	6	8
34	.8290	8281	8271	8261	8251	8241	8231	8221	8211	8202	2	3	5	7	8
35	.8192	8181	8171	8161	8151	8141	8131	8121	8111	8100	2	3	5	7	8

N.B.—Subtract Mean Differences.

NATURAL COSINES (Continued)

Deg.	0′	6′	12′	18′	24′	30′	36′	42′	48′	54′	Mean Differences 1′	2′	3′	4′	5′
36	.8090	8080	8070	8059	8049	8039	8028	8018	8007	7997	2	3	5	7	9
37	.7986	7976	7965	7955	7944	7934	7923	7912	7902	7891	2	4	5	7	9
38	.7880	7869	7859	7848	7837	7826	7815	7804	7793	7782	2	4	5	7	9
39	.7771	7760	7749	7738	7728	7716	7705	7694	7683	7672	2	4	6	7	9
40	.7660	7649	7638	7627	7615	7604	7593	7581	7570	7559	2	4	6	8	9
41	.7547	7536	7524	7513	7501	7490	7478	7466	7455	7443	2	4	6	8	10
42	.7431	7420	7408	7396	7385	7373	7361	7349	7337	7325	2	4	6	8	10
43	.7314	7302	7290	7278	7266	7254	7242	7230	7218	7206	2	4	6	8	10
44	.7193	7181	7169	7157	7145	7133	7120	7108	7096	7083	2	4	6	8	10
45	.7071	7059	7046	7034	7022	7009	6997	6984	6972	6959	2	4	6	8	10
46	.6947	6934	6921	6909	6896	6884	6871	6858	6845	6833	2	4	6	8	11
47	.6820	6807	6794	6782	6769	6756	6743	6730	6717	6704	2	4	6	9	11
48	.6691	6678	6665	6652	6639	6626	6613	6600	6587	6574	2	4	7	9	11
49	.6561	6547	6534	6521	6508	6494	6481	6468	6455	6441	2	4	7	9	11
50	.6428	6414	6401	6388	6374	6361	6347	6334	6320	6307	2	4	7	9	11
51	.6293	6280	6266	6252	6239	6225	6211	6198	6184	6170	2	5	7	9	11
52	.6157	6143	6129	6115	6101	6088	6074	6060	6046	6032	2	5	7	9	12
53	.6018	6004	5990	5976	5962	5948	5934	5920	5906	5892	2	5	7	9	12

N.B.—Subtract Mean Differences.

NATURAL COSINES (Continued)

Deg.	0′	6′	12′	18′	24′	30′	36′	42′	48′	54′	Mean Differences 1′	2′	3′	4′	5′
54	.5878	5864	5850	5835	5821	5807	5793	5779	5764	5750	2	5	7	9	12
55	.5736	5721	5707	5693	5678	5664	5650	5635	5621	5606	2	5	7	10	12
56	.5592	5577	5563	5548	5534	5519	5505	5490	5476	5461	2	5	7	10	12
57	.5446	5432	5417	5402	5388	5373	5358	5344	5329	5314	2	5	7	10	12
58	.5299	5284	5270	5255	5240	5225	5210	5195	5180	5165	2	5	7	10	12
59	.5150	5135	5120	5105	5090	5075	5060	5045	5030	5015	3	5	8	10	13
60	.5000	4985	4970	4955	4939	4924	4909	4894	4879	4863	3	5	8	10	13
61	.4848	4833	4818	4802	4787	4772	4756	4741	4726	4710	3	5	8	10	13
62	.4695	4679	4664	4648	4633	4617	4602	4586	4571	4555	3	5	8	10	13
63	.4540	4524	4509	4493	4478	4462	4446	4431	4415	4399	3	5	8	10	13
64	.4384	4368	4352	4337	4321	4305	4289	4274	4258	4242	3	5	8	11	13
65	.4226	4210	4195	4179	4163	4147	4131	4115	4099	4083	3	5	8	11	13
66	.4067	4051	4035	4019	4003	3987	3971	3955	3939	3923	3	5	8	11	14
67	.3907	3891	3875	3859	3843	3827	3811	3795	3778	3762	3	5	8	11	14
68	.3746	3730	3714	3697	3681	3665	3649	3633	3616	3600	3	5	8	11	14
69	.3584	3567	3551	3535	3518	3502	3486	3469	3453	3437	3	5	8	11	14
70	.3420	3404	3387	3371	3355	3338	3322	3305	3289	3272	3	5	8	11	14
71	.3256	3239	3223	3206	3190	3173	3156	3140	3123	3107	3	6	8	11	14

N.B.—Subtract Mean Differences.

NATURAL COSINES (Continued)

Deg.	0′	6′	12′	18′	24′	30′	36′	42′	48′	54′	Mean Differences 1′	2′	3′	4′	5′
72	.3090	3074	3057	3040	3024	3007	2990	2974	2957	2940	3	6	8	11	14
73	.2924	2907	2890	2874	2857	2840	2823	2807	2790	2773	3	6	8	11	14
74	.2756	2740	2723	2706	2689	2672	2656	2639	2622	2605	3	6	8	11	14
75	.2588	2571	2554	2538	2521	2504	2487	2470	2453	2436	3	6	8	11	14
76	.2419	2402	2385	2368	2351	2334	2317	2300	2284	2267	3	6	8	11	14
77	.2250	2233	2215	2198	2181	2164	2147	2130	2113	2096	3	6	9	11	14
78	.2079	2062	2045	2028	2011	1994	1977	1959	1942	1925	3	6	9	11	14
79	.1908	1891	1874	1857	1840	1822	1805	1788	1771	1754	3	6	9	11	14
80	.1736	1719	1702	1685	1668	1650	1633	1616	1599	1582	3	6	9	12	14
81	.1564	1547	1530	1513	1495	1478	1461	1444	1426	1409	3	6	9	12	14
82	.1392	1374	1357	1340	1323	1305	1288	1271	1253	1236	3	6	9	12	14
83	.1219	1201	1184	1167	1149	1132	1115	1097	1080	1063	3	6	9	12	14
84	.1045	1028	1011	0993	0976	0958	0941	0924	0906	0889	3	6	9	12	14
85	.0872	0854	0837	0819	0802	0785	0767	0750	0732	0715	3	6	9	12	14
86	.0698	0680	0663	0645	0628	0610	0593	0576	0558	0541	3	6	9	12	15
87	.0523	0506	0488	0471	0454	0436	0419	0401	0384	0366	3	6	9	12	15
88	.0349	0332	0314	0297	0279	0262	0244	0227	0209	0192	3	6	9	12	15
89	.0175	0157	0140	0122	0105	0087	0070	0052	0035	0017	3	6	9	12	15

N.B.—Subtract Mean Differences

LOGARITHMIC COSINES

Deg.	0′	6′	12′	18′	24′	30′	36′	42′	48′	54′	Mean Differences 1′	2′	3′	4′	5′
0	10.0000	0000	0000	0000	0000	0000	0000	0000	0000	9.9999	0	0	0	0	0
1	9.9999	9999	9999	9999	9999	9999	9998	9998	9998	9998	0	0	0	0	0
2	9.9997	9997	9997	9996	9996	9996	9996	9995	9995	9994	0	0	0	0	0
3	9.9994	9994	9993	9993	9992	9992	9991	9991	9990	9990	0	0	0	0	0
4	9.9989	9989	9988	9988	9987	9987	9986	9985	9985	9984	0	0	0	0	0
5	9.9983	9983	9982	9981	9981	9980	9979	9978	9978	9977	0	0	0	0	1
6	9.9976	9975	9975	9974	9973	9972	9971	9970	9969	9968	0	0	0	1	1
7	9.9968	9967	9966	9965	9964	9963	9962	9961	9960	9959	0	0	1	1	1
8	9.9958	9956	9955	9954	9953	9952	9951	9950	9949	9947	0	0	1	1	1
9	9.9946	9945	9944	9943	9941	9940	9939	9937	9936	9935	0	0	1	1	1
10	9.9934	9932	9931	9929	9928	9927	9925	9924	9922	9921	0	0	1	1	1
11	9.9919	9918	9916	9915	9913	9912	9910	9909	9907	9906	0	1	1	1	1
12	9.9904	9902	9901	9899	9897	9896	9894	9892	9891	9889	0	1	1	1	1
13	9.9887	9885	9884	9882	9880	9878	9876	9875	9873	9871	0	1	1	1	2
14	9.9869	9867	9865	9863	9861	9859	9857	9855	9853	9851	0	1	1	1	2

N.B.—Subtract Mean Differences.

LOGARITHMIC COSINES (Continued)

Deg.	0′	6′	12′	18′	24′	30′	36′	42′	48′	54′	Mean Differences 1′	2′	3′	4′	5′
15	9.9849	9847	9845	9843	9841	9839	9837	9835	9833	9831	0	1	1	1	2
16	9.9828	9826	9824	9822	9820	9817	9815	9813	9811	9808	0	1	1	2	2
17	9.9806	9804	9801	9799	9797	9794	9792	9798	9787	9785	0	1	1	2	2
18	9.9782	9780	9777	9775	9772	9770	9767	9764	9762	9759	0	1	1	2	2
19	9.9757	9754	9751	9749	9746	9743	9741	9738	9735	9733	0	1	1	2	2
20	9.9730	9727	9724	9722	9719	9716	9713	9710	9707	9704	0	1	1	2	2
21	9.9702	9699	9696	9693	9690	9687	9684	9681	9678	9675	0	1	1	2	2
22	9.9672	9669	9666	9662	9659	9656	9653	9650	9647	9643	1	1	2	2	3
23	9.9640	9637	9634	9631	9627	9624	9621	9617	9614	9611	1	1	2	2	3
24	9.9607	9604	9601	9597	9594	9590	9587	9583	9580	9576	1	1	2	2	3
25	9.9573	9569	9566	9562	9558	9555	9551	9548	9544	9540	1	1	2	2	3
26	9.9537	9533	9529	9525	9522	9518	9514	9510	9507	9503	1	1	2	3	3
27	9.9499	9495	9491	9487	9483	9479	9475	9471	9467	9463	1	1	2	3	3
28	9.9459	9455	9451	9447	9443	9439	9435	9431	9427	9422	1	1	2	3	3

N.B.—Subtract Mean Differences.

LOGARITHMIC COSINES (Continued)

Deg.	0′	6′	12′	18′	24′	30′	36′	42′	48′	54′	Mean Differences 1′	2′	3′	4′	5′
29	9.9418	9414	9410	9406	9401	9397	9393	9388	9384	9380	1	1	2	3	4
30	9.9375	9371	9367	9362	9358	9353	9349	9344	9340	9335	1	1	2	3	4
31	9.9331	9326	9322	9317	9312	9308	9303	9298	9294	9289	1	2	2	3	4
32	9.9284	9279	9275	9270	9265	9260	9255	9251	9246	9241	1	2	2	3	4
33	9.9236	9231	9226	9221	9216	9211	9206	9201	9196	9191	1	2	3	3	4
34	9.9186	9181	9175	9170	9165	9160	9155	9149	9144	9139	1	2	3	3	4
35	9.9134	9128	9123	9118	9112	9107	9101	9096	9091	9085	1	2	3	4	5
36	9.9080	9074	9069	9063	9057	9052	9046	9041	9035	9029	1	2	3	4	5
37	9.9023	9018	9012	9006	9000	8995	8989	8983	8977	8971	1	2	3	4	5
38	9.8965	8959	8953	8947	8941	8935	8929	8923	8917	8911	1	2	3	4	5
39	9.8905	8899	8893	8887	8880	8874	8868	8862	8855	8849	1	2	3	4	5
40	9.8843	8836	8830	8823	8817	8810	8804	8797	8791	8784	1	2	3	4	5
41	9.8778	8771	8765	8758	8751	8745	8738	8731	8724	8718	1	2	3	5	6
42	9.8711	8704	8697	8690	8683	8676	8669	8662	8655	8648	1	2	3	5	6
43	9.8641	8634	8627	8620	8613	8606	8598	8591	8584	8577	1	2	4	5	6
44	9.8569	8562	8555	8547	8540	8532	8525	8517	8510	8502	1	2	4	5	6

N.B.—Subtract Mean Differences.

LOGARITHMIC COSINES (Continued)

Deg.	0′	6′	12′	18′	24′	30′	36′	42′	48′	54′	Mean Differences				
											1′	2′	3′	4′	5′
45	9.8495	8487	8480	8472	8464	8457	8449	8441	8433	8426	1	3	4	5	6
46	9.8418	8410	8402	8394	8386	8378	8370	8362	8354	8346	1	3	4	5	7
47	9.8338	8330	8322	8313	8305	8297	8289	8280	8272	8264	1	3	4	6	7
48	9.8255	8247	8238	8230	8221	8213	8204	8195	8187	8178	1	3	4	6	7
49	9.8169	8161	8152	8143	8134	8125	8117	8108	8099	8090	1	3	4	6	7
50	9.8081	8072	8063	8053	8044	8035	8026	8017	8007	7998	2	3	5	6	8
51	9.7989	7979	7970	7960	7951	7941	7932	7922	7913	7903	2	3	5	6	8
52	9.7893	7884	7874	7864	7854	7844	7835	7825	7815	7805	2	3	5	7	8
53	9.7795	7785	7774	7764	7754	7744	7734	7723	7713	7703	2	3	5	7	9
54	9.7692	7682	7671	7661	7650	7640	7629	7618	7607	7597	2	4	5	7	9
55	9.7586	7575	7564	7553	7542	7531	7520	7509	7498	7487	2	4	6	7	9
56	9.7476	7464	7453	7442	7430	7419	7407	7396	7384	7373	2	4	6	8	10
57	9.7361	7349	7338	7326	7314	7302	7290	7278	7266	7254	2	4	6	8	10
58	9.7242	7230	7218	7205	7193	7181	7168	7156	7144	7131	2	4	6	8	10
59	9.7118	7106	7093	7080	7068	7055	7042	7029	7016	7003	2	4	6	9	11

N.B.—Subtract Mean Differences.

LOGARITHMIC COSINES (Continued)

Deg.	0′	6′	12′	18′	24′	30′	36′	42′	48′	54′	Mean Differences 1′	2′	3′	4′	5′
60	9.6990	6977	6963	6950	6937	6923	6910	6896	6883	6869	2	4	7	9	11
61	9.6856	6842	6828	6814	6801	6787	6773	6759	6744	6730	2	5	7	9	12
62	9.6716	6702	6687	6673	6659	6644	6629	6615	6600	6585	2	5	7	10	12
63	9.6570	6556	6541	6526	6510	6495	6480	6465	6449	6434	3	5	8	10	13
64	9.6418	6403	6387	6371	6356	6340	6324	6308	6292	6276	3	5	8	11	13
65	9.6259	6243	6227	6210	6194	6177	6161	6144	6127	6110	3	6	8	11	14
66	9.6093	6076	6059	6042	6024	6007	5990	5972	5954	5937	3	6	9	12	15
67	9.5919	5901	5883	5865	5847	5828	5810	5792	5773	5754	3	6	9	12	15
68	9.5736	5717	5698	5679	5660	5641	5621	5602	5583	5563	3	6	10	13	16
69	9.5543	5523	5504	5484	5463	5443	5423	5402	5382	5361	3	7	10	14	17
70	9.5341	5320	5299	5278	5256	5235	5213	5192	5170	5148	4	7	11	14	18
71	9.5126	5104	5082	5060	5037	5015	4992	4969	4946	4923	4	8	11	15	19
72	9.4900	4876	4853	4829	4805	4781	4757	4733	4709	4684	4	8	12	16	20
73	9.4659	4634	4609	4584	4559	4533	4508	4482	4456	4430	4	9	13	17	21
74	9.4403	4377	4350	4323	4296	4269	4242	4214	4186	4158	5	9	14	18	23

N.B.—Subtract Mean Differences.

LOGARITHMIC COSINES (Continued)

Deg.	0′	6′	12′	18	24′	30′	36′	42′	48′	54′	Mean Differences 1′	2′	3′	4′	5′
75	9.4130	4102	4073	4044	4015	3986	3957	3927	3897	3867	5	10	15	20	24
76	9.3837	3806	3775	3745	3713	3682	3650	3618	3586	3554	5	11	16	21	26
77	9.3521	3488	3455	3421	3387	3353	3319	3284	3250	3214	6	11	17	23	28
78	9.3179	3143	3107	3070	3034	2997	2959	2921	2883	2845	6	12	19	25	31
79	9.2806	2767	2727	2687	2647	2606	2565	2524	2482	2439	7	14	20	27	34
80	9.2397	2353	2310	2266	2221	2176	2131	2085	2038	1991	8	15	23	30	38
81	9.1943	1895	1847	1797	1747	1697	1646	1594	1542	1489	8	17	25	34	42
82	9.1436	1381	1326	1271	1214	1157	1099	1040	0981	0920	10	19	29	38	48
83	9.0859	0797	0734	0670	0605	0539	0472	0403	0334	0264	11	22	33	44	55
84	9.0192	0120	0046	9970	9894	9816	9736	9655	9573	9489	13	26	39	52	65
85	8.9403	9315	9226	9135	9042	8946	8849	8749	8647	8543	16	32	48	64	80
86	8.8436	8326	8213	8098	7979	7857	7731	7602	7468	7330	*Mean* differences no longer sufficiently accurate.				
87	8.7188	7041	6889	6731	6567	6397	6220	6035	5842	5640					
88	8.5428	5206	4971	4723	4459	4179	3880	3558	3210	2832					
89	8.2419	1961	1450	0870	0200	9408	8439	7190	5429	2419					

N.B.—Subtract Mean Differences.

NATURAL TANGENTS

Deg.	0′	6′	12′	18′	24′	30′	36′	42′	48′	54′	Mean Differences 1′	2′	3′	4′	5′
0	.0000	0017	0035	0052	0070	0087	0105	0122	0140	0157	3	6	9	12	15
1	.0175	0192	0209	0227	0244	0262	0279	0297	0314	0332	3	6	9	12	15
2	.0349	0367	0384	0402	0419	0437	0454	0472	0489	0507	3	6	9	12	15
3	.0524	0542	0559	0577	0594	0612	0629	0647	0664	0682	3	6	9	12	15
4	.0699	0717	0734	0752	0769	0787	0805	0822	0840	0857	3	6	9	12	15
5	.0875	0892	0910	0928	0945	0963	0981	0998	1016	1033	3	6	9	12	15
6	.1051	1069	1086	1104	1122	1139	1157	1175	1192	1210	3	6	9	12	15
7	.1228	1246	1263	1281	1299	1317	1334	1352	1370	1388	3	6	9	12	15
8	.1405	1423	1441	1459	1477	1495	1512	1530	1548	1566	3	6	9	12	15
9	.1584	1602	1620	1638	1655	1673	1691	1709	1727	1745	3	6	9	12	15
10	.1763	1781	1799	1817	1835	1853	1871	1890	1908	1926	3	6	9	12	15
11	.1944	1962	1980	1998	2016	2035	2053	2071	2089	2107	3	6	9	12	15
12	.2126	2144	2162	2180	2199	2217	2235	2254	2272	2290	3	6	9	12	15
13	.2309	2327	2345	2364	2382	2401	2419	2438	2456	2475	3	6	9	12	15
14	.2493	2512	2530	2549	2568	2586	2605	2623	2642	2661	3	6	9	12	16
15	.2679	2698	2717	2736	2754	2773	2792	2811	2830	2849	3	6	9	13	16
16	.2867	2886	2905	2924	2943	2962	2981	3000	3019	3038	3	6	9	13	16
17	.3057	3076	3096	3115	3134	3153	3172	3191	3211	3230	3	6	10	13	16

NATURAL TANGENTS (Continued)

Deg.	0′	6′	12′	18′	24′	30′	36′	42′	48′	54′	Mean Differences 1′	2′	3′	4′	5′
18	.3249	3269	3288	3307	3327	3346	3365	3385	3404	3424	3	6	10	13	16
19	.3443	3463	3482	3502	3522	3541	3561	3581	3600	3620	3	7	10	13	16
20	.3640	3659	3679	3699	3719	3739	3759	3779	3799	3819	3	7	10	13	17
21	.3839	3859	3879	3899	3919	3939	3959	3979	4000	4020	3	7	10	13	17
22	.4040	4061	4081	4101	4122	4142	4163	4183	4204	4224	3	7	10	14	17
23	.4245	4265	4286	4307	4327	4348	4369	4390	4411	4431	3	7	10	14	17
24	.4452	4473	4494	4515	4536	4557	4578	4599	4621	4642	4	7	11	14	18
25	.4663	4684	4706	4727	4748	4770	4791	4813	4834	4856	4	7	11	14	18
26	.4877	4899	4921	4942	4964	4986	5008	5029	5051	5073	4	7	11	15	18
27	.5095	5117	5139	5161	5184	5206	5228	5250	5272	5295	4	7	11	15	18
28	.5317	5340	5362	5384	5407	5430	5452	5475	5498	5520	4	8	11	15	19
29	.5543	5566	5589	5612	5635	5658	5681	5704	5727	5750	4	8	12	15	19
30	.5774	5797	5820	5844	5867	5890	5914	5938	5961	5985	4	8	12	16	20
31	.6009	6032	6056	6080	6104	6128	6152	6176	6200	6224	4	8	12	16	20
32	.6249	6273	6297	6322	6346	6371	6395	6420	6445	6469	4	8	12	16	20
33	.6494	6519	6544	6569	6594	6619	6644	6669	6694	6720	4	8	13	17	21
34	.6745	6771	6796	6822	6847	6873	6899	6924	6950	6976	4	9	13	17	21
35	.7002	7028	7054	7080	7107	7133	7159	7186	7212	7239	4	9	13	18	22

NATURAL TANGENTS (Continued)

Deg.	0′	6′	12′	18′	24′	30′	36′	42′	48′	54′	Mean Differences 1′	2′	3′	4′	5′
36	.7265	7292	7319	7346	7373	7400	7427	7454	7481	7508	5	9	14	18	23
37	.7536	7563	7590	7618	7646	7673	7701	7729	7757	7785	5	9	14	18	23
38	.7813	7841	7869	7898	7926	7954	7983	8012	8040	8069	5	9	14	19	24
39	.8098	8127	8156	8185	8214	8243	8273	8302	8332	8361	5	10	15	20	24
40	.8391	8421	8451	8481	8511	8541	8571	8601	8632	8662	5	10	15	20	25
41	.8693	8724	8754	8785	8816	8847	8878	8910	8941	8972	5	10	16	21	26
42	.9004	9036	9067	9099	9131	9163	9195	9228	9260	9293	5	11	16	21	27
43	.9325	9358	9391	9424	9457	9490	9523	9556	9590	9623	6	11	17	22	28
44	.9657	9691	9725	9759	9793	9827	9861	9896	9930	9965	6	11	17	23	29
45	1.0000	0035	0070	0105	0141	0176	0212	0247	0283	0319	6	12	18	24	30
46	1.0355	0392	0428	0464	0501	0538	0575	0612	0649	0686	6	12	18	25	31
47	1.0724	0761	0799	0837	0875	0913	0951	0990	1028	1067	6	13	19	25	32
48	1.1106	1145	1184	1224	1263	1303	1343	1383	1423	1463	7	13	20	27	33
49	1.1504	1544	1585	1626	1667	1708	1750	1792	1833	1875	7	14	21	28	34
50	1.1918	1960	2002	2045	2088	2131	2174	2218	2261	2305	7	14	22	29	36
51	1.2349	2393	2437	2482	2527	2572	2617	2662	2708	2753	8	15	23	30	38
52	1.2799	2846	2892	2938	2985	3032	3079	3127	3175	3222	8	16	24	31	39
53	1.3270	3319	3367	3416	3465	3514	3564	3613	3663	3713	8	16	25	33	41

NATURAL TANGENTS (Continued)

Deg.	0′	6′	12′	18′	24′	30′	36′	42′	48′	54′	Mean Differences 1′	2′	3′	4′	5′
54	1.3764	3814	3865	3916	3968	4019	4071	4124	4176	4229	9	17	26	34	43
55	1.4281	4335	4388	4442	4496	4550	4605	4659	4715	4770	9	18	27	36	45
56	1.4826	4882	4938	4994	5051	5108	5166	5224	5282	5340	10	19	29	38	48
57	1.5399	5458	5517	5577	5637	5697	5757	5818	5880	5941	10	20	30	40	50
58	1.6003	6066	6128	6191	6255	6319	6383	6447	6512	6577	11	21	32	43	53
59	1.6643	6709	6775	6842	6909	6977	7045	7113	7182	7251	11	23	34	45	56
60	1.7321	7391	7461	7532	7603	7675	7747	7820	7893	7966	12	24	36	48	60
61	1.8040	8115	8190	8265	8341	8418	8495	8572	8650	8728	13	26	38	51	64
62	1.8807	8887	8967	9047	9128	9210	9292	9375	9458	9542	14	27	41	55	68
63	1.9626	9711	9797	9883	9970	0057	0145	0233	0323	0413	15	29	44	58	73
64	2.0503	0594	0686	0778	0872	0965	1060	1155	1251	1348	16	31	47	63	78
65	2.1445	1543	1642	1742	1842	1943	2045	2148	2251	2355	17	34	51	68	85
66	2.2460	2566	2673	2781	2889	2998	3109	3220	3332	3445	18	37	55	73	92
67	2.3559	3673	3789	3906	4023	4142	4262	4383	4504	4627	20	40	60	79	99
68	2.4751	4876	5002	5129	5257	5386	5517	5649	5782	5916	22	43	65	87	108
69	2.6051	6187	6325	6464	6605	6746	6889	7034	7179	7326	24	47	71	95	119
70	2.7475	7625	7776	7929	8083	8239	8397	8556	8716	8878	26	52	78	104	131
71	2.9042	9208	9375	9544	9714	9887	0061	0237	0415	0595	29	58	87	116	145

NATURAL TANGENTS (Continued)

Deg.	0′	6′	12′	18′	24′	30′	36′	42′	48′	54′	Mean Differences 1′	2′	3′	4′	5′
72	3.0777	0961	1146	1334	1524	1716	1910	2106	2305	2506	32	64	96	129	161
73	3.2709	2914	3122	3332	3544	3759	3977	4197	4420	4646	36	72	108	144	180
74	3.4874	5105	5339	5576	5816	6059	6305	6554	6806	7062	41	81	122	163	204
75	3.7321	7583	7848	8118	8391	8667	8947	9232	9520	9812	46	93	139	186	232
76	4.0108	0408	0713	1022	1335	1653	1976	2303	2635	2972					
77	4.3315	3662	4015	4374	4737	5107	5483	5864	6252	6646					
78	4.7046	7453	7867	8288	8716	9152	9594	0045	0504	0970					
79	5.1446	1929	2422	2924	3435	3955	4486	5026	5578	6140					
80	5.6713	7297	7894	8502	9124	9758	0405	1066	1742	2432					
81	6.3138	3859	4596	5350	6122	6912	7720	8548	9395	0264	*Mean* differences no longer sufficiently accurate.				
82	7.1154	2066	3002	3962	4947	5958	6996	8062	9158	0285					
83	8.1443	2636	3863	5126	6427	7769	9152	0579	2052	3572					
84	9.5144	9.677	9.845	10.02	10.20	10.39	10.58	10.78	10.99	11.20					
85	11.430	11.66	11.91	12.16	12.43	12.71	13.00	13.30	13.62	13.95					
86	14.301	14.67	15.06	15.46	15.89	16.35	16.83	17.34	17.89	18.46					
87	19.081	19.74	20.45	21.20	22.02	22.90	23.86	24.90	26.03	27.27					
88	28.636	30.14	31.82	33.69	35.80	38.19	40.92	44.07	47.74	52.08					
89	57.290	63.66	71.62	81.85	95.49	114.6	143.2	191.0	286.5	573.0					

LOGARITHMIC TANGENTS

Deg.	0′	6′	12′	18′	24′	30′	36′	42′	48′	54′	Mean Differences 1′	2′	3′	4′	5′
0	—∞	7.2419	5429	7190	8439	9409	0200	0870	1450	1962					
1	8.2419	2833	3211	3559	3881	4181	4461	4725	4973	5208					
2	8.5431	5643	5845	6038	6223	6401	6571	6736	6894	7046					
3	8.7194	7337	7475	7609	7739	7865	7988	8107	8223	8336					
4	8.8446	8554	8659	8762	8862	8960	9056	9150	9241	9331	16	32	48	64	81
5	8.9420	9506	9591	9674	9756	9836	9915	9992	0068	0143	13	26	40	53	66
6	9.0216	0289	0360	0430	0499	0567	0633	0699	0764	0828	11	22	34	45	56
7	9.0891	0954	1015	1076	1135	1194	1252	1310	1367	1423	10	20	29	39	49
8	9.1478	1533	1587	1640	1693	1745	1797	1848	1898	1948	9	17	26	35	43
9	9.1997	2046	2094	2142	2189	2236	2282	2328	2374	2419	8	16	23	31	39
10	9.2463	2507	2551	2594	2637	2680	2722	2764	2805	2846	7	14	21	28	35
11	9.2887	2927	2967	3006	3046	3085	3123	3162	3200	3237	6	13	19	26	32
12	9.3275	3312	3349	3385	3422	3458	3493	3529	3564	3599	6	12	18	24	30
13	9.3634	3668	3702	3736	3770	3804	3837	3870	3903	3935	6	11	17	22	28
14	9.3968	4000	4032	4064	4095	4127	4158	4189	4220	4250	5	10	16	21	26
15	9.4281	4311	4341	4371	4400	4430	4459	4488	4517	4546	5	10	15	20	25
16	9.4575	4603	4632	4660	4688	4716	4744	4771	4799	4826	5	9	14	19	23
17	9.4853	4880	4907	4934	4961	4987	5014	5040	5066	5092	4	9	13	18	22

T

LOGARITHMIC TANGENTS (Continued)

Deg.	0′	6′	12′	18′	24′	30′	36′	42′	48′	54′	Mean Differences 1′	2′	3′	4′	5′
18	9.5118	5143	5169	5195	5220	5245	5270	5295	5320	5345	4	8	13	17	21
19	9.5370	5394	5419	5443	5467	5491	5516	5539	5563	5587	4	8	12	16	20
20	9.5611	5634	5658	5681	5704	5727	5750	5773	5796	5819	4	8	12	15	19
21	9.5842	5864	5887	5909	5932	5954	5976	5998	6020	6042	4	7	11	15	19
22	9.6064	6086	6108	6129	6151	6172	6194	6215	6236	6257	4	7	11	14	18
23	9.6279	6300	6321	6341	6362	6383	6404	6424	6445	6465	3	7	10	14	17
24	9.6486	6506	6527	6547	6567	6587	6607	6627	6647	6667	3	7	10	13	17
25	9.6687	6706	6726	6746	6765	6785	6804	6824	6843	6863	3	7	10	13	16
26	9.6882	6901	6920	6939	6958	6977	6996	7015	7034	7053	3	6	9	13	16
27	9.7072	7090	7109	7128	7146	7165	7183	7202	7220	7238	3	6	9	12	15
28	9.7257	7275	7293	7311	7330	7348	7366	7384	7402	7420	3	6	9	12	15
29	9.7438	7455	7473	7491	7509	7526	7544	7562	7579	7597	3	6	9	12	15
30	9.7614	7632	7649	7667	7684	7701	7719	7736	7753	7771	3	6	9	12	14
31	9.7788	7805	7822	7839	7856	7873	7890	7907	7924	7941	3	6	9	11	14
32	9.7958	7975	7992	8008	8025	8042	8059	8075	8092	8109	3	6	8	11	14
33	9.8125	8142	8158	8175	8191	8208	8224	8241	8257	8274	3	5	8	11	14
34	9.8290	8306	8323	8339	8355	8371	8388	8404	8420	8436	3	5	8	11	14
35	9.8452	8468	8484	8501	8517	8533	8549	8565	8581	8597	3	5	8	11	13

LOGARITHMIC TANGENTS (Continued)

Deg.	0′	6′	12′	18′	24′	30′	36′	42′	48′	54′	Mean Differences 1′	2′	3′	4′	5′
36	9.8613	8629	8644	8660	8676	8692	8708	8724	8740	8755	3	5	8	11	13
37	9.8771	8787	8803	8818	8834	8850	8865	8881	8897	8912	3	5	8	10	13
38	9.8928	8944	8959	8975	8990	9006	9022	9037	9053	9068	3	5	8	10	13
39	9.9084	9099	9115	9130	9146	9161	9176	9192	9207	9223	3	5	8	10	13
40	9.9238	9254	9269	9284	9300	9315	9330	9346	9361	9376	3	5	8	10	13
41	9.9392	9407	9422	9438	9453	9468	9483	9499	9514	9529	3	5	8	10	13
42	9.9544	9560	9575	9590	9605	9621	9636	9651	9666	9681	3	5	8	10	13
43	9.9697	9712	9727	9742	9757	9773	9788	9803	9818	9833	3	5	8	10	13
44	9.9848	9864	9870	9894	9909	9924	9939	9955	9970	9985	3	5	8	10	13
45	10.0000	0015	0030	0045	0061	0076	0091	0106	0121	0136	3	5	8	10	13
46	10.0152	0167	0182	0197	0212	0228	0243	0258	0273	0288	3	5	8	10	13
47	10.0303	0319	0334	0349	0364	0379	0395	0410	0425	0440	3	5	8	10	13
48	10.0456	0471	0486	0501	0517	0532	0547	0562	0578	0593	3	5	8	10	13
49	10.0608	0624	0639	0654	0670	0685	0700	0716	0731	0746	3	5	8	10	13
50	10.0762	0777	0793	0808	0824	0839	0854	0870	0885	0901	3	5	8	10	13
51	10.0916	0932	0947	0963	0978	0994	1010	1025	1041	1056	3	5	8	10	13
52	10.1072	1088	1103	1119	1135	1150	1166	1182	1197	1213	3	5	8	10	13
53	10.1229	1245	1260	1276	1292	1308	1324	1340	1356	1371	3	5	8	11	13

LOGARITHMIC TANGENTS (Continued)

Deg.	0′	6	12′	18′	24′	30′	36′	42′	48′	54′	Mean Differences. 1′	2′	3′	4′	5′
54	10.1387	1403	1419	1435	1451	1467	1483	1499	1516	1532	3	5	8	11	13
55	10.1548	1564	1580	1596	1612	1629	1645	1661	1677	1694	3	5	8	11	14
56	10.1710	1726	1743	1759	1776	1792	1809	1825	1842	1858	3	5	8	11	14
57	10.1875	1891	1908	1925	1941	1958	1975	1992	2008	2025	3	6	8	11	14
58	10.2042	2059	2076	2093	2110	2127	2144	2161	2178	2195	3	6	9	11	14
59	10.2212	2229	2247	2264	2281	2299	2316	2333	2351	2368	3	6	9	12	14
60	10.2386	2403	2421	2438	2456	2474	2491	2509	2527	2545	3	6	9	12	15
61	10.2562	2580	2598	2616	2634	2652	2670	2689	2707	2725	3	6	9	12	15
62	10.2743	2762	2780	2798	2817	2835	2854	2872	2891	2910	3	6	9	12	15
63	10.2928	2947	2966	2985	3004	3023	3042	3061	3080	3099	3	6	9	13	16
64	10.3118	3137	3157	3176	3196	3215	3235	3254	3274	3294	3	6	10	13	16
65	10.3313	3333	3353	3373	3393	3413	3433	3453	3473	3494	3	7	10	13	17
66	10.3514	3535	3555	3576	3596	3617	3638	3659	3679	3700	3	7	10	14	17
67	10.3721	3743	3764	3785	3806	3828	3849	3871	3892	3914	4	7	11	14	18
68	10.3936	3958	3980	4002	4024	4046	4068	4091	4113	4136	4	7	11	15	19
69	10.4158	4181	4204	4227	4250	4273	4296	4319	4342	4366	4	8	12	15	19
70	10.4389	4413	4437	4461	4484	4509	4533	4557	4581	4606	4	8	12	16	20
71	10.4630	4655	4680	4705	4730	4755	4780	4805	4831	4857	4	8	13	17	21

LOGARITHMIC TANGENTS—(Continued)

Deg.	0′	6′	12′	18′	24′	30′	36′	42′	48	54′	Mean Differences 1′	2′	3′	4′	5′
72	10.4882	4908	4934	4960	4986	5013	5039	5066	5093	5120	4	9	13	18	22
73	10.5147	5174	5201	5229	5256	5284	5312	5430	5368	5397	5	9	14	19	23
74	10.5425	5454	5483	5512	5541	5570	5600	5629	5659	5689	5	10	15	20	25
75	10.5719	5750	5780	5811	5842	5873	5905	5936	5968	6000	5	10	16	21	26
76	10.6032	6065	6097	6130	6163	6196	6230	6264	6298	6332	6	11	17	22	28
77	10.6366	6401	6436	6471	6507	6542	6578	6615	6651	6688	6	12	18	24	30
78	10.6725	6763	6800	6838	6877	6915	6954	6994	7033	7073	6	13	19	26	32
79	10.7113	7154	7195	7236	7278	7320	7363	7406	7449	7493	7	14	21	28	35
80	10.7537	7581	7626	7672	7718	7764	7811	7858	7906	7954	8	16	23	31	39
81	10.8003	8052	8102	8152	8203	8255	8307	8360	8413	8467	9	17	26	35	43
82	10.8522	8577	8633	8690	8748	8806	8865	8924	8985	9046	10	20	29	39	49
83	10.9109	9172	9236	9302	9367	9433	9501	9570	9640	9711	11	22	34	45	56
84	10.9784	9857	9932	0008	0085	0164	0244	0326	0409	0494	13	26	40	53	66
85	11.0580	0669	0759	0850	0944	1040	1138	1238	1341	1446	16	32	48	64	81
86	11.1554	1664	1777	1893	2012	2135	2261	2391	2525	2663	*Mean* differences no longer sufficiently accurate.				
87	11.2806	2954	3106	3264	3429	3599	3777	3962	4155	4357					
88	11.4569	4792	5027	5275	5539	5189	6119	6441	6789	7167					
89	11.7581	8038	8550	9130	9800	0591	1561	2810	4571	7581					

LOGARITHMS

	0	1	2	3	4	5	6	7	8	9	1	2	3	4	5	6	7	8	9
10	0000	0043	0086	0128	0170	0212	0253	0294	0334	0374	4	8	12	17	21	25	29	33	37
11	0414	0453	0492	0531	0569	0607	0645	0682	0719	0755	4	8	11	15	19	23	26	30	34
12	0792	0828	0864	0899	0934	0969	1004	1038	1072	1106	3	7	10	14	17	21	24	28	31
13	1139	1173	1206	1239	1271	1303	1335	1367	1399	1430	3	6	10	13	16	19	23	26	29
14	1461	1492	1523	1553	1584	1614	1644	1673	1703	1732	3	6	9	12	15	18	21	24	27
15	1761	1790	1818	1847	1875	1903	1931	1959	1987	2014	3	6	8	11	14	17	20	22	25
16	2041	2068	2095	2122	2148	2175	2201	2227	2253	2279	3	5	8	11	13	16	18	21	24
17	2304	2330	2355	2380	2405	2430	2455	2480	2504	2529	2	5	7	10	12	15	17	20	22
18	2553	2577	2601	2625	2648	2672	2695	2718	2742	2765	2	5	7	9	12	14	16	19	21
19	2788	2810	2833	2856	2878	2900	2923	2945	2967	2989	2	4	7	9	11	13	16	18	20
20	3010	3032	3054	3075	3096	3118	3139	3160	3181	3201	2	4	6	8	11	13	15	17	19
21	3222	3243	3263	3284	3304	3324	3345	3365	3385	3404	2	4	6	8	10	12	14	16	18
22	3424	3444	3464	3483	3502	3522	3541	3560	3579	3598	2	4	6	8	10	12	14	15	17
23	3617	3636	3655	3674	3692	3711	3729	3747	3766	3784	2	4	6	7	9	11	13	15	17
24	3802	3820	3838	3856	3874	3892	3909	3927	3945	3962	2	4	5	7	9	11	12	14	16
25	3979	3997	4014	4031	4048	4065	4082	4099	4116	4133	2	3	5	7	9	10	12	14	15
26	4150	4166	4183	4200	4216	4232	4249	4265	4281	4298	2	3	5	7	8	10	11	13	15
27	4314	4330	4346	4362	4378	4393	4409	4425	4440	4456	2	3	5	6	8	9	11	13	14
28	4472	4487	4502	4518	4533	4548	4564	4579	4594	4609	2	3	5	6	8	9	11	12	14
29	4624	4639	4654	4669	4683	4698	4713	4728	4742	4757	1	3	4	6	7	9	10	12	13

LOGARITHMS—(*continued*)

	0	1	2	3	4	5	6	7	8	9	1	2	3	4	5	6	7	8	9
30	4771	4786	4800	4814	4829	4843	4857	4871	4886	4900	1	3	4	6	7	9	10	11	13
31	4914	4928	4942	4955	4969	4983	4997	5011	5024	5038	1	3	4	6	7	8	10	11	12
32	5051	5065	5079	5092	5105	5119	5132	5145	5159	5172	1	3	4	5	7	8	9	11	12
33	5185	5198	5211	5224	5237	5250	5263	5276	5289	5302	1	3	4	5	6	8	9	10	12
34	5315	5328	5340	5353	5366	5378	5391	5403	5416	5428	1	3	4	5	6	8	9	10	11
35	5441	5453	5465	5478	5490	5502	5514	5527	5539	5551	1	2	4	5	6	7	9	10	11
36	5563	5575	5587	5599	5611	5623	5635	5647	5658	5670	1	2	4	5	6	7	8	10	11
37	5682	5694	5705	5717	5729	5740	5752	5763	5775	5786	1	2	3	5	6	7	8	9	10
38	5798	5809	5821	5832	5843	5855	5866	5877	5888	5899	1	2	3	5	6	7	8	9	10
39	5911	5922	5933	5944	5955	5966	5977	5988	5999	6010	1	2	3	4	5	7	8	9	10
40	6021	6031	6042	6053	6064	6075	6085	6096	6107	6117	1	2	3	4	5	6	8	9	10
41	6128	6138	6149	6160	6170	6180	6191	6201	6212	6222	1	2	3	4	5	6	7	8	9
42	6232	6243	6253	6263	6274	6284	6294	6304	6314	6325	1	2	3	4	5	6	7	8	9
43	6335	6345	6355	6365	6375	6385	6395	6405	6415	6425	1	2	3	4	5	6	7	8	9
44	6435	6444	6454	6464	6474	6484	6493	6503	6513	6522	1	2	3	4	5	6	7	8	9
45	6532	6542	6551	6561	6571	6580	6590	6599	6609	6618	1	2	3	4	5	6	7	8	9
46	6628	6637	6646	6656	6665	6675	6684	6693	6702	6712	1	2	3	4	5	6	7	7	8
47	6721	6730	6739	6749	6758	6767	6776	6785	6794	6803	1	2	3	4	5	5	6	7	8
48	6812	6821	6830	6839	6848	6857	6866	6875	6884	6893	1	2	3	4	4	5	6	7	8
49	6902	6911	6920	6928	6937	6946	6955	6964	6972	6981	1	2	3	4	4	5	6	7	8
50	6990	6998	7007	7016	7024	7033	7042	7050	7059	7067	1	2	3	3	4	5	6	7	8
51	7076	7084	7093	7101	7110	7118	7126	7135	7143	7152	1	2	3	3	4	5	6	7	8
52	7160	7168	7177	7185	7193	7202	7210	7218	7226	7235	1	2	2	3	4	5	6	7	7
53	7243	7251	7259	7267	7275	7284	7292	7300	7308	7316	1	2	2	3	4	5	6	6	7
54	7324	7332	7340	7348	7356	7364	7372	7380	7388	7396	1	2	2	3	4	5	6	6	7

LOGARITHMS—(*continued*)

	0	1	2	3	4	5	6	7	8	9	1	2	3	4	5	6	7	8	9
55	7404	7412	7419	7427	7435	7443	7451	7459	7466	7474	1	2	2	3	4	5	5	6	7
56	7482	7490	7497	7505	7513	7520	7528	7536	7543	7551	1	2	2	3	4	5	5	6	7
57	7559	7566	7574	7582	7589	7597	7604	7612	7619	7627	1	2	2	3	4	5	5	6	7
58	7634	7642	7649	7657	7664	7672	7679	7686	7694	7701	1	1	2	3	4	4	5	6	7
59	7709	7716	7723	7731	7738	7745	7752	7760	7767	7774	1	1	2	3	4	4	5	6	7
60	7782	7789	7796	7803	7810	7818	7825	7832	7839	7846	1	1	2	3	4	4	5	6	6
61	7853	7860	7868	7875	7882	7889	7896	7903	7910	7917	1	1	2	3	4	4	5	6	6
62	7924	7931	7938	7945	7952	7959	7966	7973	7980	7987	1	1	2	3	3	4	5	6	6
63	7993	8000	8007	8014	8021	8028	8035	8041	8048	8055	1	1	2	3	3	4	5	5	6
64	8062	8069	8075	8082	8089	8096	8102	8109	8116	8122	1	1	2	3	3	4	5	5	6
65	8129	8136	8142	8149	8156	8162	8169	8176	8182	8189	1	1	2	3	3	4	5	5	6
66	8195	8202	8209	8215	8222	8228	8235	8241	8248	8254	1	1	2	3	3	4	5	5	6
67	8261	8267	8274	8280	8287	8293	8299	8306	8312	8319	1	1	2	3	3	4	5	5	6
68	8325	8331	8338	8344	8351	8357	8363	8370	8376	8382	1	1	2	3	3	4	4	5	6
69	8388	8395	8401	8407	8414	8420	8426	8432	8439	8445	1	1	2	2	3	4	4	5	6
70	8451	8457	8463	8470	8476	8482	8488	8494	8500	8506	1	1	2	2	3	4	4	5	6
71	8513	8519	8525	8531	8537	8543	8549	8555	8561	8567	1	1	2	2	3	4	4	5	5
72	8573	8579	8585	8591	8597	8603	8609	8615	8621	8627	1	1	2	2	3	4	4	5	5
73	8633	8639	8645	8651	8657	8663	8669	8675	8681	8686	1	1	2	2	3	4	4	5	5
74	8692	8698	8704	8710	8716	8722	8727	8733	8739	8745	1	1	2	2	3	4	4	5	5

LOGARITHMS—(*continued*)

	0	1	2	3	4	5	6	7	8	9	1	2	3	4	5	6	7	8	9
75	8751	8756	8762	8768	8774	8779	8785	8791	8797	8802	1	1	2	2	3	3	4	5	5
76	8808	8814	8820	8825	8831	8837	8842	8848	8854	8859	1	1	2	2	3	3	4	5	5
77	8865	8871	8876	8882	8887	8893	8899	8904	8910	8915	1	1	2	2	3	3	4	4	5
78	8921	8927	8932	8938	8943	8949	8954	8960	8965	8971	1	1	2	2	3	3	4	4	5
79	8976	8982	8987	8993	8998	9004	9009	9015	9020	9025	1	1	2	2	3	3	4	4	5
80	9031	9036	9042	9047	9053	9058	9063	9069	9074	9079	1	1	2	2	3	3	4	4	5
81	9085	9090	9096	9101	9106	9112	9117	9122	9128	9133	1	1	2	2	3	3	4	4	5
82	9138	9143	9149	9154	9159	9165	9170	9175	9180	9186	1	1	2	2	3	3	4	4	5
83	9191	9196	9201	9206	9212	9217	9222	9227	9232	9238	1	1	2	2	3	3	4	4	5
84	9243	9248	9253	9258	9263	9269	9274	9279	9284	9289	1	1	2	2	3	3	4	4	5
85	9294	9299	9304	9309	9315	9320	9325	9330	9335	9340	1	1	2	2	3	3	4	4	5
86	9345	9350	9355	9360	9365	9370	9375	9380	9385	9390	1	1	2	2	3	3	4	4	5
87	9395	9400	9405	9410	9415	9420	9425	9430	9435	9440	0	1	1	2	2	3	3	4	4
88	9445	9450	9455	9460	9465	9469	9474	9479	9484	9489	0	1	1	2	2	3	3	4	4
89	9494	9499	9504	9509	9513	9518	9523	9528	9533	9538	0	1	1	2	2	3	3	4	4
90	9542	9547	9552	9557	9562	9566	9571	9576	9581	9586	0	1	1	2	2	3	3	4	4
91	9590	9595	9600	9605	9609	9614	9619	9624	9628	9633	0	1	1	2	2	3	3	4	4
92	9638	9643	9647	9652	9657	9661	9666	9671	9675	9680	0	1	1	2	2	3	3	4	4
93	9685	9689	9694	9699	9703	9708	9713	9717	9722	9727	0	1	1	2	2	3	3	4	4
94	9731	9736	9741	9745	9750	9754	9759	9763	9768	9773	0	1	1	2	2	3	3	4	4
95	9777	9782	9786	9791	9795	9800	9805	9809	9814	9818	0	1	1	2	2	3	3	4	4
96	9823	9827	9832	9836	9841	9845	9850	9854	9859	9863	0	1	1	2	2	3	3	4	4
97	9868	9872	9877	9881	9886	9890	9894	9899	9903	9908	0	1	1	2	2	3	3	4	4
98	9912	9917	9921	9926	9930	9934	9939	9943	9948	9952	0	1	1	2	2	3	3	4	4
99	9956	9961	9965	9969	9974	9978	9983	9987	9991	9996	0	1	1	2	2	3	3	3	4

ANTILOGARITHMS

	0	1	2	3	4	5	6	7	8	9	1	2	3	4	5	6	7	8	9
.00	1000	1002	1005	1007	1009	1012	1014	1016	1019	1021	0	0	1	1	1	1	2	2	2
.01	1023	1026	1028	1080	1033	1035	1038	1040	1042	1045	0	0	1	1	1	1	2	2	2
.02	1047	1050	1052	1054	1057	1059	1062	1064	1067	1069	0	0	1	1	1	1	2	2	2
.03	1072	1074	1076	1079	1081	1084	1086	1089	1091	1094	0	0	1	1	1	1	2	2	2
.04	1096	1099	1102	1104	1107	1109	1112	1114	1117	1119	0	1	1	1	1	2	2	2	2
.05	1122	1125	1127	1130	1132	1135	1138	1140	1143	1146	0	1	1	1	1	2	2	2	2
.06	1148	1151	1153	1156	1159	1161	1164	1167	1169	1172	0	1	1	1	1	2	2	2	2
.07	1175	1178	1180	1183	1186	1189	1191	1194	1197	1199	0	1	1	1	1	2	2	2	2
.08	1202	1205	1208	1211	1213	1216	1219	1222	1225	1227	0	1	1	1	1	2	2	2	3
.09	1230	1233	1236	1239	1242	1245	1247	1250	1253	1256	0	1	1	1	1	2	2	2	3
.10	1259	1262	1265	1268	1271	1274	1276	1279	1282	1285	0	1	1	1	1	2	2	2	3
.11	1288	1291	1294	1297	1300	1303	1306	1309	1312	1315	0	1	1	1	2	2	2	2	3
.12	1318	1321	1324	1327	1330	1334	1337	1340	1343	1346	0	1	1	1	2	2	2	2	3
.13	1349	1352	1355	1358	1361	1365	1368	1371	1374	1377	0	1	1	1	2	2	2	3	3
.14	1380	1384	1387	1390	1393	1396	1400	1403	1406	1409	0	1	1	1	2	2	2	3	3
.15	1413	1416	1419	1422	1426	1429	1432	1435	1439	1442	0	1	1	1	2	2	2	3	3
.16	1445	1449	1452	1455	1459	1462	1466	1469	1472	1476	0	1	1	1	2	2	2	3	3
.17	1479	1483	1486	1489	1493	1496	1500	1503	1507	1510	0	1	1	1	2	2	2	3	3
.18	1514	1517	1521	1524	1528	1531	1535	1538	1542	1545	0	1	1	1	2	2	2	3	3
.19	1549	1552	1556	1560	1563	1567	1570	1574	1578	1581	0	1	1	1	2	2	3	3	3
.20	1585	1589	1592	1596	1600	1603	1607	1611	1614	1618	0	1	1	1	2	2	3	3	3
.21	1622	1626	1629	1633	1637	1641	1644	1648	1652	1656	0	1	1	2	2	2	3	3	3
.22	1660	1663	1667	1671	1675	1679	1683	1687	1690	1694	0	1	1	2	2	2	3	3	3
.23	1698	1702	1706	1710	1714	1718	1722	1726	1730	1734	0	1	1	2	2	2	3	3	4

ANTILOGARITHMS—(*continued*).

	0	1	2	3	4	5	6	7	8	9	1	2	3	4	5	6	7	8	9
.24	1738	1742	1746	1750	1754	1758	1762	1766	1770	1774	0	1	1	2	2	2	3	3	4
.25	1778	1782	1786	1791	1795	1799	1803	1807	1811	1816	0	1	1	2	2	2	3	3	4
.26	1820	1824	1828	1832	1837	1841	1845	1849	1854	1858	0	1	1	2	2	3	3	3	4
.27	1862	1866	1871	1875	1879	1884	1888	1892	1897	1901	0	1	1	2	2	3	3	3	4
.28	1905	1910	1914	1919	1923	1928	1932	1936	1941	1945	0	1	1	2	2	3	3	4	4
.29	1950	1954	1959	1963	1968	1972	1977	1982	1986	1991	0	1	1	2	2	3	3	4	4
.30	1995	2000	2004	2009	2014	2018	2023	2028	2032	2037	0	1	1	2	2	3	3	4	4
.31	2042	2046	2051	2056	2061	2065	2070	2075	2080	2084	0	1	1	2	2	3	3	4	4
.32	2089	2094	2099	2104	2109	2113	2118	2123	2128	2133	0	1	1	2	2	3	3	4	4
.33	2138	2143	2148	2153	2158	2163	2168	2173	2178	2183	0	1	1	2	2	3	3	4	4
.34	2188	2193	2198	2203	2208	2213	2218	2223	2228	2234	1	1	2	2	3	3	4	4	5
.35	2239	2244	2249	2254	2259	2265	2270	2275	2280	2286	1	1	2	2	3	3	4	4	5
.36	2291	2296	2301	2307	2312	2317	2323	2328	2333	2339	1	1	2	2	3	3	4	4	5
.37	2344	2350	2355	2360	2366	2371	2377	2382	2388	2393	1	1	2	2	3	3	4	4	5
.38	2399	2404	2410	2415	2421	2427	2432	2438	2443	2449	1	1	2	2	3	3	4	4	5
.39	2455	2460	2466	2472	2477	2483	2489	2495	2500	2506	1	1	2	2	3	3	4	5	5
.40	2512	2518	2523	2529	2535	2541	2547	2553	2559	2564	1	1	2	2	3	4	4	5	5
.41	2570	2576	2582	2588	2594	2600	2606	2612	2618	2624	1	1	2	2	3	4	4	5	5
.42	2630	2636	2642	2649	2655	2661	2667	2673	2679	2685	1	1	2	2	3	4	4	5	6
.43	2692	2698	2704	2710	2716	2723	2729	2735	2742	2748	1	1	2	3	3	4	4	5	6
.44	2754	2761	2767	2773	2780	2786	2793	2799	2805	2812	1	1	2	3	3	4	4	5	6
.45	2818	2825	2831	2838	2844	2851	2858	2864	2871	2877	1	1	2	3	3	4	5	5	6
.46	2884	2891	2897	2904	2911	2917	2924	2931	2938	2944	1	1	2	3	3	4	5	5	6
.47	2951	2958	2965	2972	2979	2985	2992	2999	3006	3013	1	1	2	3	3	4	5	5	6
.48	3020	3027	3034	3041	3048	3055	3062	3069	3076	3083	1	1	2	3	4	4	5	6	6
.49	3090	3097	3105	3112	3119	3126	3133	3141	3148	3155	1	1	2	3	4	4	5	6	6

ANTILOGARITHMS—(*continued*).

	0	1	2	3	4	5	6	7	8	9	1	2	3	4	5	6	7	8	9
.50	3162	3170	3177	3184	3192	3199	3206	3214	3221	3228	1	1	2	3	4	4	5	6	7
.51	3236	3243	3251	3258	3266	3273	3281	3289	3296	3304	1	2	2	3	4	5	5	6	7
.52	3311	3319	3327	3334	3342	3350	3357	3365	3373	3381	1	2	2	3	4	5	5	6	7
.53	3388	3396	3404	3412	3420	3428	3436	3443	3451	3459	1	2	2	3	4	5	6	6	7
.54	3467	3475	3483	3491	3499	3508	3516	3524	3532	3540	1	2	2	3	4	5	6	6	7
.55	3548	3556	3565	3573	3581	3589	3597	3606	3614	3622	1	2	2	3	4	5	6	7	7
.56	3631	3639	3648	3656	3664	3673	3681	3690	3698	3707	1	2	3	3	4	5	6	7	8
.57	3715	3724	3733	3741	3750	3758	3767	3776	3784	3793	1	2	3	3	4	5	6	7	8
.58	3802	3811	3819	3828	3837	3846	3855	3864	3873	3882	1	2	3	4	4	5	6	7	8
.59	3890	3899	3908	3917	3926	3936	3945	3954	3963	3972	1	2	3	4	5	5	6	7	8
.60	3981	3990	3999	4009	4018	4027	4036	4046	4055	4064	1	2	3	4	5	6	6	7	8
.61	4074	4083	4093	4102	4111	4121	4130	4140	4150	4159	1	2	3	4	5	6	7	8	9
.62	4169	4178	4188	4198	4207	4217	4227	4236	4246	4256	1	2	3	4	5	6	7	8	9
.63	4266	4276	4285	4295	4305	4315	4325	4335	4345	4355	1	2	3	4	5	6	7	8	9
.64	4365	4375	4385	4395	4406	4416	4426	4436	4446	4457	1	2	3	4	5	6	7	8	9
.65	4467	4477	4487	4498	4508	4519	4529	4539	4550	4560	1	2	3	4	5	6	7	8	9
.66	4571	4581	4592	4603	4613	4624	4634	4645	4656	4667	1	2	3	4	5	6	7	9	10
.67	4677	4688	4699	4710	4721	4732	4742	4753	4764	4775	1	2	3	4	5	7	8	9	10
.68	4786	4797	4808	4819	4831	4842	4853	4864	4875	4887	1	2	3	4	6	7	8	9	10
.69	4898	4909	4920	4932	4943	4955	4966	4977	4989	5000	1	2	3	5	6	7	8	9	10
.70	5012	5023	5035	5047	5058	5070	5082	5093	5105	5117	1	2	4	5	6	7	8	9	11
.71	5129	5140	5152	5164	5176	5188	5200	5212	5224	5236	1	2	4	5	6	7	8	10	11
.72	5248	5260	5272	5284	5297	5309	5321	5333	5346	5358	1	2	4	5	6	7	9	10	11
.73	5370	5383	5395	5408	5420	5433	5445	5458	5470	5483	1	3	4	5	6	8	9	10	11
.74	5495	5508	5521	5534	5546	5559	5572	5585	5598	5610	1	3	4	5	6	8	9	10	12

ANTILOGARITHMS—(*continued*).

	0	1	2	3	4	5	6	7	8	9	1	2	3	4	5	6	7	8	9
.75	5623	5636	5649	5662	5675	5689	5702	5715	5728	5741	1	3	4	5	7	8	9	10	12
.76	5754	5768	5781	5794	5808	5821	5834	5848	5861	5875	1	3	4	5	7	8	9	11	12
.77	5888	5902	5916	5929	5943	5957	5970	5984	5998	6012	1	3	4	5	7	8	10	11	12
.78	6026	6039	6053	6067	6081	6095	6109	6124	6138	6152	1	3	4	6	7	8	10	11	13
.79	6166	6180	6194	6209	6223	6237	6252	6266	6281	6295	1	3	4	6	7	9	10	11	13
.80	6310	6324	6339	6353	6368	6383	6397	6412	6427	6442	1	3	4	6	7	9	10	12	13
.81	6457	6471	6486	6501	6516	6531	6546	6561	6577	6592	2	3	5	6	8	9	11	12	14
.82	6607	6622	6637	6653	6668	6683	6699	6714	6730	6745	2	3	5	6	8	9	11	12	14
.83	6761	6776	6792	6808	6823	6839	6855	6871	6887	6902	2	3	5	6	8	9	11	13	14
.84	6918	6934	6950	6966	6982	6998	7015	7031	7047	7063	2	3	5	6	8	10	11	13	15
.85	7079	7096	7112	7129	7145	7161	7178	7194	7211	7228	2	3	5	7	8	10	12	13	15
.86	7244	7261	7278	7295	7311	7328	7345	7362	7379	7396	2	3	5	7	8	10	12	13	15
.87	7413	7430	7447	7464	7482	7499	7516	7534	7551	7568	2	4	5	7	9	10	12	14	16
.88	7586	7603	7621	7638	7656	7674	7691	7709	7727	7745	2	4	5	7	9	11	12	14	16
.89	7762	7780	7798	7816	7834	7852	7870	7889	7907	7925	2	4	5	7	9	11	13	14	16
.90	7943	7962	7980	7998	8017	8035	8054	8072	8091	8110	2	4	6	7	9	11	13	15	17
.91	8128	8147	8166	8185	8204	8222	8241	8260	8279	8299	2	4	6	8	9	11	13	15	17
.92	8318	8337	8356	8375	8395	8414	8433	8453	8472	8492	2	4	6	8	10	12	14	15	17
.93	8511	8531	8551	8570	8590	8610	8630	8650	8670	8690	2	4	6	8	10	12	14	16	18
.94	8710	8730	8750	8770	8790	8810	8831	8851	8872	8892	2	4	6	8	10	12	14	16	18
.95	8913	8933	8954	8974	8995	9016	9036	9057	9078	9099	2	4	6	8	10	12	15	17	19
.96	9120	9141	9162	9183	9204	9226	9247	9268	9290	9311	2	4	6	8	11	13	15	17	19
.97	9333	9354	9376	9397	9419	9441	9462	9484	9506	9528	2	4	7	9	11	13	15	17	20
.98	9550	9572	9594	9616	9638	9661	9683	9705	9727	9750	2	4	7	9	11	13	16	18	20
.99	9772	9795	9817	9840	9863	9886	9908	9931	9954	9977	2	5	7	9	11	14	16	18	20

INDEX